# Enabling Technologies for Next Generation Wireless Communications

# Artificial Intelligence (AI): Elementary to Advanced Practices

*Series Editors: Vijender Kumar Solanki, Zhongyu (Joan) Lu, and Valentina E Balas*

In emerging smart city technology and industries, the role of artificial intelligence is becoming more prominent. This AI book series aims to cover the latest AI work, which will help the naïve user obtain support in solving existing problems and for experienced AI practitioners, to shed light on new avenues in the AI domains. The series covers the recent work carried out in AI and its associated domains, including Logic, Pattern Recognition, NLP, Expert Systems, Machine Learning, Block-Chain, and Big Data. The vast work domain of AI covers the latest trends and it will be helpful to those new to the field, practitioners, students, and researchers, to gain some new insights.

**Cyber Defense Mechanisms**
Security, Privacy, and Challenges
*Gautam Kumar, Dinesh Kumar Saini, and Nguyen Ha Huy Cuong*

**Artificial Intelligence Trends for Data Analytics Using Machine Learning and Deep Learning Approaches**
*K. Gayathri Devi, Mamata Rath and Nguyen Thi Dieu Linh*

**Transforming Management Using Artificial Intelligence Techniques**
Vikas Garg and Rashmi Agrawal

**AI and Deep Learning in Biometric Security**
Trends, Potential, and Challenges
*Gaurav Jaswal, Vivek Kanhangad, and Raghavendra Ramachandra*

**Enabling Technologies for Next Generation Wireless Communications**
*Edited by Mohammed Usman, Mohd Wajid, and Mohd Dilshad Ansari*

For more information on this series, please visit: https://www.crcpress.com/Artificial-Intelligence-AI-Elementary-to-Advanced-Practices/book-series/CRCAIEAP

# Enabling Technologies for Next Generation Wireless Communications

Edited by

Mohammed Usman
Mohd Wajid
Mohd Dilshad Ansari

CRC Press is an imprint of the
Taylor & Francis Group, an **informa** business

First edition published 2021
by CRC Press
6000 Broken Sound Parkway NW, Suite 300, Boca Raton, FL 33487-2742

and by CRC Press
2 Park Square, Milton Park, Abingdon, Oxon, OX14 4RN

CRC Press is an imprint of Taylor & Francis Group, LLC

© 2021 Taylor & Francis Group, LLC

The right of Mohammed Usman, Mohd Wajid, and Mohd Dilshad Ansari to be identified as the authors of the editorial material, and of the authors for their individual chapters, has been asserted in accordance with sections 77 and 78 of the Copyright, Designs and Patents Act 1988.

Reasonable efforts have been made to publish reliable data and information, but the author and publisher cannot assume responsibility for the validity of all materials or the consequences of their use. The authors and publishers have attempted to trace the copyright holders of all material reproduced in this publication and apologize to copyright holders if permission to publish in this form has not been obtained. If any copyright material has not been acknowledged please write and let us know so we may rectify in any future reprint.

Except as permitted under U.S. Copyright Law, no part of this book may be reprinted, reproduced, transmitted, or utilized in any form by any electronic, mechanical, or other means, now known or hereafter invented, including photocopying, microfilming, and recording, or in any information storage or retrieval system, without written permission from the publishers.

For permission to photocopy or use material electronically from this work, access www.copyright.com or contact the Copyright Clearance Center, Inc. (CCC), 222 Rosewood Drive, Danvers, MA 01923, 978-750-8400. For works that are not available on CCC please contact mpkbookspermissions@tandf.co.uk

*Trademark notice*: Product or corporate names may be trademarks or registered trademarks and are used only for identification and explanation without intent to infringe.

Library of Congress Cataloging-in-Publication Data

Names: Usman, Mohammed (Electrical engineering professor), editor. | Wajid, Mohd, editor. | Ansari, Mohd Dilshad, editor.
Title: Enabling technologies for next generation wireless communications / edited by Mohammed Usman, Mohd Wajid, and Mohd Dilshad Ansari.
Description: First edition. | Boca Raton, FL : CRC Press/Taylor & Francis Group, LLC, 2021. | Series: Artificial intelligence (AI). Elementary to advanced practices | Includes bibliographical references and index.
Identifiers: LCCN 2020036945 (print) | LCCN 2020036946 (ebook) | ISBN 9780367422493 (hbk) | ISBN 9781003003472 (ebk)
Subjects: LCSH: Wireless communication systems--Technological innovations.
Classification: LCC TK5103.2 .E525 2021 (print) | LCC TK5103.2 (ebook) | DDC 621.384--dc23
LC record available at https://lccn.loc.gov/2020036945
LC ebook record available at https://lccn.loc.gov/2020036946

ISBN: 978-0-367-42249-3 (hbk)
ISBN: 978-1-003-00347-2 (ebk)

Typeset in Times
by SPi Global, India

# Dedication

*To my parents, wife, and children*
*-Mohammed Usman*

*To my family*
*-Mohd Wajid*

*To my son Shahrul*
*-Mohd Dilshad Ansari*

# Contents

Preface .................................................................................................................. ix
Editors .................................................................................................................. xi
Contributors ...................................................................................................... xiii

**Chapter 1** Technology Evolution of Wireless Communications: A Survey and Look Forward ............................................................................................ 1
*M. Usman and M. Wajid*

**Chapter 2** Enabling Technologies and Enabling Business Models for Next Generation Wireless Communications ............................................... 13
*M. Mzyece*

**Chapter 3** Enabling Technologies for Internet of Everything ........................... 33
*M. Rezwanul Mahmood and Mohammad Abdul Matin*

**Chapter 4** Power Allocation Techniques for Visible Light: Nonorthogonal Multiple Access Communication Systems ........................................ 45
*C. E. Ngene, Prabhat Thakur, and Ghanshyam Singh*

**Chapter 5** Multiantenna Systems: Large-Scale MIMO and Massive MIMO ..... 79
*Bhasker Gupta*

**Chapter 6** Channel Estimation Techniques in MIMO-OFDM System ............ 101
*Asif Alam Joy, Mohammed Nasim Faruq, and Mohammad Abdul Matin*

**Chapter 7** Localization Protocols for Wireless Sensor Networks .................... 119
*Ash Mohammad Abbas and Hamzah Ali Abdul Rahman Qasem*

**Chapter 8** Distributed Intelligent Networks: Convergence of 5G, AI and IoT ........................................................................................... 137
*M. Z. Shamim, M. Parayangat, V. P. Thafasal Ijyas, and S. J. Ali*

**Chapter 9** Antenna Design Challenges for 5G: Assessing Future Direction .... 149
*S. Arif Ali, M. Wajid, and M. Shah Alam*

**Chapter 10** Design and Simulation of New Beamforming: Based Cognitive Radio for 5G Networks ................................................... 177
*Tanzeela Ashraf, Javaid A. Sheikh, Sadaf Ajaz Khan, and Mehboob-ul-Amin*

**Chapter 11** Image Transmission Analysis Using MIMO-OFDM Systems ......... 195
*Akanksha Sharma, Lavish Kansal, Gurjot Singh Gaba, and Mohamed Mounir*

**Chapter 12** Physical Layer Security in Two-Way Wireless Communication System ................................................................................................. 211
*Shashibhushan Sharma, Sanjay Dhar Roy, and Sumit Kundu*

**Chapter 13** Design and Simulation of Bio-Inspired Algorithm: Based Cognitive Radio for 5G Networks ....................................................... 239
*Sadaf Ajaz Khan, Javaid A. Sheikh, Tanzeela Ashraf, and Mehboob-ul-Amin*

**Chapter 14** Evaluating The Performance of Quasi and Rotated Quasi OSTBC System with Advanced Detection Techniques for 5G and IoT Applications ................................................................. 259
*Priyanka Mishra and Mehboob-ul Amin*

**Index** ......................................................................................................................... 277

# Preface

Wireless communication systems have evolved by leaps and bounds ever since their inception. Starting with the first generation (1G) wireless systems, which only catered to voice calls with limited support for mobility to the present-day fifth-generation (5G) systems, they have made mobile broadband a reality, providing a host of applications and services for end users. Such an evolution and development of mobile wireless communication systems has been possible due to the underlying technologies that cater to the demands and requirements in each generation of wireless systems, providing the capabilities to deliver the services and applications of utility to end users/customers. These enabling technologies span a wide range of domains/topics, such as signal processing, communication theory, information theory, antenna design, spectrum management, channel modeling, artificial intelligence, and security. Each of these enabling technologies comes under the purview of diverse, yet interrelated research specializations, and this book shall provide latest developments of these enabling technologies in the context of and application to future wireless systems. In addition, novel business models, applications, and services are required to commercialize and generate revenue from these ever-evolving wireless communications systems, as end users pay for services that benefit them rather than for the technology itself. The chapters in this book cover some of the important enabling technologies that will be key components of next generation wireless systems while also addressing the key technologies in present-day 4G/5G systems. It is hoped that this book will provide up-to-date information on emerging trends in wireless systems, their enabling technologies, and their evolving application paradigms to researchers, technologists, developers, engineers, and policy decision makers, as well as graduate students.

**Mohammed Usman**

**Mohd Wajid**

**Mohd Dilshad Ansari**

# Editors

**Mohammed Usman, PhD,** is an assistant professor in the Department of Electrical Engineering at King Khalid University, Abha, Saudi Arabia. He earned his BE in Electronics and Communication Engineering from Madras University, India, in 2002. He joined the University of Strathclyde in Glasgow, United Kingdom, as a graduate student and earned his Master of Science in Communications, Control and Digital Signal Processing in 2003. In 2008, he earned his PhD degree for his research on Fountain codes for wireless communications. He is a senior member of IEEE, USA, and Member of IET, UK. He served as editor for several books and as Organizing/TPC Chair for IEEE/Springer International conferences. His research interests are in mathematical modeling, signal processing, and AI techniques for biomedical applications, next generation wireless systems, and their applications.

**Mohd Wajid, PhD,** is an assistant professor in the Department of Electronics Engineering, Aligarh Muslim University (AMU), Aligarh, India. Dr. Wajid earned his B Tech (Electronics) from AMU – Aligarh and M.Tech (VLSI and Embedded Systems) from IIIT Hyderabad. He earned his PhD in Signal Processing from Indian Institute of Technology Delhi, India. Before joining AMU, he was associated with Jaypee University of Information Technology, Texas Instruments, Xilinx India Technology Services Private Limited, and BlueStar Limited. He is also a senior member of IEEE, USA.

**Mohd Dilshad Ansari, PhD,** is an assistant professor in the Department of Computer Science & Engineering at CMR College of Engineering & Technology, Hyderabad, India. He earned his Tech in Information Technology from Uttar Pradesh Technical University, Lucknow, UP, India, in 2009. He earned his M Tech and PhD in Computer Science & Engineering from Jaypee University of Information Technology, Waknaghat, Solan, HP, India, in 2011 and 2018, respectively. He has more than 8 years of Academic/Research Experience; Dr. Ansari has published more than 45 papers in International Journals (SCIE/Scopus) and conferences (IEEE/Springer). He is a member of various technical/professional societies such as IEEE, UACEE, and IACSIT. He has been appointed as a member of the Editorial/Reviewer Board and Technical Program Committee in numerous reputed journals/conferences. He is also serving as guest editor in reputed journals and Organized Special Sessions in IEEE/Springer Conferences. His research interests include digital and fuzzy image processing, machine learning, IoT, and cloud computing.

xi

# Contributors

**Ash Mohammad Abbas**
Department of Computer Engineering
Aligarh Muslim University
Aligarh, India

**M. Shah Alam**
Department of Electrical Engineering
College of Engineering
Riyadh, Saudi Arabia

**S. Arif Ali**
Department of Electronics Engineering
Aligarh Muslim University
Aligarh, India

**S. J. Ali**
Department of Computer Engineering
King Khalid University
Abha, Saudi Arabia

**Mehboob-ul-Amin**
Department of Electronics and
  Instrumentation Technology
University of Kashmir
Srinagar, India

**Tanzeela Ashraf**
Department of Electronics and
  Instrumentation Technology
University of Kashmir
Srinagar, India

**Mohammed Nasim Faruq**
Department of Electrical and Computer
  Engineering
North South University
Dhaka, Bangladesh

**Gurjot Singh Gaba**
Department of Wireless Communications
Lovely Professional University
Punjab, India

**Bhasker Gupta**
Department of Electronics and
  Communication Engineering
Chandigarh College of Engineering &
  Technology (CCET)
Punjab, India

**T. Ijyas V. P.**
Department of Electrical Engineering
King Khalid University
Abha, Saudi Arabia

**Asif Alam Joy**
Department of Electrical and Computer
  Engineering
North South University
Dhaka, Bangladesh

**Lavish Kansal**
Department of Electronics and
  Communication Engineering
Lovely Professional University
Punjab, India

**Sadaf Ajaz Khan**
Department of Electronics and
  Instrumentation Technology
University of Kashmir
Srinagar, India

**Sumit Kundu**
Department of Electronics and
  Communication Engineering
National Institute of Technology
Durgapur, India

**M. Rezwanul Mahmood**
Department of Electrical and Computer
  Engineering
North South University
Dhaka, Bangladesh

**Mohammad Abdul Matin**
Department of Electrical and Computer
 Engineering
North South University
Dhaka, Bangladesh

**Priyanka Mishra**
Department of Electronics and
 Communication Engineering
Noida International University
Uttar Pradesh, India

**Mohamed Mounir**
Communication and Electronics
 Department
El-Gazeera High Institute for
 Engineering and Technology
Cairo, Egypt

**M. Mzyece**
Graduate School of Business
 Administration (Wits Business School)
Wits University
Johannesburg, South Africa

**C. E. Ngene**
Department of Electrical and
 Electronics Engineering Science
Auckland Park Kingsway Campus
University of Johannesburg
Johannesburg, South Africa

**M. Parayangat**
Department of Electrical Engineering
King Khalid University
Abha, Saudi Arabia

**Hamzah Ali Abdul Rahman Qasem**
Department of Computer Engineering
Aligarh Muslim University
Aligarh, India

**Sanjay Dhar Roy**
Department of Electronics and
 Communication Engineering
National Institute of Technology
Durgapur, India

**M. Z. Shamim**
Center for Artificial Intelligence
King Khalid University
Abha, Saudi Arabia

**Akanksha Sharma**
Department of Electronics and
 Communication Engineering
Lovely Professional University
Punjab, India

**Shashibhushan Sharma**
Department of Electronics and
 Communication Engineering
National Institute of Technology
Durgapur, India

**Javaid A. Sheikh**
Department of Electronics and
 Instrumentation Technology
University of Kashmir
Srinagar, India

**Ghanshyam Singh**
Department of Electrical and
 Electronics Engineering Science
Auckland Park Kingsway Campus
University of Johannesburg
Johannesburg, South Africa

**Prabhat Thakur**
Department of Electrical and
 Electronics Engineering Science
Auckland Park Kingsway Campus
University of Johannesburg
Johannesburg, South Africa

**M. Usman**
Department of Electrical Engineering
King Khalid University
Abha, Saudi Arabia

**M. Wajid**
Department of Electronics Engineering
Aligarh Muslim University
Aligarh, India

# 1 Technology Evolution of Wireless Communications
## A Survey and Look Forward

M. Usman
King Khalid University

M. Wajid
Aligarh Muslim University

## CONTENTS

1.1 Introduction ..................................................................................................1
1.2 Historical Background and Evolution of Wireless Systems .......................2
1.3 Application Scenarios of Next Generation Wireless Systems ....................5
1.4 Requirements of Next Generation Wireless Systems ..................................6
1.5 Need for 5G and Beyond ..............................................................................7
1.6 Enabling Technologies of Next Generation Wireless Systems ...................7
    1.6.1 Spectrum .............................................................................................8
    1.6.2 mmWave and Terahertz Spectral Bands .........................................8
    1.6.3 Visible Light Communication (VLC) ..............................................9
    1.6.4 Massive MIMO (Large-Scale MIMO) ............................................9
    1.6.5 Ultra-Dense Small-Cell Networks .................................................10
    1.6.6 Network Slicing ...............................................................................10
    1.6.7 Artificial Intelligence .......................................................................10
1.7 Conclusions .................................................................................................11
References ...........................................................................................................11

## 1.1 INTRODUCTION

Communication networks and systems have been evolving at a rapid pace over the last decade. Ever since their inception, their scope and utility have also evolved tremendously – from basic speech communication to short texts to high-speed Internet connectivity to multimedia applications. At the heart of this evolution has been a transition from a carrier-centric approach toward a user-centric approach. The next generation of wireless systems will be a heterogeneous mix of multiple technologies, which will not just be competing with but also complementing one another, providing end users immersive experiences in real time along with innumerable applications. To provide such experiences and utility, the next generation of communication systems must satisfy stringent requirements such as high speed (10's of gigabits per second), submillisecond latency, and high density of connected devices while ensuring high quality of experience (QoE) for end users.

The next generation of wireless systems will be a heterogeneous mix of several communication technologies, such as millimeter wave (mmWave) communication, terahertz (THz) communication, and visible light communication (VLC), in addition to evolved versions of existing systems, which are expected to complement and cooperate with one another. Several wireless systems are emerging as candidate technologies for satisfying the requirements of next generation communication systems, which are required to provide connectivity not just in the traditional sense, but also in the rapidly emerging area of Internet of Things (IoT). To fulfill the requirements of next generation communication systems many different techniques need to be implemented. These techniques which fall into the scope of signal processing, coding theory, spectrum management, multiantenna systems (MIMO and massive MIMO), cloud computing, artificial intelligence, machine learning/deep learning, and security will be the enablers for future wireless systems.

## 1.2 HISTORICAL BACKGROUND AND EVOLUTION OF WIRELESS SYSTEMS

Subscriber numbers of mobile and wireless communication systems have witnessed enormous growth since their introduction. After the launch of Nordic Mobile Telephone (NMT), the first-generation analogue mobile system, the first digital cellular network was Advanced Mobile Phone Service (AMPS). Global System for Mobile (GSM) Communications was launched in 1991 and cellular subscriber numbers grew to 100 million in 1998 with figures reaching one billion in 2002 and two billion in 2005. By the end of 2007, cellular subscriptions exceeded three billion and network coverage extended over 80% of the world population. Mobile communication is therefore considered to be the enabler to bridge the so-called Digital Divide between the nations of the world. (Usman 2015). Such growth was possible due to mobile telephony and a vast array of affordable data services, enabled with the launch of packet-switched systems such as General Packet Radio Service (GPRS) in 2000 followed by 3G systems such as Wideband Code Division Multiple Access (WCDMA) in 2001. The 3GPP (Third Generation Partnership Project) family of standards gave a clear, cost-efficient road map from basic GSM voice and GPRS data services to true mobile broadband services based on Enhanced Data rates for GSM Evolution (EDGE) and WCDMA/HSPA technologies.

Second-generation (2G), systems which were based on time division multiple access (TDMA) offered low-rate data services in addition to conventional speech communication. GSM was the most popular 2G system holding 81.2% of global market share of active digital mobile subscriptions. GSM was used in most parts of the world except in Japan, where Personal Digital Cellular (PDC) was the 2G system used. GPRS, the packet-switched solution to GSM, provided data rates of up to 20

# Technology Evolution of Wireless Communications

kbps per time slot. By using multiple time-slots per user in the downlink, attractive services were offered. GPRS was an important evolutionary step in realizing the idea of always connected, ubiquitous Internet with multimedia and real-time traffic. GPRS evolved to EDGE, which provided data rates three times that of GPRS. EDGE utilized higher-order modulation together with Link Adaptation (LA) and Incremental Redundancy (IR) to improve the data rate. (Usman, 2015). Such has been the rapid pace of development of wireless communication systems, that pre-3G systems which were state of the art only a couple of decades back, seem ancient in the present times.

While mobile communication systems continued to evolve and grow, the Internet, which brought multimedia applications, also developed at a staggering rate. The convergence of wireless systems and the Internet became necessary to realize mobile data communications that would enable new services such as location-based services. Such services are not meaningful in a fixed network, and the first attempt to realize such a convergence was 3G, referred to as Universal Mobile Telecommunication System (UMTS). Broadband connectivity helped make the Internet a richer experience with a higher utility factor. Much of the developed world relied on fixed-line telecom networks to deliver broadband services, but emerging markets leapfrogged to using new mobile-based broadband technologies. High Speed Packet Access (HSPA), an upgrade to UMTS has by far been the most successful technology to deliver mobile broadband services. With around 900 million mobile broadband subscribers in 2012, more than 70% were being served by HSPA networks (Ericsson 2007). The mobile broadband system evolving from CDMA-2000 is called CDMA EV-DO (Evolution – Data Only). HSPA had the advantage that it could be built using the existing GSM networks and was a software upgrade of installed WCDMA networks. Another technology that competed to provide mobile broadband is IEEE 802.16e mobile Worldwide Interoperability for Microwave Access (WiMAX). HSPA and mobile WiMAX both used similar techniques to satisfy certain criteria required of mobile broadband systems – high data rates, low latency, good quality of service (QoS), good coverage, and high capacity. While HSPA and mobile WiMAX have comparable performance in certain areas such as peak data rates and spectral efficiency, the coverage range of HSPA was superior. Mobile WiMAX therefore, didn't have much technology advantage over HSPA and HSPA became the clear choice for mobile broadband services (Ericsson 2009). A major challenge is to keep the price of mobile broadband services low by making it imperative to utilize the available resources efficiently. Spectrum is a scarce and expensive resource in wireless communications and wireless channels are also the most challenging part of the radio network due to their inherent characteristics.

A rich variety of services has become possible with the advent of mobile broadband. Broadcast/multicast of high-resolution multimedia content, real-time streaming, virtual reality, online multiplayer gaming, real-time services in the cloud, and Internet of Things are examples of such services. Some of these service impose stringent requirements on the wireless communication systems in terms of data rate, latency, user density, and mobility to satisfy the end user's QoS) and QoE. To cater to these requirements the transition from 3G to 4G was made via an intermediate step called HSPA+, which could be referred to as 3.5G/3.75G. HSPA+ is an incremental upgrade to existing 3G networks as a migratory path toward achieving 4G speeds.

Average real-world download speed of 3G is around 2–3 Mbps while that of HSPA+ is around 6 Mbps with upload speeds of 0.4 Mbps and 3 Mbps, respectively ("How Fast Is 4G?" 2020). Of course, these vary across operators. The underlying technology in 3G UMTS is Wideband Code Division Multiple Access (WCDMA) and frequency division duplexing (FDD) for downlink and uplink transmissions. HSPA+ is also based on WCDMA but uses additional technologies – higher-order modulation schemes such as 16-quadrature amplitude modulation (QAM) and 64-QAM, multiple-input multiple-output (MIMO) schemes with beam-forming techniques with a further improvement to support dual-carrier (dual-cell) implementation (Rysavy 2006).

Although HSPA+ delivered significantly higher bit rates than 3G technologies, the opportunity remained for wireless operators to capitalize on the ever-increasing demand for wireless broadband, even lower latency, and multimegabit throughput. The solution is LTE (3GPP Long Term Evolution), the next generation network beyond 3G, which needs to satisfy the International Mobile Telecommunications – Advanced (IMT – Advanced) specifications to qualify as 4G technology. LTE encompasses the pillars of next generation networks:

- Broadband wireless as the new access reality with download speed of at least 100 Mbps, upload speed of at least 50 Mbps, and latency of less than 10 milliseconds.
- Achieving the aforementioned data rate even at high mobility (even traveling in cars and trains up to 350 km/h)
- Convergence of technology and networks.
- Technology shift to all-IP (Internet protocol).
- Download date rate of 1 Gbps for stationary and pedestrian users.
- Support device-to-device (D2D) or machine-to-machine (M2M) communication in which the transmitter and receiver are both moving (even relative to each other).

Two key enabling technologies that enabled "beyond-3G" networks to achieve their objectives are orthogonal frequency division multiple access (OFDMA) and MIMO, the combined use of which improved spectral efficiency and capacity of wireless networks (Nortel Networks, 2008).

While several operators marketed HSPA+ as 4G, it could be designated as 3.75G at most. The first commercial launch of 4G LTE was in 2009 with OFDMA as the underlying technology, and it was completely IP based. Typical real-world download and upload speeds of 4G LTE are 20 Mbps and 10 Mbps, respectively. A trend that could be noticed in the download and upload speeds across different generation of wireless systems is the asymmetry between download and upload speeds, which became smaller and smaller moving toward newer generations of wireless systems. While 4G LTE does not satisfy the requirements of IMT – Advanced, a consensus was reached in 2010 to nevertheless consider LTE as 4G, based on other factors.

Some of the key technologies that help 4G achieve its performance metrics are MIMO-OFDM, frequency domain equalization techniques, transmit/receive and

# Technology Evolution of Wireless Communications

spatial diversity, smart antennas with beam-forming, IPv6 support, advanced and adaptive modulation and coding schemes, software-defined radio (SDR) to support diverse wireless standards, and multiuser MIMO (MU-MIMO). The initial version of LTE supported 2x2 MIMO for both downlink and uplink and later versions extended support for up to 4x4 MIMO (downlink). The number of antennas on the user equipment (UE) is constrained by the size of the devices. To exploit some of the benefits of MIMO, the antennas need to be spaced half a wavelength apart. The main LTE bands are centered around 1.8, 2.1, and 2.6 GHz. At these frequencies, a half-wavelength ranges from 5 to 8 cm, which should be the separation between antennas to get good spatial separation of the signals.

While LTE performance is significantly improved compared to 3G-HSPA, it still does not satisfy the requirements of IMT-A. An enhanced version of LTE called LTE-Advanced (LTE-A) was released in 2011, which achieves performance closer to IMT-A requirements. The download and upload speed on real-world LTE-A deployments are 42 Mbps and 25 Mbps, respectively, which still fall short of IMT-A requirements. LTE-A improves on LTE by adopting 8x8 MIMO for downlink, 4x4 MIMO for uplink, carrier aggregation, and higher-order modulation up to 256 QAM. LTE-A further supports heterogeneous networks with macrocells, picocells, and femtocells configuration. The limitations of LTE-A (for the kind of applications being envisioned) are latency, which averages 50 ms (can be around 5–10 ms in best-case scenario), and user density (2,000 devices per sq. km). A summary of performance metrics for different wireless systems is listed in Table 1.1 ("4G LTE Advanced," 2020). It should be noted that these are metrics for best-case scenario and are poorer for typical real-world scenarios.

It becomes clear from the previous discussion that a host of different technologies act as enablers in approaching the performance requirements of wireless systems to support a variety of different applications for end users.

## 1.3 APPLICATION SCENARIOS OF NEXT GENERATION WIRELESS SYSTEMS

For applications envisioned such as virtual reality (VR) based immersive experience, real-time services, autonomous driving, connected vehicles, Internet of Things (IoT), e-Health, video surveillance, cloud-based applications and data analytics, smart cities, smart homes, smart factories (Industry 4.0), smart offices, and smart "everything," the performance metrics delivered up to 4G are not sufficient. In order to cater

### TABLE 1.1
**Performance metrics (best case) of different wireless systems**

| Performance metrics | WCDMA (UMTS) | HSPA | HSPA+ | LTE | LTE-A |
|---|---|---|---|---|---|
| Max. downlink data rate | 384 kbps | 14 Mbps | 28 Mbps | 100 Mbps | 1 Gbps |
| Max. uplink data rate | 128 kbps | 5.7 Mbps | 11 Mbps | 50 Mbps | 500 Mbps |
| Latency | 150 ms | 100 ms | 50 ms | 10 ms | <5 ms |
| Access technology | CDMA | CDMA | CDMA | OFDMA | OFDMA |

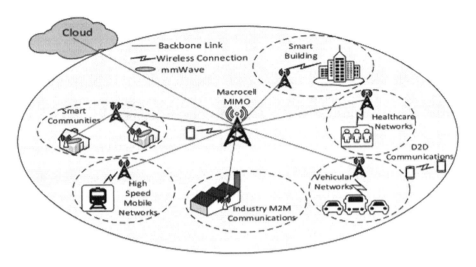

**FIGURE 1.1** Application scenarios of next generation wireless communications.

to the requirements of future applications, the current generation of wireless systems needs to be enhanced and radical new technologies are required to address the requirements, constraints, challenges, and applications of next generation wireless systems. This paves the road to 5G and beyond (Figure 1.1).

## 1.4 REQUIREMENTS OF NEXT GENERATION WIRELESS SYSTEMS

The requirements for 5G set forth by the International Telecommunications Union (ITU) to support future applications are ambitious: 20 Gbps per cell up to 1 Gbps per user, submillisecond latency, and the ability to serve in excess of 1 million devices per sq. km ("The Forces of 5G – The Next Generation of Wireless Technology Is Ready for Take-Off," 2018). To get a perspective of why such requirements on performance metrics are necessary, consider the following example on latency requirement.

The reaction time of a human between detecting a hazard and pressing the brake pedal is 1 second. An automobile doing 100 km/h travels around 28 m in that time. For an autonomously driven vehicle, this would be too late. An autonomous vehicle, which can analyze data with a latency of 1 millisecond, reacts 1,000 times faster than a human, and the distance covered between the instant of hazard detection and application of brakes is under 1 centimeter. For such applications the latency requirement is submillisecond as a guarantee rather than a best-case scenario. Current 5G deployments claim to have achieved a download speed of 1 Gbps, but the average download and upload data rates in typical real-world scenarios are 200 Mbps and 100 Mbps, respectively. The average real-world latency is about 21–26 ms, but expected to get below 1 ms in the future. ("How Fast Is 4G?" 2020). Many of these features may not be required for end users, as most application requirements of end users, such as

# Technology Evolution of Wireless Communications

streaming, videoconferencing, and online gaming, are satisfied by 4G networks. The real value of 5G will be realized in automation, smart cities, IoT applications, etc. The evolution from 5G to 6G imposes even more stringent requirements such as (NTT-Docomo, 2020):

- Extremely high capacity – exceeding 100 Gbps.
- Extremely low latency – end-to-end latency to be always less than 1 ms (not just for best-case scenario).
- Extremely massive device density – 10 million devices per sq. km with location-based services accurate in the order centimeters.
- Extremely high coverage – land, sky (up to 10,000 m altitude), sea (200 nautical miles from the coast) with Gbps data rate everywhere.
- Extremely high reliability – guaranteed QoS with high level of security.
- Extremely low energy – long battery life of devices and even devices free from charging requirements.
- Extremely low cost – affordable end-user devices as well as network components.

## 1.5 NEED FOR 5G AND BEYOND

5G has had its share of controversies, in particular, with respect to health hazards due to electromagnetic radiations. Advocacy groups opposed to 5G cite the disappearance of fauna in 5G coverage areas. While several studies have investigated the adverse effects of electromagnetic radiations of wireless systems, no concrete scientific evidence of adverse effects has been found that would stymie the rollout of 5G networks.

In an unexpected twist of events, the situation due to COVID-19 pandemic, with all its disastrous consequences, has become an unlikely proponent for speedy rollout of 5G networks across the world. Due to the pandemic, people are having to work from home, online education and e-learning have become the order of the day, online shopping has increased manifold and in many places it's the only means of purchasing even basic daily necessities, e-health is gaining traction, and with lockdown imposed in several parts of the world, people are staying in touch with family and friends via videoconferencing. People are also having to go online for leisure and entertainment. All of this has caused an order of magnitude increase in the global data traffic demand, saturating existing 4G network capacities, thereby making 5G (and beyond), which has improved capabilities in terms of capacity and device density as compared to 4G, all the more necessary.

## 1.6 ENABLING TECHNOLOGIES OF NEXT GENERATION WIRELESS SYSTEMS

Several different technologies will need to work together to fulfill the requirements of 5G and beyond 5G networks. In addition to technologies used in 4G or their advanced versions, the following techniques are being explored, some of which have already been adopted as candidate technologies for future wireless systems.

### 1.6.1 SPECTRUM

Conventionally, spectrum is considered to be a scarce and expensive resource. Existing terrestrial wireless systems typically utilize frequencies up to about 6 GHz. Legally,

> Terrestrial wireless system(s) means any terrestrial wireless communication system or equipment not incidental to a space-based commercial satellite communication system and any service provided using such a system or equipment. For the avoidance of doubt, terrestrial wireless system(s) specifically includes any equipment compatible with air interfaces or standards/protocols associated with any of the following terrestrial wireless communication systems: IS-95 (CDMA), IS-136 (US TDMA), GSM, WCDMA, CDMA2000, CDMA EVDO, iDEN systems, GPRS, UMTS, WiMax, LTE, IEEE 802.xx (including 802.16 and 802.11), OFDM/OFDMA based cellular communication systems, and Land Mobile Radio, including P25, DMR, dPMR, and TETRA, and future generations or evolutions of such systems.
> 
> ("Definition of Terrestrial Wireless System(S)," 2020)

The spectrum beyond 6 GHz is vastly vacant and not utilized primarily due to lack of technological capabilities. Thus scarcity of spectrum arises mainly due to limitations of existing technologies, which includes electronics, antennas, etc. to handle and process signals at these frequencies. The characteristic requirements of 5G and beyond 5G networks necessitate the use of higher frequencies to accommodate higher-order MIMO configurations called massive MIMO (much higher than the 8x8 MIMO supported in 4G networks), support wide channel bandwidths up to 20 GHz as compared to 20 MHz supported in LTE-A, improve beam-forming to reduce interference, and also support several types of antenna arrays – such as linear, planar, spherical, and cylindrical with arbitrary polarization (Gyasi-Agyei, 2020).

### 1.6.2 MMWAVE AND TERAHERTZ SPECTRAL BANDS

Frequencies in mmWave band, i.e., 30–300 GHz, have wavelengths of 10 mm to 1 mm and terahertz band extending up to 10 THz, also called the sub-mm band, as the wavelength is often less than 1 mm considered for the next generation of wireless communication systems. For the initial 5G releases and impending launches, the mmWave spectrum standardized by ITU extends from 24.25 to 52.6 GHz. Terahertz frequencies are likely to be utilized in 6G networks, which are expected to be deployed by 2030. Frequency band ranging from 95 GHz to 3 THz has been recommended by the U.S. FCC and is being investigated to achieve a data rate in excess of 100 Gbps (NTT-Docomo, 2020). Starting from 4G, radio access technologies (RATs) have been based on OFDM, which have already achieved performance close to the Shannon limit. Radically different and novel RATs need to be developed for the next generation of wireless systems, and as of now, there is no clarity on the definition of RAT for 6G.

### 1.6.3 Visible Light Communication (VLC)

To achieve the high data rate, high bandwidth, high security, low latency, and low interference requirements of future wireless systems, the visible range of electromagnetic spectrum is also being considered for 5G and beyond 5G systems. Communication in this band, which extends from 430 to 790 THz (wavelength 750–380 nm), is called VLC. The replacement of florescent lamps with light-emitting diodes (LEDs) for lighting in homes, offices, and other premises along with the development of solid-state technology for high-speed (fast response) LEDs and photodetectors, enables the doubling up of lighting infrastructure as communication infrastructure as well. VLC also addresses the health hazards associated with radio frequencies (RF). Communication protocols and technologies (Usman and Al-Rayif, 2018) for VLC are being investigated and developed by several research groups, and high-speed prototype VLC systems have been demonstrated. VLC systems are also called light fidelity (LiFi) systems, a term introduced by Harald Haas of the University of Edinburgh, a pioneer in the research on LiFi. As light does not penetrate through opaque objects such as brick walls and metals, LiFi signals cannot travel outside the room where the transmitter is placed. This makes VLC particularly suited for communication in aircraft cabins, hospitals, military zones etc., where prevention of interference with flight electronic systems, medical equipment, and security is of utmost importance. Another advantage of VLC is that it operates in nonlicensed bands (Khan, 2017).

### 1.6.4 Massive MIMO (Large-Scale MIMO)

While multiple-antenna technology has matured over the years and has been incorporated into 4G networks that support up to 8x8 MIMO configurations, massive MIMO uses a very large number of antennas in the hundreds or even thousands. Some of the advantages of having a very large number of antennas are as follows.

*Capacity gain* – Channel capacity increases proportional to the number of antennas (smaller of the number of transmit and receive antennas). This is called as multiplexing gain, achieved by transmitting a different stream of data from each antenna.

*Diversity gain* – The same data stream is transmitted over each of the transmit antennas in an orthogonal or near orthogonal manner. This exploits multipath fading to its advantage, thereby improving the reliability of the received signal. With a large number of transmit and receive antennas, diversity gain improves with the number of antennas. Of course, there is a trade-off between multiplexing gain and diversity gain with massive MIMO just as in MIMO.

*Improved beam-forming* – Having a large number of antennas allows directing the transmitted signal to very narrow regions in space, referred to as pencil beam-forming, mitigating the effects of interference.

Massive MIMO also offers other advantages such as better energy efficiency, lower latency, and better immunity to intentional jamming. Massive MIMO has opened up several new research problems that were not relevant in small-scale MIMO systems ("Massive MIMO: News, Commentary, Myth-Busting," 2020). To implement massive MIMO, the size of antennas has to be made small to accommodate a large number of antennas onto a single device, as mobile devices have limited real estate. Smaller antenna dimensions also complement communication at higher-frequency bands such as mmWave and THz bands as antenna dimensions are required to be proportional to wavelength (or some fraction of it) for efficient radiation.

### 1.6.5 Ultra-Dense Small-Cell Networks

The higher-frequency bands, i.e., mmWave and THz, being considered for next generation wireless suffer from larger attenuation or path loss. This is due to higher free space loss as well as atmospheric absorption and blockage by walls and thick foliage, as compared to lower frequencies used in existing systems up to 4G. This limits the useful range of these frequency bands to about 1 km, and real-world deployments have even smaller range (100–400 m) due to the typical power densities used in practical networks. Thus, ultra-dense small-cell networks are required for network deployments operating at these frequencies. Careful design of such small-cell networks is important, as initial deployments of 5G will coexist with existing macro-cell 4G networks for several years. While stand-alone mmWave small cells may be present in some locations, most 5G mmWave small cells will overlay existing 4G macro cells at least in the initial years of 5G rollout. The 5G small cells will cater to higher capacity requirements, while the 4G macro cells will cater to coverage and mobility requirements (Athanasiadou et al., 2020). It is therefore important to carefully plan 5G network deployment to maximize efficiency and minimize cost.

### 1.6.6 Network Slicing

Another important feature in 5G and beyond is network slicing, which allows virtualization for optimal utilization of physical resources by various applications and services. This allows that each user/service/application gets only the physical resources it needs, thereby maximizing energy efficiency and minimizing cost. Each user/service/application is allocated resources based on its requirements, thus allowing all of them to operate on the same physical hardware. In other words, network slicing is the multiplexing of several virtual networks onto the same physical network, where each virtual network caters to the end-to-end requirements of a service or application (Foukas et al., 2017; Zhang et al., 2017). This is a paradigm shift from the one-size-fits-all concept in earlier and existing wireless networks to a tailored fit depending on the requirements of each service/application.

### 1.6.7 Artificial Intelligence

Artificial intelligence (AI) has matured considerably over the last decade and is providing solutions to previously intractable problems. AI will be a vital technology component in the next generation of wireless systems that will address many of the expected problems in the development of beyond 5G networks. These networks are being designed to support an extremely large number of connected devices that will generate an extremely large amount of data – data that will predominantly be unstructured. Future networks will have a dense network architecture, with each base station site involving a large number of antennas (massive MIMO), which make channel modeling, mobility management, optimization of physical resource utilization, etc. an extremely challenging task using conventional methods. AI techniques such as machine learning (ML) and deep learning (DL) have the capabilities of handling such tasks efficiently and reliably by dynamically adapting to varying scenarios and predicting the behavior of networks and users in real time. The relation between various network settings and parameters that needs to be determined and optimized is expected to be highly nonlinear. Simultaneous optimal configuration of these settings with adaptability in real time depending on the end user/application/service requirements can be performed using sophisticated AI algorithms. Some of the potential use cases of AI in beyond 5G networks are channel modeling, channel sensing, channel estimation, signal processing with massive MIMO, network configuration, management and optimization, and network slicing design (Wang et al., 2020).

## 1.7 CONCLUSIONS

In this chapter, an introduction to the evolution of wireless systems starting from the first-generation systems to present-day 4G/5G systems is provided. The evolution beyond 5G has also been discussed. The requirements, features, applications, and underlying technologies that have been and are the enablers of each generation of wireless systems are presented. Current generation systems as well as the next generation of wireless systems require a host of different technologies working together to fulfill the demands and requirements of these networks. While an overview of some of the important features, requirements, and enabling technologies has been presented, the list is neither complete nor comprehensive. The next generation of wireless systems will be such a complex concoction of several disparate technologies cooperating and collaborating with each other that it is neither possible nor justifiable to present all of them without a context in a single book, let alone a chapter.

In the chapters that follow, a discussion of some of these enabling technologies, application scenarios, etc. of the next generation of wireless systems, in some limited context is presented.

## REFERENCES

4G LTE Advanced. 2020. Electronics Notes. https://www.electronics-notes.com/articles/connectivity/4g-lte-long-term-evolution/what-is-lte-advanced.php.

Athanasiadou, Georgia E., Panagiotis Fytampanis, Dimitra A. Zarbouti, George V. Tsoulos, Panagiotis K. Gkonis, and Dimitra I. Kaklamani. 2020. "Radio Network Planning towards 5g mmWave Standalone Small-Cell Architectures." *Electronics (Switzerland)* 9: 1–10. doi:10.3390/electronics9020339

Definition of Terrestrial Wireless System(S). 2020. *Law Insider*. https://www.lawinsider.com/dictionary/terrestrial-wireless-systems.

Ericsson. 2007. "Long Term Evolution (LTE): An Introduction." *Ericsson Whitepaper*.

Ericsson. 2009. "Technical Overview and Performance of HSPA and Mobile WiMAX." *Ericsson Whitepaper*.

Foukas, Xenofon, Georgios Patounas, Ahmed Elmokashfi, and Mahesh K. Marina. 2017. "Network Slicing in 5G: Survey and Challenges." *IEEE Communications Magazine* 55 (5): 94–100. doi:10.1109/MCOM.2017.1600951

Gyasi-Agyei, Amoakoh. 2020. *Wireless Internet of Things – Principles and Practice*. Singapore: World Scientific. doi:10.1142/11308

How Fast Is 4G? 2020. 4G.Co.UK. https://www.4g.co.uk/how-fast-is-4g/.

Khan, Latif Ullah. 2017. "Visible Light Communication: Applications, Architecture, Standardization and Research Challenges." *Digital Communications and Networks* 3 (2): 78–88. doi:10.1016/j.dcan.2016.07.004.

Massive MIMO: News, Commentary, Myth-Busting. 2020. FP7 Project MAMMOET. http://www.massive-mimo.net/.

Nortel Networks. 2008. "Long-Term Evolution (LTE): The Vision beyond 3G." Nortel Networks White Paper.

NTT-Docomo. 2020. "White Paper – 5G Evolution and 6G."

Rysavy, Peter. 2006. "Mobile Broadband – EDGE, HSPA and LTE." www.rysavy.com.

The Forces of 5G – "The Next Generation of Wireless Technology Is Ready for Take-Off." 2018. *The Economist*. https://www.economist.com/business/2018/02/08/the-next-generation-of-wireless-technology-is-ready-for-take-off.

Usman, Mohammed. 2015. *Fountain Codes for Mobile Wireless Channels*. Saarbrücken: LAP Lambert Academic Publishing.

Usman, Mohammed, and Mohammed Ibrahim Al-Rayif. 2018. "Threshold Detection for Visible Light Communication Using Parametric Distribution Fitting." *International Journal of Numerical Modelling: Electronic Networks, Devices and Fields* 31 (3): 1–10. doi:10.1002/jnm.2305

Wang, Cheng Xiang, Marco Di Renzo, Slawomir Stańczak, Sen Wang, and Erik G. Larsson. 2020. "Artificial Intelligence Enabled Wireless Networking for 5G and Beyond: Recent Advances and Future Challenges." *IEEE Wireless Communications* 27 (1): 16–23. doi:10.1109/MWC.001.1900292

Zhang, Haijun, Na Liu, Xiaoli Chu, Keping Long, Abdol Hamid Aghvami, and Victor C.M. Leung. 2017. "Network Slicing Based 5G and Future Mobile Networks: Mobility, Resource Management, and Challenges." *IEEE Communications Magazine* 55 (8): 138–145.doi:10.1109/MCOM.2017.1600940

# 2 Enabling Technologies and Enabling Business Models for Next Generation Wireless Communications

M. Mzyece
Wits University

## CONTENTS

2.1 Introduction ........................................................................................................ 13
2.2 Main Contributions and Related Works ........................................................ 15
2.3 Enabling Technologies and Enabling Business Models for Wireless Communications .............................................................................................. 17
2.4 Assessment of Enabling Technologies and Enabling Business Models for Previous-Generation, Current-Generation, and Emerging-Generation Wireless Communications ................................................................................ 17
2.5 Assessment of Enabling Technologies and Enabling Business Models for Next Generation Wireless Communications .......................................... 22
2.6 Integrated Framework for Enabling Technologies and Enabling Business Models for Next Generation Wireless Communications ............ 25
2.7 Conclusions ........................................................................................................ 28
Notes ............................................................................................................................ 29
References .................................................................................................................... 29

## 2.1 INTRODUCTION

This chapter explores the connections between enabling technologies and enabling business models for next generation wireless communications. A business model defines the explicit or implicit architecture and design of organizational mechanisms and processes used to create, deliver, and capture value (Teece 2010). At the heart of all business models are value propositions: the bundles of products and services that provide value for specified customer segments (Osterwalder and Pigneur 2010). Each value proposition is a specific set of benefits that designated customers derive from the particular products and services offered to them by organizations. Business model innovation is about creating, delivering, and capturing such value in *novel ways*, where the novelty can be in how benefits are generated, how benefits are delivered to customers, how benefits are monetized, or how organizational resources are built and used to generate, deliver, or monetize benefits to customers (Afuah 2018).

Technical research on telecommunications generally and wireless communications specifically tends to focus almost exclusively on technology-related issues; in-depth considerations of business model–related issues are comparatively rare. For instance, Huurdeman (2003), in his comprehensive worldwide overview of telecommunications covering the period 1750–2000, delves into various key technological, historical, and policy issues to explain the rich history of the global telecommunications industry. However, he says relatively little about the business model aspects of this history, even though they are clearly relevant in such a market-oriented industry. Similarly, Molisch (2011) very ably surveys wireless communications in carefully selected technical and technological detail, but, once again, with very little regard for how all the technicalities and technologies are connected to wireless communications business models. These are just two well-known sources that are representative of the larger corpus of published technical works in the field. It is worth noting that this is only an observation of an important omission in the literature, which this chapter will attempt to begin to fill. It is *not* a criticism of these two excellent works on that basis as such. After all, no single research work that aims to be doable and useful can – nor should – attempt to be universal in its scope. Delimiting the research scope is essential for both tractability and utility.

Several classic real-world examples demonstrate the importance of considering enabling technologies alongside enabling business models. In the PC era, IBM's adoption and design of open architecture personal computers in the early 1980s had a profound long-term impact on the commercial and strategic evolution of the computing industry (Greenstein 1998). In the Internet era (the mid- to late-1990s onward), the development of Google and earlier search engines was underpinned by the sponsored search business model (Jansen and Mullen 2008). In the smartphone era (2007 onward), the dominant rise of Apple's iPhone technology ecosystem was based on the reverse razor/blade business model, whereby demand for relatively low-margin "blades" (music, mobile apps, and other digital content) drove sales and usage of high-margin "razors" (iPhones) (Ovans 2015).

Among the best real-world examples of this phenomenon in wireless communications specifically are prepaid ("pay-as-you-go") business models, which account for over 50% of the world's wireless subscribers, and historically over 90% of subscribers in most African countries as well as other parts of the world like Italy (Conroy and Hill 2009; Layton 2014). By contrast, otherwise technically promising wireless communications technologies such as wireless application protocol (WAP) (Mzyece 2001) ultimately failed because they lacked viable business models.

The technological and commercial development of wireless communications highlights the crucial role of business models and business innovation across all wireless communications generations. The prepaid business models of the 2G era used emerging technologies to fulfill customer requirements that could not be met by the postpaid (contract-based) business models of 1G wireless communications. Subsequently, new business models emerged in the 3G and 4G eras to support voice over IP (VoIP), over-the-top (OTT), and other new services, based on technological innovation and convergence with data communications networks like the Internet. Ongoing technological developments in 5G and beyond 5G wireless communications will continue to drive emerging and future services based on new business models.

For these reasons, this chapter seeks to explore the intersection and interaction of enabling technologies and enabling business models for next generation wireless communications. Section 2.2 outlines the chapter's main contributions and clearly contextualizes and differentiates them with respect to previous contributions in related works in the literature. Section 2.3 defines enabling technologies and enabling business models for wireless communications and examines the vital relationship between them. Section 2.4 focuses on enabling technologies and enabling business models for previous-generation (1G–3G), current-generation (4G), and emerging-generation (5G) wireless communications. Section 2.5 focuses on enabling technologies and enabling business models for next generation (beyond 5G) wireless communications. Section 2.6 proposes an integrated framework for enabling technologies and enabling business models for next generation wireless communications. Finally, Section 2.7 concludes the chapter by distilling key findings, offering recommendations, and highlighting potential areas for future research.

## 2.2 MAIN CONTRIBUTIONS AND RELATED WORKS

The purpose of this section is to describe this chapter's three main contributions in the context of related works from the literature. In summary, the chapter's three main contributions are

1. A holistic assessment of specific aspects of enabling technologies and enabling business models for previous-generation, current-generation, and emerging-generation wireless communications;
2. A holistic assessment of generic aspects of enabling technologies and enabling business models for next generation wireless communications; and
3. An integrated framework for enabling technologies and enabling business models for next generation wireless communications.

The first main contribution assesses intersections and interactions between *specific aspects* of enabling technologies and enabling business models in previous-generation, current-generation, and emerging-generation wireless communications. As noted in Section 2.2, studies that explicitly investigate the connections between wireless technologies and business models are relatively rare, but some do exist. (Camponovo and Pigneur 2003) applied business model analysis to the wireless communications industry of the early 2000s. They identified the industry's unique business characteristics (mobility, network effects, and proprietary assets) and classified the primary and secondary actors in the industry into five major categories: technology providers (device and network equipment vendors), services providers (content, applications, and payments services vendors), network providers (including mobile network operators [MNOs] and Internet service providers [ISPs]), regulation-related players (governments, regulatory authorities and standardization groups), and users (corporates, corporate groups, consumers, and consumer groups). Their analysis still has some validity at a high level in terms of certain major categories, but the exact roles and role players in each of the categories may have changed over time. Additionally, new categories have emerged, and some of the current actors occupy multiple categories. For example, Google is simultaneously a technology provider, a services provider, and a network provider in some scenarios. At roughly the same time (the early 2000s), Li and Whalley (2002) analyzed emerging business models in the telecommunications industry, but with a much wider focus than wireless communications and with less emphasis on underlying technological characteristics than is the case with this chapter's work. Their analysis is still insightful, although some of the key features of the then telecommunications industry that they emphasized have fundamentally changed. For instance, mobile portals are not as relevant as they were in the early 2000s. In the early 2010s, Dhar and Varshney (2011) examined a specific wireless communications use case and its associated business models, namely location-based services, but generalizability to other use cases and both previous and subsequent generations of wireless communications is somewhat limited. More recently, Yaghoubi et al. (2018) and Ahokangas et al. (2019) assessed the viability of various alternative business model scenarios for 5G transport networks and 5G micro operators, respectively, but they did not cover various other application scenarios or generations of wireless communications.

The second main contribution assesses intersections and interactions between *generic aspects* of enabling technologies and enabling business models in next generation wireless communications. Despite the growing number of publications on 6G (Dang et al. 2020; David et al. 2019; Giordani et al. 2020; Rappaport et al. 2019; Yuan et al. 2020), it is worth noting that because substantive research and standardization work on 6G has barely even begun, many specific technical details presented in these works are necessarily highly speculative. Furthermore, some scholars have argued that 5G may be the last generation of wireless communications in the conventional sense of a package of large-scale technological advancements delivered at a given point or period in time (Dohler 2018). Consequently, this chapter focuses instead on the most likely *generic aspects* of enabling technologies and enabling business models for next generation wireless communications. These generic aspects include *heterogeneous wireless technologies* (Bosch et al. 2020), *artificial intelligence and machine learning* (AI/ML) (Wang et al. 2020), *digital platforms* (Cusumano, Yoffie, and Gawer 2020), and *digital business models* (Cusumano, Yoffie, and Gawer 2020; Hanafizadeh, Hatami, and Bohlin 2019; Li 2020; Veit et al. 2014; Verhoef and Bijmolt 2019).

The third and final main contribution is an integrated framework for enabling technologies and enabling business models for next generation wireless communications. The proposed framework incorporates six key technology elements whose business implications are typically omitted from technical research: *technology creation* (Arthur 2007), *technology architecture and modularity* (Baldwin and Clark 2006; Clark 2018), *technology S-curves* (Scillitoe 2013), *technology cycles* (Anderson and Tushman 1990), *technology standardization* (Biddle 2017), and *technology profitability* (Teece 2018). The proposed framework relates these six key technology elements to a seventh element: *business model innovation* (Afuah 2018; Osterwalder and Pigneur 2010; Skog, Wimelius, and Sandberg 2018; Taran et al. 2016). The proposed framework differs significantly from existing frameworks. For example, unlike the lean start-up framework (Blank 2013), which applies an iterative process of business model hypothesis (the business model canvas), customer development, and agile product/service development to general business scenarios, the proposed framework applies the six key technology-related elements and the business model innovation element holistically to next generation wireless communications. While Glisic (2016) has used game theory to model some business aspects of advanced wireless networks, using game theoretic modeling necessarily restricts the scope and depth of how the related business scenarios can be investigated (MacKenzie and DaSilva 2006). By contrast, the proposed framework has no such constraints.

## 2.3 ENABLING TECHNOLOGIES AND ENABLING BUSINESS MODELS FOR WIRELESS COMMUNICATIONS

What are "enabling technologies" and "enabling business models" in wireless communications? A starting point for understanding these two concepts and terms is their common word *enabling*. This word can be used in several possible general senses: as the noun (i.e., name), adjective (i.e., description), or verb (i.e., action) form of the root word *enable*. The various specific meanings of the word *enable* include "to empower; to make adequate or proficient; to make competent or capable; to supply with the requisite means or opportunities to an end or for an object; to make possible; to give effectiveness to; to make operational" (OED Online 2020). All of these senses and meanings are broadly applicable in the context of this chapter.

More concretely, however, "enabling technologies" and "enabling business models" are simply those respective aspects of wireless communications that make it work, technically and commercially. Both aspects are essential and complementary in wireless communications. On the one hand, enabling technologies for wireless communications are the various technology principles, techniques, resources, components, devices, subsystems, systems, and standards that make wireless communications *technically feasible*. On the other hand, enabling business models for wireless communications describe how organizations individually and collectively use these enabling technologies and other organizational resources to create, deliver, and capture value, for, to, and from their customers *commercially in specific business environments and ecosystems*. The successful development and deployment of GSM provides an excellent case study of the symbiotic relationship between enabling technologies and enabling business models in wireless communications (Temple 2010).

## 2.4 ASSESSMENT OF ENABLING TECHNOLOGIES AND ENABLING BUSINESS MODELS FOR PREVIOUS-GENERATION, CURRENT-GENERATION, AND EMERGING-GENERATION WIRELESS COMMUNICATIONS

Table 2.1 compares enabling technologies and enabling business models across previous, current, and emerging generations of wireless communications, from

# TABLE 2.1
## Comparison of enabling technologies and business models across wireless generations

| Generation (years) | Services & devices | Data rates | Latency | Spectrum allocation(s) | Enabling technologies | Enabling business models |
|---|---|---|---|---|---|---|
| 1G (1970s-1980s) | - voice services<br>- fax<br>- car phones & large handsets | - 2.4 kb/s | - up to 400 ms for TACS handover latency alone | - 450 / 800 / 900 MHz | - TACS, AMPS, NMT, C450<br>- cells / frequency reuse<br>- analogue techniques<br>- FDMA | - technology-driven (e.g., Bell Labs)<br>- contracts/licensing (vendors)<br>- postpaid subscriptions (MNOs) |
| 2G (1990s) | - voice services<br>- text messaging (SMS)<br>- value-added services (VAS)<br>- handheld phones | - up to 9.6 kb/s (GSM), 171.3 kb/s (GPRS, 8 timeslots), 473.6 kb/s (EDGE, 8 timeslots) | - 300-1,000 ms | - 450 / 800 / 900 MHz<br>- 1.8 / 1.9 GHz | - GSM / GPRS / EDGE, IS-95, D-AMPS, PDC<br>- cells / frequency reuse<br>- digital techniques<br>- FDMA, TDMA, CDMA<br>- network virtualisation | - politically & commercially driven (e.g., EU & other governments, deregulation, new private MNOs)<br>- contracts/licensing (vendors)<br>- postpaid subscriptions (M(V)NOs)<br>- prepaid subscriptions (M(V)NOs) |
| 3G (2000s) | - voice services<br>- text messaging (SMS) multimedia messaging (MMS)<br>- Internet / data / mobile apps<br>- video<br>- handheld & portable devices | - up to 8-10 Mb/s (HSDPA) | - 100-500 ms | - 800 / 850 / 900 MHz<br>- 1.7 / 1.9 / 2.1 GHz | - UMTS, CDMA2000<br>- cells / frequency reuse<br>- digital techniques + all-IP<br>- W-CDMA<br>- network virtualisation<br>- managed services | - contracts/licensing (vendors, ISPs)<br>- postpaid subscriptions (M(V)NOs)<br>- prepaid subscriptions (M(V)NOs)<br>- free / freemium (Internet firms)<br>- disintermediation (hyperscale platforms) |

# Enabling Technologies and Enabling Business Models

| | | | | | |
|---|---|---|---|---|---|
| **4G** (2010s) | - voice services<br>- messaging (SMS/MMS)<br>- Internet / data / mobile apps<br>- video streaming<br>- OTT services (e.g., WhatsApp, Skype, Hangouts, Zoom, etc.)<br>- smartphones, wearables, smart speakers, tablets, notebooks, etc. | - up to 300 Mb/s (downlink) | - 50 ms | - 700 MHz<br>- 1.8 / 2.1 / 2.3 / 2.6 GHz | - LTE, LTE-Advanced, WiMAX +<br>Heterogeneous wireless networks (e.g., Wi-Fi)<br>- cells / frequency reuse<br>- digital techniques + all-IP<br>- OFDM, MIMO<br>- network virtualisation managed services | - contracts/licensing (vendors, ISPs, integrators, specialists, generalists)<br>- postpaid subscriptions (M(V)NOs)<br>- prepaid subscriptions (M(V)NOs)<br>- free / freemium (Internet firms)<br>- disintermediation (hyperscale platforms, OTT players)<br>- fees / ads (digital platforms) |
| **5G** (2020s) | generic use cases:<br>- eMBB l mMTC l URLLC<br>specific use cases (among others):<br>- Live ultra-HD video, AR/VR<br>- Internet of Things (IoT) & Industrial IoT (factories, etc.)<br>- connected/autonomous cars<br>- tactile Internet | - up to 20 Gb/s (downlink) | - down to 1 ms | - low bands (<1 GHz)<br>- mid bands (1-6 GHz)<br>- high bands (24-60 GHz) | - non-Standalone (NSA) & standalone (SA) 5G NR<br>- digital techniques + all-IP<br>- OFDM, massive MIMO, mmWave<br>- SDR/SDN l network slicing (NFV/VNF) l network orchestration & automation | Existing business models from the 4G era, plus the following emerging / experimental business models:<br>- micro operator models<br>- co-creation models<br>- partnership & ecosystem models<br>- brokerage models<br>- digital platform models |

*Source:* The author's own analysis and synthesis based on multiple sources, including research publications, reports, standards bodies, industry associations, websites, and various other references. For example, Huurdeman (2003), Temple (2010), Molisch (2011), Ahokangas et al. (2019), ITU, 3GPP, GSMA, etc.

first-generation (1G) wireless communications to fifth-generation (5G) wireless communications.

Several important insights can be deduced from Table 2.1. The first deduction that can be drawn is that generational developments in the services and technical characteristics of wireless communications, such as data rates, latencies, and spectrum allocations, have been closely linked with the features of the associated enabling technologies. The largest technological shift over the generations has been in the implementation of digital techniques and all-IP networks, which has in turn facilitated high-quality IP-based services and convergence with the Internet and other IP-based networks. This trend is set to continue in emerging 5G wireless communications, which will, accordingly, consist of heterogeneous, IP-based digital networks as the key enabling technologies. New 5G enabling technologies include software-defined radio (SDR) and software-defined networking (SDN), network slicing, and network orchestration and automation. Data, in its various forms, is arguably the single most important resource in this highly interconnected, IP-based network of networks. These 5G enabling technologies must therefore support a large variety of generic and specific use cases with different service quality characteristics, quality of service (QoS) metrics, and quality of experience (QoE) requirements. This constitutes a considerable network management challenge. It also interesting to note the evolution in devices used to access services across the generations from the cumbersome car phones and bulky handsets of 1G to the array of 5G wireless devices currently available in a wide range of form factors, sizes, and specifications.

The second deduction that can be made based on Table 2.1 is that, over the past four to five decades, as wireless communications has developed and become increasingly sophisticated in its services, technical features, and enabling technologies, so too have its enabling business models. In the 1G era of the 1970s and 1980s, when wireless communications was only used to provide mobile access to fixed-line telecommunications type services like voice or fax, the postpaid business model based on contract subscriptions was sufficient (and, for the most part, was all that was possible and available). From the 1990s onward, the prepaid business model based on pay-as-you-go subscriptions was created, and it drove the spectacular worldwide success of 2G wireless communications, especially based on the GSM/GPRS/EDGE standards. The current 5G wireless communications era is characterized by a plethora of different business models, some inherited from 4G and earlier wireless eras, others still emerging and being experimented with, including new digital platform business models (Cusumano, Yoffie, and Gawer 2020). Overall, therefore, wireless communications' enabling business models have become increasingly complex and diverse over time.

Take the example of WhatsApp, the OTT smartphone messaging app that was acquired by Facebook in 2014 for $19 billion (Facebook Investor Relations 2014). According to one estimate (Brodsky 2020), WhatsApp and other OTT messaging apps carry about 2.3 times as much international voice traffic as traditional telecommunications carriers. And yet WhatsApp is apparently provided free of charge to its customers, and they are not exposed to any unwanted advertising. So what is WhatsApp's business model, and in particular, how does it possibly make any money to justify its huge price tag? In two ways: from businesses and consumers.

On the business side, WhatsApp uses a freemium business model: the free WhatsApp Business App, which is targeted at small businesses, and the premium WhatsApp Business API, which is designed for medium and large businesses (Kudritskiy 2019). The consumer side business model is more subtle. WhatsApp consumers do not pay directly for using its services. Although Facebook does not "require people to use a common identifier or link their accounts to use multiple products in our Family [Facebook, Instagram, Messenger, and WhatsApp]," it does "seek to attribute multiple user accounts within and across products to individual people. To calculate these metrics, we rely upon complex techniques, algorithms and machine learning models that seek to count the individual people behind user accounts, including by matching multiple user accounts within an individual product and across multiple products when we believe they are attributable to a single person, and counting such group of accounts as one person" (Facebook Investor Relations 2020). Because many people have multiple accounts across Facebook's "Family" of products, these accounts can still be linked back to them as individuals, and hence monetized. Furthermore, some experts have commented on the immense commercial value available in the vast amounts of behavioral data that WhatsApp collects on its customers (Murphy 2019). In other words, even though WhatsApp consumers are not charged directly for using WhatsApp, Facebook can, and apparently does, still make a great deal of money from WhatsApp.

Third, it can be deduced from Table 2.1 that there are very wide differences in the impact on the actors in wireless communications caused by the evolution of enabling technologies and enabling business models. For instance, third-party value-added services (VAS) providers of premium mobile services (voicemail, missed call alerts, etc.) and premium mobile content (news, ring tones, etc.) were significant players in the 2G era. They have since largely disappeared, having been overtaken by technological and business model developments such as mobile apps. While the general roles of the technology providers and networks providers has remained largely intact in the evolution from 1G to 5G, the specific identities of many of the actors have changed, as have the technical details of the work performed in these roles. Some actors have disappeared altogether: AT&T was broken up in the 1980s due to antitrust action; Nortel went bankrupt in 2009; and Microsoft discontinued development of Windows Phone and withdrew from the mobile operating systems space for a combination of technology and business reasons. New actors have come onto the scene: Virgin Mobile became the world's first mobile virtual network operator (MVNO) in 1999; Safaricom/Vodafone launched M-Pesa in 2007, which went on to become the world's most successful mobile money service; and Huawei became the largest vendor in the global telecommunications software and professional services market in 2017 (Cooperson 2018), building on its earlier ascent to the top of the global telecommunications network equipment market in 2012 (Lee 2012).

Entirely new digital roles and digital role players have emerged that have completely disrupted the "dominant logic" (Skog, Wimelius, and Sandberg 2018) of wireless communications, including hyperscalers (e.g., Google), OTT players (e.g., WhatsApp), and digital platforms (e.g., Apple). In this regard, it is also worth noting the ever-increasing fragmentation and reconfiguration of value ecosystems in wireless communications over time.

The fourth and final deduction from Table 2.1 is that there are *intragenerational* and *intergenerational* dynamics with respect to the interchange and interplay between enabling technologies being the main driver versus enabling business models being the main driver. In 1G wireless communications, the fundamental underlying research and development (R&D) on the enabling technologies was done by AT&T Bell Labs in the United States, NTT Electrical Communications Labs in Japan, and other corporate research entities (Huurdeman 2003), making enabling technologies the main driver. In 2G wireless communications, the main impetus was based on political and commercial considerations that were later translated into technological outcomes, making enabling business models the main driver, at least initially (Temple 2010). We observe this constant interchange and interplay between enabling technologies and enabling business models being the main driver throughout the successive generations of wireless communications. The most important point to note is that long-term viability and sustainability of wireless communications generations depends crucially on achieving synergies between enabling technologies and enabling business models both *within* individual organizations and *across* ecosystems of organizations and other relevant actors.

## 2.5 ASSESSMENT OF ENABLING TECHNOLOGIES AND ENABLING BUSINESS MODELS FOR NEXT GENERATION WIRELESS COMMUNICATIONS

What exactly is meant by "next generation wireless communications" in the context of this chapter? Any such designations, of course, face the obvious risk of quickly becoming obsolete as they are overtaken by technological and marketplace developments. The first peer-reviewed IEEE publication that contained the exact term "next generation wireless communications" was in 1994 (Cheung, Beach, and McGeehan 1994), in the 2G era, when it was used in reference to the 3G UMTS standard.[1] By the early 2000s, with the first commercial deployments of UMTS networks, it was no longer appropriate to speak of UMTS as "next generation wireless communications," and the technical community shifted the use of the term to 4G technologies, and then later on to 5G technologies. In 2020, a strong uptick in commercial 5G network deployments was projected following conceptual standardization of 5G being completed in 2018 and initial 5G network launches in 2019 (GSMA 2020). Therefore, for the purposes of this chapter, "next generation wireless communications" will be defined to mean what comes after 5G, i.e., the wireless communications networks of 2030 and beyond (based on the approximately 10-year intergenerational spans in Table 2.1).

Modifying and updating the categories and definitions of (Camponovo and Pigneur 2003), Figure 2.1 shows the ecosystem of key actors in next generation wireless communications.[2] The following observations can be made about the key actors in this ecosystem:

1. *Users and user communities* may or may not be formal or informal customers, aren't tethered (literally or metaphorically) to any of the organizational actors

# Enabling Technologies and Enabling Business Models

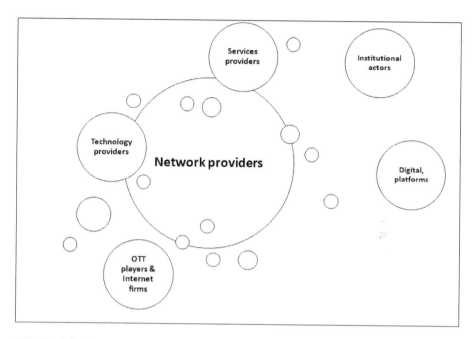

**FIGURE 2.1** Ecosystem of key actors in next generation wireless communications.

in the ecosystem, can roam freely and associate at will with any of these actors (or, indeed, among themselves), and are not only consumers but also important providers of wireless technologies and services in some instances (von Hippel 2017).

2. *Network providers* include MNOs, MVNOs, ISPs, WISPs, Wi-Fi operators, fixed-network operators (e.g., FTTX operators), and various other communications infrastructure providers and operators like data centers and satellite operators.

3. *Technology providers* include wireless component, access device, and network equipment manufacturers (e.g., Qualcomm, Samsung, and Ericsson, respectively) and other network infrastructure manufacturers (e.g., Cisco for IP networks). There are other types of technology providers as well, such as systems software, design, and intellectual property firms (e.g., Microsoft, Arm, and InterDigital, respectively); consultants, system integrators, and resellers (e.g., Accenture); and various downstream distribution channel players (e.g., wholesalers and retailers).

4. *Services providers* include various content providers (creators, developers, owners, aggregators, syndicators, distributors, portals, and platforms), application providers (e.g., applications software, mobile app, and games developers), and payments providers (e.g., banks, credit card companies, and digital payments platforms).

5. *OTT players and Internet firms* like Skype (now part of Microsoft) and Netflix

which are able to connect and offer products and services directly to users without going through the organizational intermediaries that have traditionally "owned" such users.
6. *Digital platforms* include hyperscale platforms (e.g., Google), social media platforms (e.g., Facebook), and marketplace platforms (e.g., Amazon), which have become the most powerful and profitable commercial actors not just in wireless communications, but also in many other sectors, by exploiting well-integrated enabling technologies and enabling business models (digital technologies and platform business models).
7. *Institutional actors* set, influence, or maintain the overall institutional context in which wireless communications takes place and include governments, regulators, standards organizations, universities, industry associations and lobby groups, and customer associations and lobby groups.

It is interesting and important to note that some actors occupy multiple categories in this ecosystem and play these multiple roles simultaneously. It has already been mentioned that users and user communities can be consumers *and* technology providers *and* services providers at the same time. They can also be institutional actors through consumer associations, for example. Alphabet, Google's parent company, wields enormous influence precisely because it arguably occupies all seven categories.

It is also evident from reflecting on Figure 2.1 that next generation wireless communications will not only have a complex ecosystem of key actors, but also an even more complex associated ecosystem of enabling technologies and enabling business models. It is far too early to be definitive about the specific details of the enabling technologies and enabling business models of next generation wireless communications. However, there are a few commonly agreed generic features to extrapolate from. In terms of enabling technologies, next generation wireless communications will certainly require managing heterogeneous wireless technologies and deploying AI/ML. Associated enabling business models will definitely involve digital platforms and a large variety of digital business models.

Next generation wireless communications will require orchestration of heterogeneous wireless technologies. Traditionally, management and control of these technologies has been isolated and uncoordinated, resulting in poor resource management, performance, and service quality. There are currently no complete solutions to this problem. A complete solution would have to include four key features (Bosch et al. 2020): technology integration, load balancing latency management, technology coexistence, and intelligent global network management. Technology abstraction and machine learning are potential components of a complete solution. Unless the wireless heterogeneity problem is resolved, the enabling business models of next generation wireless communications are likely to be extremely varied, fragmented, complex, and therefore suboptimal in performance. Fully orchestrated heterogeneous wireless technology management is thus a key enabling technology of next generation wireless communications.

AI/ML will be another essential enabling technology in next generation wireless communications. Current applications of AI/ML in wireless communications include (Wang et al. 2020): channel measurements, modeling, and estimation; physical layer research; and network management and optimization. Future applications of AI/ML will require new distributed, ultrafast, and lightweight machine learning algorithms for various use cases. Furthermore, as noted earlier, AI/ML is a key part of solving the wireless heterogeneity problem. Finally, because AI is a general-purpose technology (GPT) (Agrawal, Gans, and Goldfarb 2019), there are many other potentially significant impacts it could have on enabling technologies and enabling business models in next generation wireless communication that are impossible to foresee.

The enabling business models of next generation wireless communications will necessarily involve digital platforms, either directly or indirectly. Digital platforms use digital technologies to bring together two or more market actors and grow through network effects. Four major trends are predicted for digital platforms over the next 20 years (Cusumano, Yoffie, and Gawer 2020): *Hybrid platforms* (combinations of innovation platform and transaction platforms) will become the dominant strategy due to increasing competition; *AI and data analytics* will help drive more innovation; *more market power* is expected to be concentrated in a small number of large platforms; and platforms will face increasing demands for *regulation and content curation*. Overall, these trends point to heightened competition from and among digital platforms in next generation wireless communications. Once again, AI is expected to play a pivotal role.

Finally, in next generation wireless communications, traditional business models will have to compete, complement, or coordinate with a large variety of digital business models arising from different sectors and perspectives, including: digital platforms (Cusumano, Yoffie, and Gawer 2020); ISPs (Hanafizadeh, Hatami, and Bohlin 2019); creative industries (corresponding largely to services providers in Figure 2.1) (Li 2020); IT industries (Veit et al. 2014); and digital marketing (Verhoef and Bijmolt 2019).

## 2.6 INTEGRATED FRAMEWORK FOR ENABLING TECHNOLOGIES AND ENABLING BUSINESS MODELS FOR NEXT GENERATION WIRELESS COMMUNICATIONS

Figure 2.2 proposes an integrated framework for enabling technologies and enabling business for next generation wireless communications. The framework has seven elements:

1. *Technology creation*. Radically new enabling technologies for next generation wireless communications are created by a process of recursive problem solving (Arthur 2007). They therefore consist of a combination of hierarchical components arranged to fulfill some desired purpose by exploiting some base principle. This recursive, componential, purposive nature of technology creation has important implications for business model innovation – it suggests that enabling business models can shape enabling technologies just as much as enabling technologies shape enabling business models.
2. *Technology architecture and modularity*. The notion of technology architecture and modularity follows logically from the componential nature of technology creation. Technology architecture and modularity have important inherent

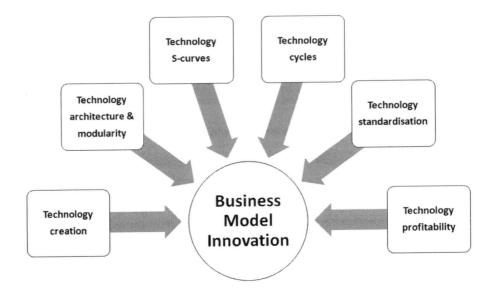

**FIGURE 2.2** Integrated framework for enabling technologies and enabling business models for next generation wireless communications.

business implications that are manifested in organizations' combined technical and business performance (Baldwin and Clark 2006; Clark 2018). In other words, the architecture and modularity of enabling technologies is inextricably linked with the organization's enabling business model.

3. *Technology S-curves.* The theory of technology S-curves explains how an enabling technology's market-related performance improves though the collective R&D efforts of multiple actors over time with a given context (Scillitoe 2013). Eventually, the maturity of the technology is reached, and additional R&D efforts produce diminishing returns, at which point the focal firm should redirect its efforts to a new enabling technology with a new technology S-curve. This theory ties in with enabling business models through its focus on market-related (i.e., business model) performance.

4. *Technology cycles.* Technology cycles differ from technology S-curves in that technology cycles focus on industry-wide patterns of technology change and adoption, and use the concepts of dominant designs, technological discontinuities, and competence-destroying and competence-enhancing discontinuities to understand these patterns (Anderson and Tushman 1990). These ideas can be combined with enabling business models to assess what stage of the technology cycle the organization's industry is in, what investments to make in potential enabling technologies, and how to configure enabling business models accordingly. The concept of dominant designs is particularly important as it

shows that, frequently, the emergence of a dominant design is not because of technological superiority, but due to other key factors, including superior business models or superior business model innovation. Furthermore, technology cycles demonstrate the inherent unpredictability of the evolution of technology, its $n$th-order effects, and its long-term consequences.

5. *Technology standardization.* As noted by Biddle (2017), technology standards in the ICT industry like those that will be used in next generation wireless communications are "created, maintained and propagated in a bewildering variety of ways, by a diverse set of actors." Understanding this extremely complex phenomenon is critical to successfully matching enabling technologies and enabling business models in next generation wireless communications.

6. *Technology profitability.* Teece (2018) recently applied his classic Profiting from Innovation (PFI) framework to the latest developments in wireless communications. He recommended assessing a wireless communications enabling technology's profitability based on: the appropriability regime, complementary assets and technologies and related business model issues, standards and installed base effects, timing, and ecosystem strength. Critical weaknesses in any one of these areas suggests that a firm should seriously reconsider its choice of enabling technology.

7. *Business model innovation.* This element of the framework requires analyzing and synthesizing the various enabling technology elements of the framework and the enabling business model element to create a sustainable and viable enabling business model. The business model innovation element of the framework involves an iterative process (Afuah 2018; Osterwalder and Pigneur 2010; Skog, Wimelius, and Sandberg 2018; Taran et al. 2016). Additionally, the process should focus on one or a combination of novel ways to generate customer benefits, to deliver benefits to customers, to monetize benefits, or to build and use organizational resources to generate, deliver, or monetize benefits to customers (Afuah 2018).

This integrated framework can be used to holistically analyze the technology dynamics of any potential enabling technology for next generation wireless communications and to iteratively design potential enabling business models. For instance, because the analysis in Section 2.5 showed that AI/ML will be a key enabling technology in next generation wireless communications, the integrated framework can be used to explore the technology dynamics and business model innovation aspects of AI/ML in next generation wireless communications. Furthermore, when applying this integrated framework, it is imperative to regularly review and incorporate any relevant information from the wider business model environment, including market forces, industry forces, macroeconomic forces, and key trends (Osterwalder and Pigneur 2010).

Finally, the usefulness of this integrated framework is not restricted to next generation wireless communications. It can also be applied to current-generation and emerging-generation wireless communications, and, with a little imagination, to previous-generation wireless communications as well. It is also applicable to other technology domains.

## 2.7 CONCLUSIONS

This chapter explored the connections between enabling technologies and enabling business models for next generation wireless communications. It noted that relatively few works in the technical literature have explicitly studied these connections, despite the existence of several prominent real-world examples demonstrating their critical importance.

The chapter made three main research contributions. First, it assessed the intersections and interactions between *specific aspects* of enabling technologies and enabling business models for previous-generation, current-generation, and emerging-generation wireless communications. It showed the crucial interrelationships between the enabling technologies and enabling business models of successive wireless generations, from 1G to 5G. Second, it assessed the intersections and interactions between *generic aspects* of enabling technologies and enabling business models for next generation wireless communications, which was defined as beyond 5G. It showed that enabling technologies and enabling business models would continue to influence and affect each other in next generation wireless communications, indeed to an even greater degree than in previous generations. Third, it proposed an integrated framework for enabling technologies and enabling business models for next generation wireless communications. It showed how the framework can be used both as an analytical tool (technology dynamics) and a design tool (business model innovation) for enabling technologies and enabling business models for next generation wireless communications. The framework is extremely versatile and can be applied to other generations of wireless communications as well as to other technology domains.

The chapter also made three ancillary contributions in the form of schematics to visualize and summarize important aspects of the three main research contributions: a comparison of enabling technologies and business models across wireless generations (1G–5G), the ecosystem of key actors in next generation wireless communications, and an integrated framework for enabling technologies and enabling business models for next generation wireless communications.

Take 6G as an example. Ongoing developments of wireless communications toward 6G (Table 2.1) point to services and devices proliferating, data rates soaring, latencies plummeting, spectrum allocations multiplying, and enabling technologies and enabling business models becoming increasingly versatile and complex. This calls for entirely new approaches to wireless communications such as ambient wireless services that work seamlessly across multiple devices and interfaces, end-to-end network resource sharing, and dynamic business models. The configuration and relative sizes and roles of key actors in the 6G ecosystem (Figure 2.1) will be radically different from that of previous generations, a factor that all current actors should incorporate in their strategies and tactics. The interplay of technology dynamics and business model innovation (Figure 2.2) will be most acute in general-purpose technologies (GPTs) like AI/ML (Agrawal, Gans, and Goldfarb 2019; Wang et al. 2020), but also, potentially, in emerging physical layer technologies such as terahertz communications (Rappaport et al. 2019), meta-surfaces and meta-materials (Yuan et al. 2020), and quantum communications and computing (Dang et al. 2020). As such, 6G role players should be highly adaptable and agile with respect to 6G technology dynamics and business models.

The following specific recommendations are offered to key role players in next generation wireless communications. For practitioners, the findings demonstrate the unmistakable value of taking both enabling technology *and* enabling business model considerations into account when making both engineering and management decisions. The integrated technology dynamics and business model innovation framework is particularly useful and powerful in this regard. For policymakers, the results suggest that there are potential risks of policy failure as well as potential opportunities for policy innovation in areas such as technology sector legislation and regulation (which should be based on a well-informed, longitudinal understanding of technology, business, and other relevant issues), education and skills development (which should equip all workers with at least basic technology and business literacy). For researchers, both in wireless communications and other STEM-related disciplines, this chapter shows how and why it is so vital to *not* be restricted by traditional disciplinary boundaries. Simultaneously exploring traditionally separate research domains facilitates discovery of important new insights that are also valuable for each separate domain.

The work reported in this chapter raises many new research questions; three are mentioned for purposes of illustration. In the first place, it would be interesting to further investigate the ecosystem of key actors in next generation wireless communications to identify and analyze key relationships and subecosystems and how they specifically relate to enabling technologies and enabling business models. This may require extrapolating based on reliable data from previous and current generations of wireless communications to ground the analysis in reality. A second research question would be to apply the integrated technology dynamics and business model innovation framework to 5G. This is a potentially rich area of enquiry because of 5G's status as a newly launched generation of wireless communications. This provides a unique opportunity to generate research that can be tested and validated against a growing body of real-world data and evidence. Finally, there is a critical need to better understand the relationships between enabling technologies and enabling business models on the one side and on the other side, the enabling strategies (Magretta 2002; Veit et al. 2014) through which they are translated into market results.

## NOTES

1 Found using Google Scholar's advanced search function and cross-checked with IEEE Xplore.
2 With some minor modifications, the proposed schematic can also be applied to current-generation and emerging-generation wireless communications.

## REFERENCES

Afuah, Allan. 2018. *Business model innovation: Concepts, analysis, and cases.* 2nd ed. New York: Routledge.
Agrawal, Ajay, Joshua Gans, and Avi Goldfarb, eds. 2019. *The economics of artificial intelligence: An agenda.* Chicago, IL: The University of Chicago Press.

Ahokangas, Petri, Marja Matinmikko-Blue, Seppo Yrjölä, Veikko Seppänen, Heikki Hämmäinen, Risto Jurva, and Matti Latva-aho. 2019. "Business models for local 5G micro operators." *IEEE Transactions on Cognitive Communications and Networking* 5, no. 3: 730–740.

Anderson, Philip, and Michael L. Tushman. 1990. "Technological discontinuities and dominant designs: A cyclical model of technological change." *Administrative Science Quarterly* 35, no. 4: 604–633.

Arthur, W. Brian. 2007. "The structure of invention." *Research Policy* 36, no. 2: 274–287.

Baldwin, Carliss Y., and Kim B. Clark. 2006. "Modularity in the design of complex engineering systems." In *Complex engineered systems: Science meets technology*, edited by Dan Braha, Ali A. Minai, and Yaneer Bar-Yam, 175–205. New York: Springer.

Biddle, C. Bradford. 2017. "No standard for standards: Understanding the ICT standards-development ecosystem." In *The Cambridge handbook of technical standardization law: Competition, antitrust, and patents, Cambridge Law Handbooks*, edited by Jorge L. Contreras, 17–28. Cambridge, UK: Cambridge University Press. doi:10.1017/9781316416723.004

Blank, Steve. 2013. "Why the lean start-up changes everything." *Harvard Business Review* 9, no. 5: 63–72.

Bosch, Patrick, Tom De Schepper, Ensar Zeljković, Jeroen Famaey, and Steven Latré. 2020. "Orchestration of heterogeneous wireless networks: State of the art and remaining challenges." *Computer Communications* 149, January: 62–77.

Brodsky, Paul. 2020. Carriers down to 465 billion minutes in 2018. https://blog.telegeography.com/international-voice-market-update-2019-2020 (accessed June 13, 2020).

Camponovo, Giovanni, and Yves Pigneur. 2003. "Business model analysis applied to mobile business." In *Proceedings of the 5th International Conference on Enterprise Information Systems (ICEIS)*, April 23–26, 2003, Angers, France.

Cheung, Joseph C. S., Mark A. Beach, and Joseph P. McGeehan. 1994. "Network planning for third-generation mobile radio systems." *IEEE Communications Magazine* 32, no. 11: 54–59.

Clark. David D. 2018. *Designing an internet*. Cambridge, MA: The MIT Press.

Conroy, Victoria, and John Hill. 2009. Italy: The powerhouse of prepaid. https://www.verdict.co.uk/cards-international/special-reports/italy-the-powerhouse-of-prepaid/ (accessed June 13, 2020).

Cooperson, Dana. 2018. Huawei overtakes Ericsson to lead the USD69 billion telecoms software market in 2017. https://www.analysysmason.com/research/content/comments/market-share-comment-rma09/ (accessed June 13, 2020).

Cusumano, Michael A., David B. Yoffie, and Annabelle Gawer. 2020. "The future of platforms." *MIT Sloan Management Review* 61, no. 3: 46–54.

Dang, Shuping, Osama Amin, Basem Shihada, and Mohamed-Slim Alouini. 2020. "What should 6G be?" *Nature Electronics* 3, no. 1: 20–29. doi:10.1038/s41928-019-0355-6

David, Klaus, Jaafar Elmirghani, Harald Haas, and Xiao-Hu You, eds. 2019. "Defining 6G: Challenges and opportunities [special issue]." *IEEE Vehicular Technology Magazine* 41, no. 3: 14–16.

Dhar, Subhankar and Upkar Varshney. 2011. "Challenges and business models for mobile location-based services and advertising." *Communications of the ACM* 54, no. 5: 121–129.

Dohler, Mischa. 2018. "The future and challenges of communications – Toward a world where 5G enables synchronized reality and an internet of skills." *Internet Technology Letters* 1, no. 2: e33. doi:10.1002/itl2.33

Facebook Investor Relations. 2014. Facebook to acquire WhatsApp. https://investor.fb.com/investor-news/press-release-details/2014/Facebook-to-Acquire-WhatsApp/default.aspx (accessed June 13, 2020).

Facebook Investor Relations. January 30, 2020. Facebook 2019 annual report (Form 10-K). *SEC Filings Details*. https://investor.fb.com/financials/sec-filings-details/default.aspx?FilingId=13872030 (accessed June 13, 2020).

Giordani, Marco, Michele Polese, Marco Mezzavilla, Sundeep Rangan, and Michele Zorzi. 2020. "Toward 6G networks: Use cases and technologies." *IEEE Communications Magazine* 58, no. 3, 55–61.

Glisic, Savo G. 2016. *Advanced wireless networks: Technology and business models*. 3rd ed. Chichester, UK: John Wiley & Sons.

Greenstein, Shane. 1998. "Industrial economics and strategy: Computing platforms." *IEEE Micro* 18, no. 3: 43–53.

GSMA. 2020. 2019 saw 5G become a commercial reality – 2020 will take it to the mass market. https://www.gsma.com/futurenetworks/digest/5g-in-2019/ (accessed June 13, 2020).

Hanafizadeh, Payam, Parastou Hatami, and Erik Bohlin. 2019. "Business models of Internet service providers." *Netnomics* 20, no. 1: 55–99.

Huurdeman, Anton A. 2003. *The worldwide history of telecommunications*. Hoboken, NJ: John Wiley & Sons.

Jansen, Bernard J., and Tracy Mullen. 2008. "Sponsored search: An overview of the concept, history, and technology." *International Journal of Electronic Business* 6, no. 2: 114–131.

Kudritskiy, Iaroslav. 2019. What is WhatsApp Business? https://respond.io/blog/whatsapp-business/ (accessed June 13, 2020).

Layton, Roslyn. 2014. "The prepaid mobile market in Africa." In *The African mobile story*, edited by Knud Erik Skouby and Idongesit Williams, 51–78. Aalborg, Denmark: River Publishers.

Lee, Cyrus. 2012. Huawei surpasses Ericsson as world's largest telecom equipment vendor. https://www.zdnet.com/article/huawei-surpasses-ericsson-as-worlds-largest-telecom-equipment-vendor/ (accessed June 13, 2020).

Li, Feng, and Jason Whalley. 2002. "Deconstruction of the telecommunications industry: From value chains to value networks." *Telecommunications Policy* 26, no. 9–10: 451–472.

Li, Feng. 2020. "The digital transformation of business models in the creative industries: A holistic framework and emerging trends." *Technovation* 92–93, 102012: 1–10. doi:10.1016/j.technovation.2017.12.004

MacKenzie, Allen B., and Luiz A. DaSilva. 2006. *Game theory for wireless engineers*. San Rafael, CA: Morgan & Claypool. doi:10.2200/S00014ED1V01Y200508COM001

Magretta, Joan. 2002. "Why business models matter." *Harvard Business Review* 80, no. 5: 86–92.

Molisch, Andreas F. 2011. *Wireless communications*. 2nd ed. Chichester, UK: John Wiley & Sons.

Murphy, Hannah. March 27, 2019. "How Facebook could target ads in age of encryption." *Financial Times*. https://www.ft.com/content/0181666a-4ad6-11e9-bbc9-6917dce3dc62

Mzyece, Mjumo. 2001. "Wireless application protocol (WAP)." *IEEE Vehicular Technology Society News* 48, no. 2: 7–12. https://www.researchgate.net/publication/334746916_Wireless_Application_Protocol_WAP

Osterwalder, Alexander, and Yves Pigneur. 2010. *Business model generation: A handbook for visionaries, game changers, and challengers*. Hoboken, NJ: Wiley.

Ovans, Andrea. 2015. "What is a business model? [digital article]." *Harvard Business Review*, 23, 1–7. https://hbr.org/2015/01/what-is-a-business-model

OED Online. 2020 *Oxford english dictionary*. Oxford: Oxford University Press. https://www.oed.com/ (accessed June 13, 2020).

Rappaport, Theodore S., Yunchou Xing, Ojas Kanhere, Shihao Ju, Arjuna Madanayake, Soumyajit Mandal, Ahmed Alkhateeb, and Georgios C. Trichopoulos. 2019. "Wireless communications and applications above 100 GHz: Opportunities and challenges for 6G and beyond." *IEEE Access* 7, 78729–78757. doi:10.1109/ACCESS.2019.2921522

Scillitoe, Joanne L. 2013. "Technology S-curve." In *Encyclopedia of management theory*, edited by Eric H. Kessler, 846–849. Thousand Oaks, CA: SAGE Publications.

Skog, Daniel A., Henrik Wimelius, and Johan Sandberg. 2018. "Digital disruption." *Business & Information Systems Engineering* 60, no. 5: 431–437.

Taran, Yariv, Christian Nielsen, Marco Montemari, Peter Thomsen, and Francesco Paolone. 2016. "Business model configurations: A five-V framework to map out potential innovation routes." *European Journal of Innovation Management* 19, no. 4: 492–527. doi:10.1108/EJIM-10-2015-0099

Teece, David J. 2010. "Business models, business strategy and innovation." *Long Range Planning* 43, no. 1–2: 172–194. doi:10.1016/j.lrp.2009.07.003

Teece, David J. 2018. "Profiting from innovation in the digital economy: Enabling technologies, standards, and licensing models in the wireless world." *Research Policy* 47, no. 8: 1367–1387. doi:10.1016/j.respol.2017.01.015

Temple, Stephen. 2010. *Inside the mobile revolution: A political history of GSM*. Self-published e-book. http://www.gsmhistory.com/wp-content/uploads/2013/01/Inside-a-Mobile-Revolution-Temple-20101.pdf (accessed June 13, 2020).

Veit, Daniel, Eric Clemons, Alexander Benlian, Peter Buxmann, Thomas Hess, Dennis Kundisch, Jan Marco Leimeister, Peter Loos, and Martin Spann. 2014. "Business models: An information systems research agenda." *Business & Information Systems Engineering* 6, no. 1: 45–53.

Verhoef, Peter C., and Tammo H. A. Bijmolt. 2019. "Marketing perspectives on digital business models: A framework and overview of the special issue." *International Journal of Research in Marketing* 36, no. 3: 341–349.

von Hippel, Eric. 2017. *Free innovation*. Cambridge, MA: MIT Press.

Wang, Cheng-Xiang, Marco Di Renzo, Slawomir Stanczak, Sen Wang, and Erik G. Larsson. 2020. "Artificial intelligence enabled wireless networking for 5G and beyond: Recent advances and future challenges." *IEEE Wireless Communications* 27, no. 1: 16–23.

Yaghoubi, Forough, Mozhgan Mahloo, Lena Wosinska, Paolo Monti, Fabricio de Souza Farias, Joao Crisstomo Weyl Albuquerque Costa, and Jiajia Chen. 2018. "A techno-economic framework for 5G transport networks." *IEEE Wireless Communications* 25, no. 5: 56–63.

Yuan, Yifei, Yajun Zhao, Baiqing Zong, and Sergio Parolari. 2020. "Potential key technologies for 6G mobile communications." *Science China Information Sciences* 63, no. 183301:1–19. doi:10.1007/s11432-019-2789-y

# 3 Enabling Technologies for Internet of Everything

*M. Rezwanul Mahmood and Mohammad Abdul Matin*
North South University

## CONTENTS

3.1 Introduction .................................................................................................33
3.2 Enabling Technologies for IoE .....................................................................35
    3.2.1 Cloud Computing ..............................................................................35
    3.2.2 Fog Computing .................................................................................35
    3.2.3 Edge Computing ...............................................................................36
    3.2.4 Machine to Machine .........................................................................36
    3.2.5 Machine Learning .............................................................................36
3.3 Data Management and Security in IoE .........................................................37
3.4 System Management and Protection for IoE ................................................37
3.5 Applications of IoE .......................................................................................38
    3.5.1 Healthcare .........................................................................................38
    3.5.2 Power ................................................................................................39
    3.5.3 Education System .............................................................................39
    3.5.4 Smart Environment ...........................................................................39
3.6 Enabling IoE in Developing Countries ..........................................................40
3.7 Conclusion ....................................................................................................40
References ............................................................................................................41

## 3.1 INTRODUCTION

Internet of Things (IoT) networks are based on ubiquitous and pervasive computing, embedded devices, communication technologies, sensor networks, Internet protocols, and so on (Al-Fuqaha et al. 2015). Along with the protocols such as M2M, Cognitive Radio (CR), and IPv6, low-cost and low-power consuming devices such as sensors, RFID tags, etc. have assisted to form an economical network. The number of mobile devices and M2M connections are expected to reach the billions in the near future, as highlighted in Rawat, Singh and Bonnin (2016). Thus a number of research efforts have been dedicated toward secured and uninterrupted connectivity among

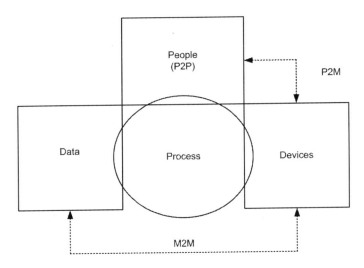

**FIGURE 3.1** Components of IoE and their interactions.

the IoT devices. The potential of IoT has been attempted to explore by extending its concept, which gave birth to the concept of Internet of Everything (IoE). IoT considers the connection among "things" or devices, whereas IoE creates interconnection among the elements, which are people, devices, data, and process (de Matos, Amaral and Hessel 2017). Figure 3.1 represents the environment of IoE formed with these elements and indicates how people-to-people (P2P), people-to-machine (P2M), and machine-to-machine (M2M) connect (Miraz et al. 2015).

In IoE networks, people are connected via social network and devices such as smartphones, smart TV, computers, and wearable devices. This allows people to be a part of IoE system and act as a node generating or receiving information. The concept of device in IoE is similar to that in IoT. It will not only generate data but also provide services according to decisions made by the IoE system. The generated data will be converted from raw form to useful information for higher-level analysis to provide better service to the end user. The process component will ensure proper interconnection among the components of IoE and utilization of processed data for decision making to transfer accurate information to the end user/devices.

IoT features have enabled building effective automation systems for homes and industries, healthcare services, smart grid, and an opportunity to make innovative applications. Such applications will be smarter and provide better facility with the help of IoE. With the effort of expanding the Internet connectivity across the world by exploring protocols such as Low Power Wide Area Network (LPWAN) (Kim and Kim 2017; Palattella and Accettura 2018), remote devices will also be connected to the Internet, and thus IoE will be realized more effectively. Paradigms such as cloud computing, edge computing, and machine learning will aid in storing and analyzing the data and thus perform decision making within short time.

This chapter presents the enabling paradigms for IoE and highlights recent progress on these enablers. Besides, the advancement in the application sector, for instance, healthcare, power sector, education system, and smart environment are also

# Enabling Technologies for Internet of Everything

presented to show the effect of implementation of IoE in these applications. A scenario of the current situation of realizing IoE in developing countries is also stated in this chapter.

The rest of the chapter has been organized as follows: Section 3.2 describes the potential technologies for enabling IoE. Data management and security is crucial in IoE, which is discussed in Section 3.3. Section 3.4 describes the security solution, which can provide IoE system protection based on policy management. The applications of IoE are summarized in Section 3.5. IoE provides an opportunity for improving human lifestyles in developing countries. Therefore, IoE needs to be enabled in developing countries, which is explained in Section 3.6. The concluding remarks are given in Section 3.7.

## 3.2 ENABLING TECHNOLOGIES FOR IoE

### 3.2.1 Cloud Computing

Cloud computing provides Internet-enabled computing resources and ensures low hardware cost, device independency, and accessibility to compute and obtain results irrespective of location and time (Wang et al. 2010). IoE combined with cloud computing supports effortless improvement of computing and storage services/applications for the developers. Security threats such as leakage of data privacy or illegal access to data exist in such systems, which can be addressed by using searchable encryption (SE) solutions. A data-sharing framework is needed that minimizes false results return, support dynamic update, fair arbitration, and multiowner setting, which is done in Miao et al. (2019).

Cyber-physical system (CPS) can create a smart artificial system by means of the improvement in remote sensing, IoT, and cloud computing. However, remote sensors have limited storage space and computing assets. A cloud computing system can be a potential solution to make system framework more adaptable and allow invigilating data (Kaur, Riaz, and Mushtaq 2020).

### 3.2.2 Fog Computing

Fog computing connects the end user and cloud data station located at distant place (Bonomi et al. 2012). It serves the purpose of reduction of service deployment delay, support for mobility, and awareness of location for IoT application. The incremental number of heterogeneous data at the network edge increases the data volume, which results in delay in data transfer to the cloud and reception of services (Kazmi et al. 2016). In such cases, fog computing acts as an extension of cloud computing and ensures faster service delivery.

The integration of cloud computing, fog computing and Internet of Everything will create opportunities for newer and improved technologies and innovations. This raises the complexity of the system, which can be managed by fog orchestrator, a coordinator of services provided between IoE and cloud computing. Different architectures and issues related to fog orchestrator are being studied in Velasquez et al. (2018) for its improvement. Fog of Everything (FoE) is another new paradigm proposed in Baccarelli et al. (2017). The authors also provided the protocols and elements of FoE and simulated a prototype of FoE enabled system. Another study on fog computing enabled smart city network demonstrates the energy efficiency of the network solution along with low latency (Naranjo et al. 2019). Fog enabled IoE network is not immune of network congestion which leads to delay in service provision. So optimization of routing algorithms can minimize such issues, which is aimed in (Nath et al. 2017).

### 3.2.3 Edge Computing

The computation performed by cloud computing at the edge of a network is challenging to carry out due to increasing data generation at that region. For efficient computation, the concept of edge computing is introduced (Shi et al. 2016). Edge computing is similar to fog computing; however, edge computing works with devices, whereas fog computing works on system infrastructure (Cao, Zhang and Shi 2018). Edge computing allows a limited amount of data to be transferred from devices located at the network edge to cloud and reduces system complexity. Besides, enabling some of the computation to edge computing instead of cloud computing can make the overall wireless network energy efficient. The uploading of data directly from end devices to cloud can consume greater bandwidth if the device-generated data is large in volume. Thus it is recommended to process the data before uploading to cloud directly to occupy less bandwidth and ensure privacy of the user/data when required, which can be possible with the help of edge computing.

### 3.2.4 Machine to Machine

IoE can be implemented in the real world with the help of IoT and automated M2M (Balfour 2015). A global standard is required to serve the purpose of IoE, which is oneM2M, as described in (Balfour 2015). The oneM2M operates on any network transport architecture and thus allows the global M2M devices to be connected over an end-to-end network. A software-defined network perimeter (SDP) technology can look after the security aspect of the M2M network.

According to (Balfour 2015), it can be said that fourDscape architecture, developed by Balfour Technologies, can enable promising implementation of oneM2M and SDP technology due to their resemblance in terms of concepts. The authors have described the components of fourDscape and shown how these components can support oneM2M. M2M common services such as addressing and identification of resources, data management and repository, network services, and security will be served by the fourDscape platform.

### 3.2.5 Machine Learning

Because IoE-enabled devices generate a huge volume of data, the management and processing of such data is a necessity for providing required services. Machine learning techniques process multidimensional data and enhance the performance of IoE applications and devices (AlSuwaidan 2019). Based on data processing tasks, different machine learning techniques have been discussed in Mahdavinejad et al. (2018), such as k-nearest neighbors, support vector machine, linear regression, and k-means. In applications such as healthcare and smart grid, machine learning techniques are chosen on the basis of processing requirements. In Sianaki et al. (2018), it is shown that the support vector machine, logistic regression, and random forest have been most studied in the healthcare system. Also, to study on health issues such as depression and anxiety, logistic regression has mostly been used. Also, smart grid data analysis requires machine learning to predict generation and consumption of electrical power. However, the computation performed with the help of machine learning models may poses the risk of undesired revelation of data. This revelation can be limited by using randomization technique (Cimato and Damiani 2018).

## 3.3 DATA MANAGEMENT AND SECURITY IN IoE

It was previously mentioned that IoE devices will generate a large amount of data within a short period of time. So the exchange of data among different components of the IoE system needs to be managed in a secured and efficient manner. In Abu-Elkheir, Hayajneh and Ali (2013), it is mentioned that data-centric and source-centric methods can be effective for allowing accumulation and processing of different types of data collected from all kinds of sources. In Charmonman and Mongkhonvanit (2015), the authors present some points regarding the privacy and security of IoE device-generated data.

In this section, some recent research on data sharing framework, and management models have been surveyed. In Pena, Sarkar and Maheshwari (2015), the authors present a big-data-centric framework that supports the security-cognition interaction-enabled architecture. The proposed framework has been discussed in terms of the cloud architecture, security, and cognition aspect. In AlSuwaidan (2019), a data-driven management model has been proposed. This model consists of three layers, namely data sources layer, master data management layer, and top-level data management services layer. The data sources layer is composed of data sources such as people, machines, and documents. The master data management layer comprises data cleaning, processing, privacy, integration, etc. Last, the top-level data management services layer consists of security, integrity, federation, and cloud services. The author describes the function of each of the components of the model in the article. The data-sharing framework (Miao et al. 2019), based on searchable encryption (SE) schemes, ensures data sharing and security from attack such as keyword guessing. Researchers have been focusing on cryptography schemes for data protection. However, classic cryptography schemes depend on mathematical computation, which limits the efficiency of data protection. Quantum cryptography is therefore considered a better approach to protect data more effectively with respect to classic cryptography (Pradeep, Rao, and Vikas 2019).

## 3.4 SYSTEM MANAGEMENT AND PROTECTION FOR IoE

Various devices such as smartphones, Internet-enabled vehicles, and household appliances are equipped with complicated computational abilities to provide intelligent services to mankind. This creates the need of communication among the IoE-enabled devices for providing effective services. The software systems can be considered as agents or multiagent system, as mentioned in Schatten, Ševa and Tomičić (2016). The increase in IoE devices will introduce multiagent systems (MASs) in large scale. So it is a must to address the management of such a system in large magnitude. In Schatten, Ševa, and Tomičić (2016), the author reviews organizational design methods for a large-scale multiagent system (LSMAS) and provides suggestion on the ontological method for organizational designing schemes for improvement of the system. The interaction among the agents occurs autonomously or with negligible human intervention, which raises the requirements of evaluation of trustable communication. In Kuada (2018), a trust management system thus has been presented for an IoE system that allows the devices to calculate trust values and make decisions on the basis of IoE applications. In an IoE system where the user can interfere for reception of services, an authentication technique is useful for secured communication between users and the system. So, a wireless authentication key has been one of the interests for researchers. In Komar, Edelev, and Koucheryavy (2016), a system comprising a wireless authentication key scheme has been studied that uses NFC tags to eliminate battery needs and reduce cost and cryptographic schemes to avoid computation on NFC tag side and provide security to data. According to Kim, Jo, and Shon (2020), devices with low performance in a system can be affected by cyber threats. In order to monitor the behavior of such devices in an industrial IoE environment, an auto-encoder-based anomaly detection method has been proposed in Kim, Jo, and Shon (2020).

## 3.5 APPLICATIONS OF IoE

This section summarizes recent research on applications such as healthcare, power systems, education, and smart environment and demonstrates the results of introducing IoE in these sectors.

### 3.5.1 HEALTHCARE

Health monitoring devices, connected to the Internet, can play important roles in healthcare service provisions. IoE-enabled wearables and applications can provide cost-effective and efficient medical services. These technologies also enable patients to consult physicians effortlessly and take medications both at hospitals and homes (Mele and Russo-Spena 2019). For example, a diabetic patient can send health vitals to a healthcare center for analysis (Ismail 2017). The patient can then receive required medical assistance according to inspection of the data. To make it happen, the data from different sensor devices is collected, preprocessed for reducing data volume, and transferred to storage devices (Honan et al. 2016). The data is then processed, and corresponding services are provided to the patient. However, such an operation is challenging to execute. So, cloud platforms (e.g., HEAL and CoCaMAAL) can act as a middleware and alleviate this kind complexity (Manashty and Thompson 2017). A collaborative service-based well-being service bus can integrate health devices, applications, and systems to a paradigm named enterprise service bus (Meridou et al. 2015) and a health-related social network transfer data to the people associated with it, which ensures better remote healthcare services (Meridou et al. 2017).

### 3.5.2 Power

Smart grid allows the merging of renewable and nonrenewable energy sources to generate electricity. Machine learning, which can act as an enabler for IoT and eventually IoE, can manage the electrical grid system if there is any sort of fault. The increased use of renewable energy source generated electricity, for instance, solar PV and wind electricity, has created the need of forecasting the electricity generation from these sources, which can be fulfilled by machine learning. Besides, it can monitor the electrical energy consumption to optimize the overall performance of an electrical generation system (Sianaki et al. 2018). Advanced metering infrastructure (AMI) permits two-way communications between the user smart meters and utility office due to which information related to power consumption, billing, and others can be transferred easily (Desai et al. 2019). Remote controlling and monitoring of loads powered by sources such as solar PV panel(s) can enhance better utilization of solar energy and generated electrical energy, which is aimed in Aruna and Venkataswamy (2018).

### 3.5.3 Education System

The components of IoE, which are people, objects, process, and data, can be explored to create a smart education system that can improve the teaching methodologies, administrative operations, research facilities, library resources, etc. Some proposals for smart educational environment, such as educational cyber-physical systems (ECPS) for University 4.0 (Bachir and Abenia 2019), IoE-enabled smart embedded system (IoE-SES) (Rathod et al. 2020), and an Internet of Everything (IoE)-based educational model using deep learning techniques (Ahad, Tripathi and Agarwal 2018) are intended to provide quality education in real and virtual classrooms, evaluate curriculum, record all the academic and administrative data, monitor power consumption of the institution and hygiene, etc.

### 3.5.4 Smart Environment

In smart homes, the home appliances are operated with the help of remote or automatic controlling devices, which improves human lifestyle (Feng, Setoodeh and Haykin 2017). Edge computing is a promising enabler for realizing smart home systems because it provides advantages of network load reduction, lower service response time, and data protection from outsiders (Cao et al. 2017). The management of smart devices, data, and services can be ensured with the help of a home operating system. Such operating systems connect IoE home devices with cloud, people, and developers to transfer and analyze data, request and receive services, and provide interface for system development, respectively (Cao et al. 2017).

IoE networks can ensure public safety during any emergency situation such as environmental hazards, accidents, and crimes. Also, for protection of places like museums and cultural heritage sites, a cost-effective smart IoE network is essential. Garzia and Sant'Andrea (2016) and Gambetti et al. (2017) utilize genetic algorithm (GA) to create integrated security systems for the Papal Basilica and Sacred Convent of Saint Francis and the World War One Commemorative Museum of Fogliano Redipuglia, both located in Italy. The implemented systems provide safety to visitors and accessibility to disabled people along with safety of the sites.

## 3.6 ENABLING IoE IN DEVELOPING COUNTRIES

In developed countries, IoE is seen as a system consisting of Internet-enabled appliances (Gubbi et al. 2013) and maintenance tools for urban management (Majeed 2017). However, for developing countries, IoE provides an opportunity for improving human lifestyle with the help of improving economic condition and investment (Majeed 2017). The evolution of information and communication technology (ICT) in a faster rate can enable the potential of IoE to make progress in an economy, which can accelerate the development of the nations in several areas. For example, the improvement of air quality of China and new methodologies for clean water in Africa can be made possible with the help of IoE-based technologies.

In Adewale et al. (2019), the authors survey Nigeria's overview on the potential of enabling IoE in the country. According to the survey, most of the residents are aware of the IoE concept. However, about 41% of the participants do not agree to the possibility of IoE implementation. The survey also shows the technical aspects of communication services, according to which it is observed that security issues and subscriber privacy need to be addressed if IoE is to be implemented.

The secured interconnection of devices and interoperability among them is an important challenge. A configuration method with bootstrapping scheme (Majeed 2017), proposed for devices located within developing regions, may provide device security and eliminates exploitation of data within networks with low protection.

## 3.7 CONCLUSION

Internet of Everything (IoE) links people, data, process, and devices and creates a smart world. It is the extension of Internet of Things (IoT), which only deals with communication among devices. IoE allows processing of data received from not only devices but also from people because people are also a part of the network. Technologies such as cloud and fog computing have enabled faster and secured computation and possibility to extend the region of service delivery. Healthcare, education, and smart home services have also evolved with the help of IoE. Recent research on enabling technologies, data management and security, system management, and applications have been explored in this chapter. Besides, current scenarios on the possibilities of implementing IoE in developing countries have been summarized. The application of IoE services is expected to contribute in the growth of human lifestyle and progress in the development of a country by considering the interoperability, scalability, and security aspects of IoE systems.

## REFERENCES

Abu-Elkheir, M., M. Hayajneh, and N. A. Ali. 2013. "Data management for the Internet of Things: Design primitives and solution." *Sensors* 13 (11): 15582–15612.

Adewale, A. A., A. S. Ibidunni, A. A. Atayero, S. N. John, O. Okesola, and R. R. Ominiabohs. 2019. "Nigeria's preparedness for Internet of Everything: A survey dataset from the work-force population." *Data in Brief* 23, 103807.

Ahad, M. A., G. Tripathi, and P. Agarwal. 2018. "Learning analytics for IoE based educational model using deep learning techniques: Architecture, challenges and applications." *Smart Learning Environments* 5 (1): 1–16.

Al-Fuqaha, A., M. Guizani, M. Mohammadi, M. Aledhari, and M. Ayyash. 2015. "Internet of things: A survey on enabling technologies, protocols, and applications." *IEEE Communications Surveys & Tutorials* 17 (4): 2347–2376.

AlSuwaidan, L. 2019. "Data management model for Internet of Everything." *International Conference on Mobile Web and Intelligent Information Systems*. Istanbul.

Aruna, S., and R. Venkataswamy. 2018. "Academic workbench for streetlight powered by solar PV system using Internet of Everything (IoE)." *2018 International Conference on Communication, Computing and Internet of Things (IC3IoT)*, Chennai.

Baccarelli, E., P. G. V. Naranjo, M. Scarpiniti, M. Shojafar, and J. H. Abawajy. 2017. "Fog of everything: Energy-efficient networked computing architectures, research challenges, and a case study." *IEEE Access* 5: 9882–9910.

Bachir, S., and A. Abenia. 2019. "Internet of Everything and educational cyber physical systems for University 4.0." *International Conference on Computational Collective Intelligence*, Hendaye.

Balfour, R. E. 2015. "Building the 'Internet of Everything' (IoE) for first responders." *2015 Long Island Systems, Applications and Technology*, New York.

Bonomi, F., R. Milito, J. Zhu, and S. Addepalli. 2012. "Fog computing and its role in the Internet of Things." *Proceedings of the First Edition of the MCC Workshop on Mobile Cloud Computing*, Helsinki, Finland, 13–16.

Cao, J., L. Xu, R. Abdallah, and W. Shi. 2017. "EdgeOS_H: A home operating system for Internet of Everything." *2017 IEEE 37th International Conference on Distributed Computing Systems (ICDCS)*, Atlanta.

Cao, J., Q. Zhang, and W. Shi. 2018. "Introduction." Chap. 1 in *Edge Computing: A Primer*, 1–8. Springer, Cham.

Charmonman, S., and P. Mongkhonvanit. 2015. "Special consideration for Big Data in IoE or Internet of Everything." *2015 13th International Conference on ICT and Knowledge Engineering (ICT & Knowledge Engineering 2015)*, Bangkok.

Cimato, S., and E. Damiani. 2018. "Some ideas on privacy – Aware data analytics in the Internet-of-Everything." In *From Database to Cyber Security*, edited by P. Samarati, I. Ray, and I. Ray, 113–124. Springer, Cham.

de Matos, E., L. A. Amaral, and F. Hessel. 2017. "Context-aware systems: technologies and challenges in Internet of Everything environments." In *Beyond the Internet of Things*, edited by J. M. Batalla, G. Mastorakis, C. X. Mavromoustakis, and E. Pallis, 1–25. Springer, Cham.

Desai, S., R. Alhadad, N. Chilamkurti, and A. Mahmood. 2019. "A survey of privacy preserving schemes in IoE enabled Smart Grid Advanced Metering Infrastructure." *Cluster Computing* 22 (1): 43–69.

Feng, S., P. Setoodeh, and S. Haykin. 2017. "Smart Home: Cognitive interactive people-centric Internet of Things." *IEEE Communications Magazine* 55 (2): 34–39.

Gambetti, M., F. Garzia, F. J. V. Bonilla, et al. 2017. "The new communication network for an Internet of Everything based security/safety/general management/visitor's services for the Papal Basilica and Sacred Convent of Saint Francis in Assisi, Italy." *2017 International Carnahan Conference on Security Technology (ICCST)*, Madrid.

Garzia, F., and L. Sant'Andrea. 2016. "The Internet of Everything based integrated security system of the World War One Commemorative Museum of Fogliano Redipuglia in Italy." *2016 IEEE International Carnahan Conference on Security Technology (ICCST)*, Orlando.

Gubbi, J., R. Buyya, S. Marusic, and M. Palaniswami. 2013. "Internet of Things (IoT): A vision, architectural elements, and future directions." *Future Generation Computer Systems* 29 (7): 1645–1660.

Honan, G., A. Page, O. Kocabas, T. Soyata, and B. Kantarci. 2016. "Internet-of-Everything oriented implementation of secure Digital Health (D-Health) systems." *2016 IEEE Symposium on Computers and Communication (ISCC)*, Messina.

Ismail, S. F. 2017. "IOE solution for a diabetic patient monitoring." *2017 8th International Conference on Information Technology (ICIT)*, Amman.

Kaur, M. J., S. Riaz, and A. Mushtaq. 2020. "Cyber-physical cloud computing systems and Internet of Everything." In *Principles of Internet of Things (IoT) Ecosystem: Insight Paradigm*, edited by Sheng-Lung Peng, Souvik Pal, and Lianfen Huang, 201–227. Springer, Cham.

Kazmi, A., Z. Jan, A. Zappa, and M. Serrano. 2016. "Overcoming the heterogeneity in the Internet of Things for Smart Cities." *International Workshop on Interoperability and Open-source Solutions*, Stuttgart, Germany, 20–35.

Kim, D. Y., and S. Kim. 2017. "Dual-channel medium access control of low power wide area networks considering traffic characteristics in IoE." *Cluster Computing* 20 (3): 2375–2384.

Kim, S., W. Jo, and T. Shon. 2020. "APAD: Autoencoder-based payload anomaly detection for industrial IoE." *Applied Soft Computing* 88: 106017.

Komar, M., S. Edelev, and Y. Koucheryavy. 2016. "Handheld wireless authentication key and secure documents storage for the Internet of Everything." *2016 18th Conference of Open Innovations Association and Seminar on Information Security and Protection of Information Technology (FRUCT-ISPIT)*, St. Petersburg.

Kuada, E. 2018. "Trust modelling and management system for a hyper-connected World of Internet of Everything." *2018 IEEE 7th International Conference on Adaptive Science & Technology (ICAST)*, Accra.

Mahdavinejad, M. S., M. Rezvan, M. Barekatain, P. Adibi, P. Barnaghi, and A. P. Sheth. 2018. "Machine learning for Internet of Things data analysis: A survey." *Digital Communications and Networks* 4 (3): 161–175.

Majeed, A. 2017. "Developing countries and Internet-of-Everything (IoE)." *2017 IEEE 7th Annual Computing and Communication Workshop and Conference (CCWC)*, Las Vegas.

Manashty, A., and J. L. Thompson. 2017. "Cloud platforms for IoE Healthcare context awareness and knowledge sharing." In *Beyond the Internet of Things*, edited by J. M. Batalla, G. Mastorakis, C. X. Mavromoustakis, and E. Pallis, 303–322. Springer, Cham.

Mele, C., and T. Russo-Spena. 2019. "Innovation in sociomaterial practices: The case of IoE in the healthcare ecosystem." In *Handbook of Service Science*, edited by P. P. Maglio, C. A. Kieliszewski, J. C. Spohrer, K. Lyons, L. Patrício, and Y. Sawatani. Springer, Cham.

Meridou, D., A. Kapsalis, P. Kasnesis, C. Patrikakis, I. Venieris, and D. T. Kaklamani. 2015. "An Event-driven Health Service Bus." *Proceedings of the 5th EAI International Conference on Wireless Mobile Communication and Healthcare*, London, 267–271.

Meridou, D. T., M. E. Ch. Papadopoulou, A. P. Kapsalis, et al. 2017. "Improving quality of life with the Internet of Everything." In *Beyond the Internet of Things*, edited by J. M. Batalla, G. Mastorakis, C. X. Mavromoustakis, and E. Pallis. New York: Springer International Publishing, 377–408.

Miao, Y., X. Liu, K. K. R. Choo, R. H. Deng, H. Wu, and H. Li. 2019. "Fair and dynamic data sharing framework in cloud-assisted Internet of Everything." *IEEE Internet of Things Journal* 6 (4): 7201–7212.

Miraz, M. H., M. Ali, P. S. Excell, and R. Picking. 2015. "A review on Internet of Things (IoT), Internet of Everything (IoE) and Internet of Nano Things (IoNT)." *2015 Internet Technologies and Applications (ITA)*, Wrexham.

Naranjo, P. G. V., Z. Pooranian, M. Shojafar, M. Conti, and R. Buyya. 2019. "FOCAN: A Fog-supported smart city network architecture for management of applications in the Internet of Everything environments." *Journal of Parallel and Distributed Computing* 132: 274–283.

Nath, S., A. Seal, T. Banerjee, and S. K. Sarkar. 2017. "Optimization using swarm intelligence and dynamic graph partitioning in IoE infrastructure: Fog computing and cloud computing." *International Conference on Computational Intelligence, Communications, and Business Analytics*, Thessaloniki, Greece.

Palattella, M. R., and N. Accettura. 2018. "Enabling Internet of Everything everywhere: LPWAN with satellite backhaul." *2018 Global Information Infrastructure and Networking Symposium (GIIS)*, Thessaloniki.

Pena, P. A., D. Sarkar, and P. Maheshwari. 2015. "A big-data centric framework for smart systems in the world of Internet of Everything." *2015 International Conference on Computational Science and Computational Intelligence (CSCI)*, Las Vegas.

Pradeep, Ch. N., M. K. Rao, and B. S. Vikas. 2019. "Quantum cryptography protocols for IOE security: A perspective." *International Conference on Advanced Informatics for Computing Research*, Shimla.

Rathod, A., P. Ayare, R. Bobhate, R. Sachdeo, S. Sarode, and J. Malhotra. 2020. "IoE-enabled smart embedded system: An innovative way of learning." In *Information and Communication Technology for Sustainable Development*, edited by M. Tuba, S. Akashe, and A. Joshi, 659–668. Springer, Singapore.

Rawat, P., K. D. Singh, and J. M. Bonnin. 2016. "Cognitive radio for M2M and Internet of Things: A survey." *Computer Communications* 94: 1–29.

Schatten, M., J. Ševa, and I. Tomičić. 2016. "A roadmap for scalable agent organizations in the Internet of Everything." *Journal of Systems and Software* 115: 31–41.

Shi, W., J. Cao, Q. Zhang, Y. Li, and L. Xu. 2016. "Edge computing: Vision and challenges." *IEEE Internet of Things Journal* 3 (5): 637–646.

Sianaki, O. A., A. Yousefi, A. R. Tabesh, and M. Mahdavi. 2018. "Internet of everything and machine learning applications: Issues and challenges." *2018 32nd International Conference on Advanced Information Networking and Applications Workshops (WAINA)*, Krakow.

Velasquez, K., D. P. Abreu, M. R. M. Assis, et al. 2018. "Fog orchestration for the Internet of Everything: State-of-the-art and research challenges." *Journal of Internet Services and Applications* 9 (1): 14.

Wang, L., G. Von Laszewski, A. Younge, et al. 2010. "Cloud computing: A perspective study." *New Generation Computing* 28 (2): 137–146.

# 4 Power Allocation Techniques for Visible Light
## Nonorthogonal Multiple Access Communication Systems

C. E. Ngene, Prabhat Thakur, and
Ghanshyam Singh
University of Johannesburg

## CONTENTS

| | |
|---|---|
| 4.1 Introduction | 46 |
| 4.2 Related Work | 46 |
|     4.2.1 Visible Light Communication | 46 |
|     4.2.2 Nonorthogonal Multiple Access | 47 |
|     4.2.3 Visible Light–Nonorthogonal Multiple Access Communication System | 49 |
| 4.3 Problem Formulation and Potential Contribution | 50 |
| 4.4 System Model for VL-NOMA Communication System | 50 |
| 4.5 Power Allocation Techniques with Decoding Order for VL-NOMA Communication Systems | 51 |
| 4.6 Conventional Power Allocation Scheme for VL-NOMA Communication System | 55 |
| 4.7 Inverse Power Allocation Scheme for VL-NOMA Communication System | 55 |
| 4.8 Adaptive Power Allocation Scheme for VL-NOMA Communication System | 58 |
| 4.9 Gain Ratio Power Allocation Scheme for VL-NOMA Communication System | 62 |
| 4.10 User Data Rate for GRPA VL-NOMA Communication System | 64 |
| 4.11 Joint Power Allocation for VL-NOMA Communication System | 65 |
| 4.12 Optimal PA Technique in VL-NOMA Communication System | 69 |

4.13 Comparison of Power Allocation Techniques ............................................. 70
4.14 Summary ........................................................................................................ 72
References ............................................................................................................. 73

## 4.1 INTRODUCTION

Due to exponential growth of smartphones with other connected devices, the need of radio frequency (RF)-based high-speed wireless data/information transfer technologies are pushing toward their limits, and rapidly growing wireless data traffic is expected. Recently, it was estimated the monthly global mobile data traffic will reach up to 77 exabytes by 2022, and annual traffic will be almost one-zettabyte as reported by Cisco VNI (visual networking index) forecast update for (2017–2022). Therefore, both the optical fiber Internet backbone speeds and RF spectrum scarcity along with channel capacity gap between these two have motivated the research fraternity to anticipate an alternative solution and spectrum resources. The visible light (VL) communication has received potential attention as an alternative or complementary technology to RF communications primarily for indoor environments. Further, due to considerable advances in the micro- and nano-electronic devices like the light-emitting diodes (LEDs)/photodiodes (PDs) as a sources/detector, respectively, the visible light supports remarkably high-speed wireless communication (transmit data in the visible light spectrum with the wavelength interval of 380–780 nm) and emerged as an auspicious technology in various applications like the Internet-of-Things (IoT), 5G/6G communication, underwater communication, vehicle-to-vehicle (V2V) communication, and so forth. The light wave is used as a communication channel without additional signals for data transmission and provides an opportunity for energy efficiency. Additionally, various potential features (wide available bandwidth, low-power consumption, unlicensed spectrum, enhanced confidentiality, and antielectromagnetic interference) make it the most attractive natural spectrum resource for future-generation communication systems (Vega et al., 2018). Further, the spectral efficient communication is the prime objective of the future-generation communication systems. Recently, the nonorthogonal multiple access (NOMA) communication technology has emerged as a prominent spectral efficient technique, which already has been well explored using the RF/microwave regime of the spectrum. The simultaneous exploitation of the visible light communication and NOMA techniques give birth to a new domain known as the VL-NOMA communication systems. The work related to the visible light communication, NOMA, and VL-NOMA techniques are presented as follows.

## 4.2 RELATED WORK

### 4.2.1 Visible Light Communication

The visible light communication technique has provided an unlicensed vast optical bandwidth (~THz) unlike the RF counterparts, and it has two potential applications such as the visibility and communication for which generally free-space medium is employed for transmission. In the laboratory environments, the pre- and postprocessing data rates up to several Gbps have been already demonstrated. The visible light communication has been integrated with the solid-state lighting technology because of its promising and constant progress for indoor communication systems (Cole and Driscoll, 2012; Pathak et al., 2015). This communication technology uses LEDs to replace the RF sources in different applications. However, the modulation bandwidth of the off-the-shelf LED that is designed mainly for illumination performance without considering the communication scenarios is a challenging task in practical implementation. LEDs are used for illuminations, communications, and modulating the LED light intensity in an indoor environment (Dimitrov and Haas, 2015) because its light intensity is modulated at a high-speed rate to avoid flickering (Kim et al., 2016). Considering the visible light communication applications in several areas, the researchers have studied visible light communication based on its unique characteristics for illumination and communication. Visible light is license-free spectrum, and high data rate make it acceptable for future wireless networks in advance technology (Feng et al., 2018). A study has been presented by Kim et al., 2016 considering the peak optical intensity when transmitting to prevent damage to eye when constantly viewed and have compared the potential characteristics of visible light over conventional RF for power controlled communication systems when transmitted using optical intensity for information communication. Further, the researchers (Gong et al., 2015; Kashef et al., 2014; Shen-Cong et al., 2016a; Shen-Hong et al., 2016b) have presented a technique to maintain the average optical intensity for illumination to be constant and allow the user to adjust if needed. The wireless access is offered as an "add-on service" of the luminary with miniaturization of visible light communication chips. With this potential characteristic and incontestable performance, the visible light is mainly positioned as an additive wireless access application to RF-based solutions for data off-loading. The visible light communication is exploited to enhance the indoor connectivity and facilitate user capacity with cost-effective deployment of extremely dense indoor network in downlink scenario. Further, it is also a fascinating alternative in the safety-critical and hostile environments in which the RF application is intolerable or partially restricted. Various health-conscious people also prefer to minimize the use of Wi-Fi, and it is even impermissible or restricted by governments in primary schools. Therefore, in addition to the indoor application, LEDs are progressively exploited in the outdoors, traffic-signals, automotive vehicle exterior lighting, and so forth, which also establish the way for visible light–based vehicle-to-vehicle (V2V) and vehicle-to-infrastructure (V2I) communication. Moreover, the RF system suffers from high interference levels to fulfill the desired latency, reliability, scalability, and capacity demand of vehicular networks, especially in congested environments. Therefore, it is expected to be a major breakthrough with the broad acceptance of visible light communication for intelligent transportation systems (ITSs).

### 4.2.2 Nonorthogonal Multiple Access

The nonorthogonal multiple access (NOMA) is an emerging communication technique that serves multiple users with the same resource at the same time, space, and frequency, within the network coverage. Various researchers (Ding et al., 2017;

Higuchi and Benjebbour, 2015; Islam et al., 2017) have proposed that low latency and ultra-high data rate will be in high demand for next generation communication systems, which have been fulfilled by employing the NOMA technique. With the power domain multiplexing access (PDMA) and code domain multiplexing access (CDMA), other NOMA techniques such as sparse code multiple access (SCMA), pattern division multiple access, and low-density spreading-CDMA (LDS-CDMA) have also been explored. However, NOMA has been explored more with PDMA techniques than other techniques as reported in Islam et al. (2017). In the PDMA technique, the multiuser's transmission and reception at the transmitter and receiver has been established using successive interference cancellation (SIC) at the receiver for better reception for multiusers, and superposition coding (SC) is performed at the transmitter for instantaneous transmission of multiusers data within the same channel. The orthogonal frequency division multiplexing (OFDM) in visible light communication has already been explored for significantly high data rate communication systems. However, due to its resistance to high spectral efficiency and intersymbol interference (Yuichi et al., 2001), the NOMA communication technique has attracted attention of various researchers as a new multiple-access technology. It also offers to work between both the throughput and system fairness in the wireless communications (Kazmi et al., 2018; Zhu et al., 2019). The RF communication networks have positioned its research to be explored by NOMA as a better candidate for future works (Xu et al., 2019; Yang et al., 2019). Further, it has been proven that NOMA performs better than orthogonal multiple access (OMA) on coverage and spectral efficiency techniques (Wei et al., 2016). OMA cannot allow multiple users simultaneously to use the entire space in the system, while NOMA can provide by "superposition coding" technique at the transmitter and "successive interference cancellation" technique at the receiver. Various researchers have studied NOMA performance in the RF communication systems. However, the power allocation and subchannel allocation techniques are proposed by (Al-Imari et al., 2014; Di et al., 2015; Thakur and Singh, 2019) for NOMA communication systems to yield the best output of the RF systems sum-rate as compare to that in a multicarrier. The consumed power of network data rate is summarized by Zhang et al. (2016, 2017b) for power distribution in the NOMA communication system that utilizes this network's energy competence. Further, NOMA has been defined as a power domain, which is recommended as a candidate for fifth-generation communication systems (Yang et al., 2019). Ding et al. (2014) has studied and observed that NOMA with RF networks provides a substantial enhancement in their throughput. It also possesses the characteristics of a spectrally efficient multiple-access (MA) system for mobile communication systems in the next generation (Ding et al., 2014; Selvam and Kumar, 2019; Sohail et al., 2018). The use of NOMA for downlink scenario has been proposed by the researchers for a 3G partnership project (Third Generation Partnership Project [3GPP], 2015) in LTE systems, which can be mentioned as multiuser superposition transmission. In addition to this, NOMA can reduce the latency and increase the system capacity as one of the characteristics of radio access technology in 5G mobile communication systems (5G radio access, 2014; Proposed solutions for new radio access; 2015). In the NOMA communication system the users can be deployed randomly in cellular downlinks (Ding et al., 2014), and it can attend to multiple users concurrently in the power domain with ease, while orthogonal frequency division multiplexing (OFDM) cannot (Timotheou and Krikidis, 2015). Further, Thakur et al. (2019) have presented a framework for simultaneous use of cognitive radio (CR) and NOMA in order to improve the spectral efficiency, which is named as CR-NOMA. Moreover, to enhance the spectral efficiency, Thakur et al. (2019) and; Thakur and Singh (2020) have also explored the simultaneous use of CR-NOMA with the multiple-input multiple-output (MIMO). The performance of the proposed framework is analyzed for the downlink as well as uplink scenarios.

### 4.2.3 VISIBLE LIGHT–NONORTHOGONAL MULTIPLE ACCESS COMMUNICATION SYSTEM

The NOMA communication systems have been established to accommodate a large number of users with different techniques in communications systems, more especially when implemented in visible light communication (VLC) with high signal-to-noise ratio (SNR) for more performance gain (Timotheou and Krikidis, 2015). In the VLC downlink systems, the NOMA communication techniques are engaged in the process to enhancement of high-rate throughput with multiuser (Tao et al., 2019). NOMA is used in VLC systems for the realistic enhancement of its throughput (Lin et al., 2019; Zhao et al., 2018), particularly when multiple users are using the transmitted signal, unlike OMA, OFDMA, and time division multiple access (TDMA). The VLC limited bandwidth efficiency has been supported by NOMA for more efficient bandwidth (Guo et al., 2019; Rodoplu et al., 2020). NOMA performs better in VLC than in RF systems in indoor communication environments (Dai et al., 2015). With direct line-of-sight (DLOS) communication in visible light, the free space makes it possible to estimate accuracy in indoor VLC networks channels and with a user's low mobility. This estimated accuracy in the VLC channel helps to reduce the errors found at the successive interference at the receiver circuit (Bawazir et al., 2018; Marshoud et al., 2016, 2017). Researchers have studied the behavior of NOMA toward VLC systems, ranging from the coverage point of view and their rate. In Li et al. (2020), the authors have obtained a larger data rate in NOMA than that of OMA, using double users in one optical cell. Further, researchers have observed within the communication system that with the sum-rate positively tuning the field of view of PDs and semiangle of LEDs (Tao et al., 2018). In Kizilirmak et al. (2015), NOMA behavior is compared in a VLC network by its illumination, and superiority was shown clearly. Early research on the behavior evaluation for multiusers in VL-NOMA communication system has been made in Shahjalal et al. (2019). Potential challenges focus on the PA in VL-NOMA communication systems and is explored based on the trade-off within user fairness, decoding order, and entire achievable data rate, which is a major concern as it affects system performance. We provide some basic different power allocation schemes with their architecture and comparison to illustrate how they will work better for particular communication systems. We explored these schemes to solve the problem of high achievable data rates, user fairness, sum rate, higher spectral efficiency, and user experiences to the end users in VL-NOMA.

## 4.3 PROBLEM FORMULATION AND POTENTIAL CONTRIBUTION

The performance of VLC and NOMA systems is the function of the power allocated to the users. The power allocation (PA) techniques are employed to enhance the performance and improve the signals to the end users in a system in terms of sum rate, high achievable data rate, user fairness, higher spectral efficiency, user experiences, and so on. Thus, it is worthwhile to study the various power allocation mechanisms with their pros and cons.

Therefore, in this chapter, we have summarized various power allocation techniques with regard to VL-NOMA communication systems. The discussed techniques are as follows: (1) conventional power allocation, (2) gain ratio power allocation (GRPA), (3) inverse power allocation, (4) adaptive power allocation (APA), (5) joint power allocation, and (6) optimal power allocation techniques in VL-NOMA communication systems. The rest of the chapter is structured as follows.

## 4.4 SYSTEM MODEL FOR VL-NOMA COMMUNICATION SYSTEM

The lighting system that comprises the end users is used to demonstrate how the proposed system model operates in Figure 4.1. The LEDs are used as a lighting access that transmits data under an indoor room environment to the end users. The signal reaches to various users accordingly with the help of various PA techniques. In the presented model, NOMA is applied in the VLC system to redistribute the signals to the users with weaker and stronger signal paths. Different users access the signal with different electronic devices from various angles under an indoor environment where the lighting can reach. The signals intended for the end users are superimposed by using power domain multiplexing at the transmitting end and distributed evenly to the users with weak and stronger paths, depending on the location of

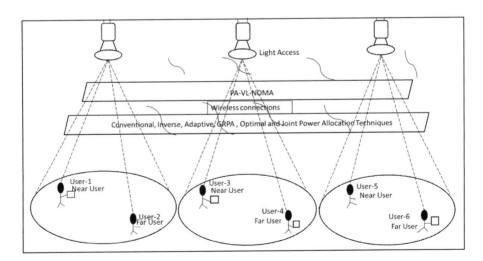

**FIGURE 4.1** Framework of lighting access using PA techniques to enhance the performance of the VL-NOMA communication system.

# Power Allocation Techniques for Visible Light

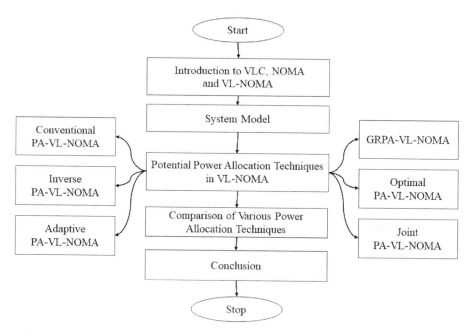

**FIGURE 4.2** Schematic of various power allocation schemes in VL-NOMA communication systems.

accessibility. The successive interference cancellation (SIC) is used to receive and decode the signals at the end users. The power to the different users in NOMA is assigned through various PA schemes to make sure that each user accesses the signal properly. The better performance and enhancement of these signals in the VL-NOMA system is performing by different PA is presented.

The use of these PA techniques in the VL-NOMA system improves the performance of the system by following ways: to enhance the system spectral efficiency, to improve the system user data rate, to better the performance of user fairness, to improve the system user data sum-rate, to improve on user experiences, and network coverage is easily accessed by multiusers when they are in operation, It works perfectly well in indoor VLC room environments, to open a research area in telecommunication industries. We have presented a flow diagram in this chapter as shown in Figure 4.2 for better understanding of the presented work. We discuss each of them in sections, providing arguments, problems, and solutions to improve the performance of the VL-NOMA system.

## 4.5 POWER ALLOCATION TECHNIQUES WITH DECODING ORDER FOR VL-NOMA COMMUNICATION SYSTEMS

The conventional PA, inverse PA, and adaptive PA techniques can be group under PA, which has the unique techniques for decoding its signals to enhance the performance of the system in VL-NOMA communication scenario to the end users

(Dong et al., 2019). The architecture of PA networks for decoding the VL-NOMA communication system is illustrated in Figure 4.3, where we perform the scenario on how the PA is utilized through decoding the signals in the system. One lighting access LED bulb is used that transmits data, and the data transmitted passes through decoding sections as the users receive the signals (Dong et al., 2019). Figure 4.3 illustrates a PA scheme for decoding signals in the VL-NOMA communication scenario. The LED is used as an optical carrier to send the signal to different users $M$, downlink indoor environment is used as a case study, and its features in this context are for illumination, converting electrical signals to optical signals, modulating the light emitted through its intensity, and sending data. Figure 4.3 demonstrates that multiple signals are superposed and transmitted using a single transmitter; however, multiple receivers are there to decode their own signal using successive interference cancellation (Dong et al., 2019). The notation $R$ is for maximum cell radius; $H$ is for the LED vertical distance to the received users; $p$ represents the PA of each user; $P_1$, $P_2$, $P_k$, $P_M$ denotes PA assigned to number of users; $S_1$, $S_2$, $S_k$, $S_M$ represents each number of signals of the users present; $r_1$, $r_k$, $r_M$ is for the separation of the LED horizontal $k$th users ($User_1U_1$, $User_kU_k$, $User_MU_M$); $f$ represents the frequency of the signal; and $t$ represents the time taken for the signal to reach to each user. The demonstrated architecture considers Lambertian radiation pattern as LED is stationed at each user to receive the signal having a FOV, which is given as $\psi_{FOV}$.

The line-of-sight (LOS) element in this scenario is observed to be greater than the diffuse components in $dB$ because of its weakness (Chen et al., 2017). Calculation of the $k$th users' direct current (DC) channel gain is done using the wideband VLC nature, effect of the shadow, and LOS channel. The equation for this given as:

$$h_k \frac{(m+1)AR_p}{2\pi d_k^2} \cos^m(\Phi_k)\cos(\psi_k)T_s(\psi_k)g(\psi_k) = \frac{AR_p(m+1)H^{m+1}T_s(\psi_k)g(\psi_k)}{2\pi(r_k^2+H^2)^{m+3/2}}, \quad (4.1)$$

where the Lambertian radiation pattern order is denoted by $m$, the LED semiangle denotes $\phi_1/2$, $\phi_1/2$ for $m = -1/\log_2(\cos(\phi_{1/2}))$, $A$, $R_p$, and $d_k$ denote the PD physical area, PD responsivity, and the distance of Euclidean within the LED and $k$th user. $\phi_k$ denotes the irradiance angel at the $k$th user, $\psi_k$ denotes the incidence angle at the $k$th user, $T_s\psi_k$ describes the optical fiber gain use at the receiver side, the vertical height represents $H$, and the visual front end receiver used denotes by $g(\psi_k)$ that is given as in Kahn and Barry (1997):

$$g(\psi_k) = \begin{cases} \dfrac{n^2}{\sin^2\psi_{FOV}} & 0 \leq \varphi_k \leq \psi_{FOV} \\ 0 & \varphi_k \geq \psi_{FOV}, \end{cases} \quad (4.2)$$

where the optical concentrator refractive index use at the receiver denotes $n$, the DC channel gain considered in this context, every user from $U_1, \ldots, U_M$ is selected in an ascending directive given as:

$$h_1 \leq h_2 \leq \ldots \leq h_k \leq \ldots \leq H_M. \quad (4.3)$$

# Power Allocation Techniques for Visible Light 53

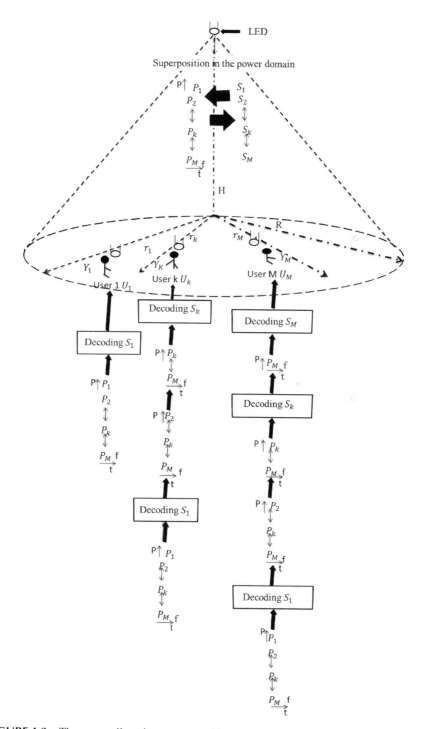

**FIGURE 4.3** The power allocation system architectures for decoding signals in a VL-NOMA communication system.

The transmitter side narrates NOMA principle from Figure 4.3, using this $\{s_i, i = 1, 2, \ldots M\}$ information at the power domain, which superimposes to various users with relative power values such as $\{P_i, i = 1, 2, \ldots M\}$ and consecutively transmitted with this equation as:

$$x = \sum_{i=1}^{M} a_i \sqrt{P_{elec}} s_i + I_{DC}, \quad (4.4)$$

where the signal superposed in $\{s_i, i = 1, 2, \ldots, M\}$ denotes $x$, the total information signals of all the electrical power denotes by $P_{elec}$. The DC bias included just before transmission for proper signal denotes $I_{DC}$, and the coefficient of the power allocation for the user $i$th user denotes $a_i$, that used to confirm these two situations with equations as:

$$\sum_{i=1}^{M} a_i^2 = 1, \quad (4.5)$$

for effective constraint of the total power, which takes the same order as in Equation (4.3). According to the basic principles of conventional NOMA in Dong et al. (2019), likewise, the coefficient power allocation, other related parameters, the factor of the power allocation $\alpha$, is given as $\alpha = a_i^2 / a_{i-1}^2, i = 1, \ldots M$, that is either variable or constant with $i$ considering the different types of power allocation methods such as GRPA or fixed power allocation (FPA). For the $k$th user, the received signal is obtained by taking the DC channel gain and the Additive White Gaussian Noise (AWGN) into consideration in the system and eliminating the DC term at the receiver, which is given as:

$$y_k = \sqrt{P_{elec}} h_k \left( \sum_{i=1}^{M} a_i s_i \right) + z_k, \quad (4.6)$$

where the signal received at the $k$th user denotes $y_k$, and the AWGN having a zero mean with $\sigma_k^2$ variance denotes $z_k$. Therefore, $\sigma_k^2 = N_0 B$, the spectral power density noise denotes $N_0$, and the bandwidth channel denotes $B$. Subsequently, to extract $s_k$ within the signal received successive interference cancellation (SIC) carried out following these procedures, such as to make sure that the first user signal information is obtained. The subsequent signals are taken as noise; hence $s_1$ is subtracted from the signal received. The outstanding interference fraction represented by $\varepsilon$ (Andrews and Meng, 2003), and the signal information from the stronger user channel gain is processed while the second user taken as noise, the information signal obtained using $s_2$ and last, the former technique follows the sequence order obtained as $s_3, \ldots, s_{k-1}, s_k$. The Theorem of Shannon is expressed based on the realizable data rate for the $k$th user, which simplified in the equation as:

$$R_k \begin{cases} \dfrac{B}{2}\log_2\left(1+\dfrac{(h_k a_k)^2}{\sum_{i=1}^{k-1}\varepsilon(h_k a_i)^2+\sum_{j=k+1}^{M}(h_k a_j)^2+1/p}\right) & k=1,\ldots,M-1 \\ \dfrac{B}{2}\log_2\left(1+\dfrac{(h_k a_k)^2}{\sum_{i=1}^{k-1}\varepsilon(h_k a_i)^2+1/p}\right) & k=M, \end{cases} \quad (4.7)$$

where $\rho = P_{elec}/(N_0 B)$ and 1/2 is the scaling factor obtained from the signal constraint real-valued, which is symmetry, and $\varepsilon$ represents residual interference fraction.

## 4.6 CONVENTIONAL POWER ALLOCATION SCHEME FOR VL-NOMA COMMUNICATION SYSTEM

Conventional PA provides high power to users with a worse channel condition to enhance the data rate of the signals in Dong et al. (2019). The decoding order is performed to better the system performance. In this scenario conventional PA is applied in VL-NOMA to enhance the performance of the system throughput. Hence, user fairness with regard to its gain is achieved as a result of the entire achievable data rate. This is proven in Equations (4.8–4.14); we combine conventional PA and inverse PA to improve the balance of user fairness and the entire achievable data rate. Therefore, we present both in Figure 4.4 and Figure 4.5 as inversely proportional to each other in terms of user fairness and the entire achievable data rate using decoding order approach on their signals.

## 4.7 INVERSE POWER ALLOCATION SCHEME FOR VL-NOMA COMMUNICATION SYSTEM

The architectures for decoding order in inverse power allocation scheme for VL-NOMA communication systems is presented in Figure 4.5. From Figure 4.5 IPA technique is established when from the input information (transmitter side), less power is assigned to users with low-quality channel condition and at the output information (receiver side), the decoding order is deployed to users with a low-quality channel condition. Inverse decoding order of signals is just the opposite of conventional decoding order of signals (Dong et al., 2019). We then illustrate the difference between the links in IPA technique and the conventional power allocation technique using similar equations. From Figures 4.4 and 4.5, two users are used for proper illustration, making $M = 2$. These equations are given as:

$$a_i \geq a_2, a_1^2 + a_2^2 = 1, \quad (4.8)$$

$$a_1' \leq a_2', \quad (4.9)$$

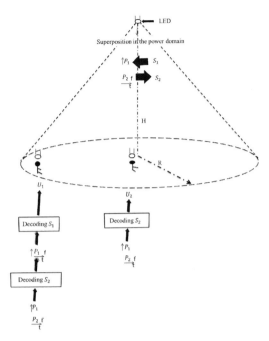

**FIGURE 4.4** Architectures for decoding order in conventional power allocation scheme for VL-NOMA communication systems.

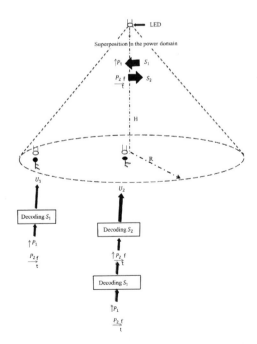

**FIGURE 4.5** Architectures for decoding order in inverse power allocation scheme for VL-NOMA communication systems.

# Power Allocation Techniques for Visible Light

and

$$a'^2_1 + a'^2_2 = 1, \qquad (4.10)$$

Where $a_i$ denotes the coefficient of the power allocation $i$ for the user within the IPA and conventional PA technique case, subsequently, then $i = 1, 2$. Hence, in Equations (4.10) and (4.11) can be obtained from the two cases below.

$$\begin{cases} a_1^2 = \dfrac{1}{1+\alpha} \\ a_2^2 = \dfrac{\alpha}{1+\alpha} \end{cases} \qquad (4.11)$$

$$\begin{cases} a'^2_1 = \dfrac{\alpha}{1+\alpha} \\ a'^2_2 = \dfrac{1}{1+\alpha} \end{cases} \qquad (4.12)$$

Moreover, $\alpha$, which is the PA factor that is assumed to be lower than 1, is given as $a^2_2/a^2_1$ or $a'^2_1/a'^2_2$ as in the typical IPA case, subsequently. After combining Equations (4.6) and (4.10), the conventional PA case for the two users regarding the entire achievable data rate is written as:

$$\begin{aligned} R_{total} &= \frac{B}{2}\log_2\left(1+\frac{(h_1 a_1)^2}{(h_1 a_2)^2+1/p}\right)+\frac{B}{2}\log_2\left(1+\frac{(h_2 a_2)^2}{\varepsilon(h_2 a_1)^2+1/p}\right) \\ &= \frac{B}{2}\log_2\left[\left(1+\frac{(h_1 a_1)^2}{h_1 a_2^2+1/p}\right)\left(1+\frac{(h_2 a_2)^2}{\varepsilon(h_2 a_1)^2+1/p}\right)\right] \\ &= \frac{B}{2}\log_2\left[\left(\frac{h_1^2}{h_1^2\alpha+(\alpha+1)/p}\right)\left(1+\frac{h_2^2\alpha}{\varepsilon h_2^2+(\alpha+1)/p}\right)\right] \end{aligned} \qquad (4.13)$$

Equally, the IPA case for the two users regarding the entire achievable data rate is written as:

$$\begin{aligned} R'_{total} &= \frac{B}{2}\log_2\left(1+\frac{(h_1 a'_1)^2}{\varepsilon(h_1 a'_2)^2+1/p}\right)+\frac{B}{2}\log_2\left(1+\frac{(h_2 a'_2)^2}{(h_2 a'_1)^2+1/p}\right) \\ &= \frac{B}{2}\log_2\left[\left(1+\frac{(h_2 a'_2)^2}{(h_2 a'_1)^2+1/p}\right)\right]=\frac{B}{2}\log_2\left[\left(1+\frac{(h_2 a'_2)^2}{(h_2 a'_1)^2+1/p}\right)\right] \\ &= \frac{B}{2}\log_2\left[\left(1+\frac{h_1^2\alpha}{\varepsilon h_1^2+(\alpha+1)/p}\right)\left(1+\frac{h_2^2}{h_2^2\alpha+(\alpha+1)/p}\right)\right] \end{aligned} \qquad (4.14)$$

## 4.8 ADAPTIVE POWER ALLOCATION SCHEME FOR VL-NOMA COMMUNICATION SYSTEM

The adaptive PA is adopted in this section based on multiattribute decision making (MADM), making the decision parameters to apply in the entire user fairness and achievable data rate. It is considered to be used for the combination of conventional PA and inverse PA. The focus is to combine the PA factor and techniques and the changes in the location of the most suitable users (Dong et al., 2019). Furthermore, $u \times v$ becomes the choice space matrix. The number of the candidate system taken to be $u$, given that is $u = 2$ and $v$ has chosen to be power allocation factors of the candidate number, that is assumed. The implementation of this process is described as the first standard deviation method; this utilizes a mathematical solution to solve the multiple attribute decision-making (MADM) variance problem (Wang, 2003). It is applied to address each decision parameter of actual weight. Hence, the order preference technique by the resemblance to the ideal solution Technique for Order Preference by Similarity to Ideal Solution (TOPSIS), that sort the best message of the real data, is also used to sort the best order combinations chosen by the candidate as reported in Xiao and Li (2018). The parameters decision on each actual weight is processed using standard deviation technique; the equations given as (Dong et al., 2019):

First, the decision matrix C is offered as:

$$C = \begin{pmatrix} C_{11} & C_{12} \\ C_{21} & C_{22} \\ \cdots & \cdots \\ \cdots & \cdots \\ C_{P1} & C_{P2} \end{pmatrix}, \quad (4.15)$$

where $p$ represents the combination of the total number of the candidates, therefore $p = u \times v$, and features $C_{k1}$, and $C_{k2}$ represents the entire achievable user fairness and data, respectively if the combination of the candidate is taken as $k$th. The given parameters are beneficiary due to their regularization in Lahby et al. (2014). The equation is written as:

$$C_{ki} = \frac{S_{ki} - \min(S_{xi}, 1 \leq x \leq p)}{\max(S_{xi}, 1 \leq x \leq p) - \min(S_{xi}, 1 \leq x \leq p)}, \quad 1 \leq k \leq p, i = 1, 2, \quad (4.16)$$

where the entire achievable value data rate combination of the candidate when the $k$th selected is denoted by $S_{k1}$, this can be calculated with Equations (4.13) and (4.14), and $S_{k2}$ represents the user fairness value combination of the candidate when the $k$th is selected, that is written as:

# Power Allocation Techniques for Visible Light

$$S_{k2} = \frac{\min(R_1|k, R_2|k)}{\max(R_1|k, R_2|k)}$$

$$= \begin{cases} \dfrac{\min\left(\dfrac{B}{2}\log_2\left(1+\dfrac{h_1^2}{h_1^2\alpha+(\alpha+1)/p}\right), \dfrac{B}{2}\log_2\left(1+\dfrac{h_2^2\alpha}{\varepsilon h_2^2+(\alpha+1)/p}\right)\right)}{\max\left(\dfrac{B}{2}\log_2\left(1+\dfrac{h_1^2}{h_1^2(\alpha+1)/p}\right), \dfrac{B}{2}\log_2\left(1+\dfrac{h_2^2\alpha}{\varepsilon h_2^2+(\alpha+1)/p}\right)\right)}, \\ \quad k \le p/2, r_1 \ge r_2, \text{ or } k > p/2, r_1 \le r_2 \\[2ex] \dfrac{\min\left(\dfrac{B}{2}\log_2\left(1+\dfrac{h_1^2\alpha}{\varepsilon h_1^2+(\alpha+1)/p}\right), \dfrac{B}{2}\log_2\left(1+\dfrac{h_2^2}{h_2^2\alpha+(\alpha+1)/p}\right)\right)}{\max\left(\dfrac{B}{2}\log_2\left(1+\dfrac{h_1^2\alpha}{\varepsilon h_1^2+(\alpha+1)/p}\right), \dfrac{B}{2}\log_2\left(1+\dfrac{h_2^2}{\varepsilon h_2^2\alpha+(\alpha+1)/p}\right)\right)}, \\ \quad k \le p/2, r_1 \le r_2, \text{ or } k > p/2, r_1 \ge r_2, \end{cases} \quad (4.17)$$

where the user-1 and user-2 achievable data rate is denoted by $R_1|k, R_2|k$, subsequently, $\varepsilon$ denotes the residual interference fraction of the channel gain, if the combination of the candidate $k$th is selected, Second, the calculation of each decision parameters of the actual weight given as:

$$\omega_j = \frac{\sqrt{\sum_{i=1}^{P}\left(C_{ij} - \frac{1}{P}\sum_{i=1}^{P}C_{ij}\right)^2 / (P-1)}}{\sum_{i=1}^{2}\sqrt{\sum_{i=1}^{P}\left(C_{ij} - \frac{1}{p}\sum_{i=1}^{P}C_{ij}\right)^2 / (p-1)}}, \quad j=1,2, \quad (4.18)$$

where $\omega_1, \omega_2$ denotes the entire achievable user fairness and data rate of the actual weights purposes, subsequently, when each parameter of the actual weight is selected, conferring to TOPSIS, then the superior candidate combination is selected due to the main step that is used, such as:

*Step 1* The weight normalized decision matrix $D$ constructed as:

$$D = \begin{pmatrix} D_{11} & D_{12} \\ D_{21} & D_{22} \\ \cdots & \cdots \\ \cdots & \cdots \\ D_{P1} & D_{P2} \end{pmatrix} = \begin{pmatrix} \omega_1 C_{11} & \omega_2 C_{12} \\ \omega_1 C_{21} & \omega_2 C_{22} \\ \cdots & \cdots \\ \cdots & \cdots \\ \omega_1 C_{P1} & \omega_2 C_{P2} \end{pmatrix}, \quad (4.19)$$

*Step II* The positive ideal solution matrix $Y^+$ is determined as:

$$Y^+ = \left(Y_1^+ Y_2^+\right) = \left(\max_k{}^{(D_{k1})} \quad \max_k{}^{(D_{k2})}\right), k = 1, 2, \ldots, p. \quad (4.20)$$

*Step III* The negative ideal solution matrix $Y^-$ is determined as:

$$Y^- = \left(Y_1^- Y_2^-\right) = \left(\max_k{}^{(D_{k1})} \quad \max_k{}^{(D_{k2})}\right), k = 1, 2, \ldots, p. \quad (4.21)$$

*Step IV* The Euclidean distance between each solution and the positive ideal solution calculated as:

$$F_k^+ = \sqrt{\sum_{i=1}^{2}\left(D_{ki} - Y_i^+\right)^2}, \ k = 1, 2, \ldots, p. \quad (4.22)$$

*Step V* The Euclidean distance between each solution and the negative ideal solution calculated as:

$$F_k^- = \sqrt{\sum_{i=1}^{2}\left(D_{ki} - Y_i^+\right)^2}, k = 1, 2, \ldots, p. \quad (4.23)$$

*Step VI* The comparative nearness of each result to the ideal solution calculated as:

$$G_k = \frac{F_k^-}{F_k^+ + F_k^-}, \quad 0 \leq G_k \leq 1, k = 1, 2, \ldots, p. \quad (4.24)$$

*Step VII* The superior combination of the power allocation factor and power allocation system is found correspondingly as:

$$\underset{k}{\mathrm{argmax}} G_k, \quad k = 1, 2, \cdots, p. \quad (4.25)$$

Further, the adaptive allocation system is examined using a real-time scenario where users $M$ and $M > 2$ exist. The definition of $\alpha_{(i-1)i}$, which refers to the power allocation factor within user $(i-1)$th and user $i$th is clarified here for quick identification. The equation $\alpha_{(i-1)}$ can be written as:

$$\alpha_{(i-1)i} = \frac{a_i^2}{a_{i-1}^2}, i = 2, \ldots M. \quad (4.26)$$

The process, which is realistic to find the optimal $\alpha_{(i-1)}$, $i = 2, \ldots M$, is given as follows. First, $\alpha_{12}$, which is now an optimized variable to be, is used for power allocation expression of the entire other factors, such as $\alpha_{23}, \ldots, \alpha(M-1)M$. Following the

# Power Allocation Techniques for Visible Light

GRPA concept, $\alpha_{(i-1)i} = (h_1/h_i)^i$, $i = 2, ..., M$. Due to this equation, the recursion relations of the power allocation factor is easily obtained as:

$$\alpha_{i(i+1)} / \alpha_{(i-1)i} = h_1 h_i^i / h_{i+1}^{i+1}, \quad i = 2, ..., M, \tag{4.27}$$

With these relations of recursion, $\alpha_{23}, ..., \alpha_{(M-1)M}$ can be easily expressed in terms of $\alpha_{12}$. Furthermore, the optimal of $\alpha_{12}$ is found by adaptive system propose, in that the Equations of (4.13), (4.14), and (4.17) are supposed to extend because the entire user's decision parameters taken into consideration, which is into account the user fairness and the entire achievable data rate. Precisely, Equations (4.13), (4.14), and (4.17) can be extending to Equations (4.28–4.30b), subsequently.

$$R_{total} = \sum_{k=1}^{M} R_k = \frac{B}{2}\log_2\left(1 + \frac{(h_M a_M)^2}{\sum_{i=1}^{M-1}\varepsilon(h_M a_i)^2 + 1/p}\right)$$

$$+ \sum_{k=1}^{M-1}\frac{B}{2}\log_2\left(1 + \frac{h_k a_k^2}{\sum_{i=1}^{k-1}\varepsilon(h_k a_i)^2 + \sum_{j=k+1}^{M}(h_k a_j)^2 + 1/p}\right) \tag{4.28}$$

$$R'_{total} = \sum_{k=1}^{M} R'_k = \frac{B}{2}\log_2\left(1 + \frac{(h_1 a'_1)^2}{\sum_{i=2}^{M}\varepsilon(h_1 a'_i)^2 + 1/p}\right)$$

$$+ \sum_{k=2}^{M}\frac{B}{2}\log_2\left(1 + \frac{(h_k a'_k)^2}{\sum_{i=k+1}^{M}\varepsilon(h_k a'_i)^2 + \sum_{j=1}^{k-1}(h_k a'_j)^2 + 1/p}\right), \tag{4.29}$$

$$S_{k2} = \frac{\min(R_1|k, R_2|k, ..., R_M|k)}{\max(R_1|k, R_2|k, ..., R_M|k)}, \quad \text{for Conventional PA,} \tag{4.30a}$$

and

$$S_{k2} = \frac{\min(R'_1|k, R'_2|k, ..., R'_M|k)}{\max(R'_1|k, R'_2|k, ..., R'_M|k)}, \quad \text{for Inverse PA,} \tag{4.30b}$$

where $a_k, a'_k$ is given as:

$$a_k \begin{cases} \sqrt{1/(1 + \alpha_{12} + \alpha_{12} \times \alpha_{23} \times ... \times \alpha_{(M-1)M})}, & k = 1 \\ \sqrt{\alpha_{12} \times \alpha_{23} \times ... \times \alpha_{(k-1)k} / \left(\begin{array}{c}1 + \alpha_{12} + \alpha_{12} \times \alpha_{23} + ... \\ + \alpha_{12} \times \alpha_{23} \times ... \times \alpha_{(M-1)M}\end{array}\right)}, & k = 2, ..., M \end{cases} \tag{4.31}$$

$$a'_k \begin{cases} \sqrt{1 / \begin{pmatrix} 1 + \alpha_{(M-1)M} + \alpha_{(M-1)M} \times \alpha_{(M-2)(M-1)} + \cdots \\ + \alpha_{12} \times \alpha_{23} \times \ldots \times \alpha_{(M-1)M} \end{pmatrix}} & , k = 1 \\ \sqrt{\alpha_{k(k+1)} \times \ldots \times \alpha_{(M-1)M} / \begin{pmatrix} 1 + \alpha_{(M-1)M} \times \alpha_{(M-2)M} \times \alpha_{(M-2)(M-1)} \\ + \ldots + \alpha_{12} \times \alpha_{23} \times \ldots \times \alpha_{(M-1)M} \end{pmatrix}} & , k = 2, \ldots, M \end{cases} \quad (4.32)$$

The expressions of $\alpha_{23}, \ldots, \alpha_{(M-1)M}$ can be substituted into Equations (4.31) and (4.32), it can be also obtain as $a_k$, $a'_k$ relations to $\alpha_{12}$. Last, when the optimal $\alpha_{12}$ is gotten, it is very easy to determine the optimal of $\alpha_{23}, \ldots, \alpha_{(M-1)M}$, respectively.

## 4.9 GAIN RATIO POWER ALLOCATION SCHEME FOR VL-NOMA COMMUNICATION SYSTEM

The gain ratio power allocation (GRPA) considers using the improvement of the user sum rate and mathematical approach to solve the problems of distributing power allocation signals power domain control evenly to the weaker and stronger users. To analyze the performance of GRPA when used in a random mobility walking prototype to simulate and coordinate the users within the network in an indoor environment has been proposed by Tao et al. (2018). This approach can be employed effectively using frequency resources in VL-NOMA systems.

A framework for GRPA in VL-NOMA is presented to enhance the entire system throughput, including the user fairness, total achievable data rate, and decoding order when in a transmission using LED as a carrier (Tao et al., 2018). The photodiode (PD) is usually used in VLC as a signal receiver. The LOS and mobility approach is not considered in this designed framework because they are not noticed; hence Komine and Nakagawa (2004) have used Lambertian radiation approach. The channel gain of VLC is considered for the $k$th user, which is given as (Tao et al., 2018):

$$h_k = \frac{(m+1)A}{2\pi d_k^2} \cdot \cos^m(\varnothing_k) \cdot T_{filter} \cdot g(\psi_k) \cdot \cos(\psi_k), \quad (4.33)$$

where $h_k$, is the channel gain of VLC for $k$th user, $m$ is the Lambertian emission order, $A$ is the area of the PD receiver detected, $d_k$ is the distance within the PD and the LED of the $k$th user, $T_{filter}$ is the optical gain filter constant, and $g(\psi_k)$ is the optical concentrator gain of the $k$th user. Hence, $m$ and $g(\psi_k)$ are given by Equations (4.34) and (4.35). From Figure 4.6, $\Phi_{1/2}$ represents the semiangle of the LED, $\Psi_{FOV}$ represents the angle field of view width, $L$ is the height of the $k$th user, and $r_k$ is the radius of the $k$th user. The Lambertian order of emission is denoted as $m$ in this equation, which is written as:

$$m = \frac{1}{\log_2(\cos(\Phi_{1/2}))}, \quad (4.34)$$

# Power Allocation Techniques for Visible Light

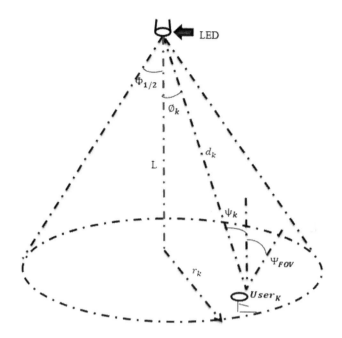

**FIGURE 4.6** The channel framework for GRPA for VL-NOMA communication system.

and the optical concentrator gain of the kth user is given as:

$$g(\psi_k) = \begin{cases} \dfrac{n^2}{\sin^2(\Psi_{FOV})}, & 0 \le \psi_k \le \Psi_{FOV} \\ 0, & \psi_k > \Psi_{FOV}, \end{cases} \quad (4.35)$$

where the $n$ denotes the refractive index constant. From Figure 4.6, the $L$ and $r_k$ can be substituted for $d_k$ for kth user in Equation (4.36), which is given as:

$$h_k = \dfrac{\left(\dfrac{A.T_{filter} \cdot g(\psi_k)}{2\pi}\right) \cdot (m+1) \cdot L^{m+1}}{\left(r_k^2 + L^2\right)^{\frac{m+3}{2}}} = \dfrac{\rho \cdot (m+1) \cdot L^{m+1}}{\left(r_k^2 + L^2\right)^{\frac{m+1}{2}}}, \quad (4.36)$$

where $\rho$ represents a constant equal to $\dfrac{A.T_{filter} \cdot g(\psi_k)}{2\pi}$. Then the received electrical power signal of the kth user is given as:

$$PR_k = y \cdot h_k \cdot PT_k. \quad (4.37)$$

where the received electrical signal power is $PR_k$, the transmitted optical signal power is $PT_k$, and the optical electrical conversion efficiency constant is $y$. From Figure 4.6, the VL-NOMA, the *k*th users enjoy the optical signal power transmitted *PT*, which is fixed, hence, the signal power is redistributed among itself and is given as:

$$a_1.PT > a_2.PT > ... > a_k.PT > ... a_k.PT, s.t. \sum_{k=1}^{K} a_k = 1, \quad (4.38)$$

where the signal power redistributed factor parameter is $a_k$, the signal transmitted from the LED to the *k*th user received is recorded in Equation (4.39), which is given as:

$$Y_k = y.PT.h_k. \sum_{i=1}^{K} a_i.s_i + n_k, \quad (4.39)$$

where the ON-OFF Keying signal denotes $s_i$, and the Gaussian noise with zero mean denotes $n_k$, which can be seen in (Jiang et al., 2018; Lapidoth et al., 2009; Shen-Hong et al., 2017a, 2017b; Yin et al., 2016). The *k*th users with weak channel quality condition are given more power signal for uniform data rate of all the *k*th users. This can lead to the ascending order of the channel gains of the users, which is given with the same order as in Equation (4.3).

## 4.10 USER DATA RATE FOR GRPA VL-NOMA COMMUNICATION SYSTEM

The GRPA uses user data rate to enhance the performance of a system in VL-NOMA scenario. The VLC system uses VLC channel gain $h_k$ where GRPA target optimization is to select the optimal parameter $a_k$ for its maximization throughput. The user data rate equation is given as (Tao et al., 2018):

$$Rate' k(h_k, a_k) = B'.\log_2(1 + \eta. SINR_k), k < K \quad (4.40)$$

where *Rate'* denotes the user data rate, $B'$ denotes the constant of the VLC bandwidth, $\eta$ denotes a constant $\frac{2}{\pi e}$ of the NOMA bandwidth common factor that can be extracted, and $SINR_k$ denotes the signal to interference noise ratio for the *k*th user, which is normally used to measure user data rate in VLC (Li et al., 2015; Yapici and Guvenc, 2019; Zhang et al., 2017a; Zhou and Zhang, 2017). The user data rate can also be expressed as:

$$SINR_k = \frac{(y.h_k. a_k.PT)^2}{\sum_{i=k+1}^{K}(y.h_k. a_k.PT)^2 + N_0 B}, k <, \quad (4.41)$$

where a constant noise power spectral density denoted $N_0$. For VL-NOMA achievable data rate for the *k*th user is given as:

$$Rate_k(h_k.a_k) = \frac{B}{2}.\log_2\left(1 + \frac{2}{\pi e}.SINR_k\right), k < K, \quad (4.42)$$

where the Euler number is $e$ and the value determined is $\pi$.

## 4.11 JOINT POWER ALLOCATION FOR VL-NOMA COMMUNICATION SYSTEM

The joint PA techniques combine two techniques and solve them jointly with the assistance of an algorithm such as: joint PA and orientation of LEDs in Obeed et al. (2018) solving their problems through algorithm, optimizing the system to consider the PA coefficients of the LED and vectors in VL-NOMA systems (Tran and Kim, 2019). PA and cell formation in Guo et al. (2009) are linked issues that have been solved jointly with the help of an algorithm. The Joint PA system came to optimally assign resources, namely, the power and channels in the multicarrier and maximize the performance of the system in VL-NOMA communication networks. It allocates resources to users in a closely optimal method to benefit NOMA technique. The artificial neural network (ANN) is deployed for channel performance assignment. It is noticed that simulation results record better performance in the system than the previous work done (He et al., 2019).

From Figure 4.7, we present a joint PA technique with LED that coordinates normal vectors in VL-NOMA indoor environment. The LED is used as an access for signal carrier to the end users. The PA is integrated with VLC lighting access (LED) to send the information to different end users (Varma, 2018; Varma et al., 2018; Wang

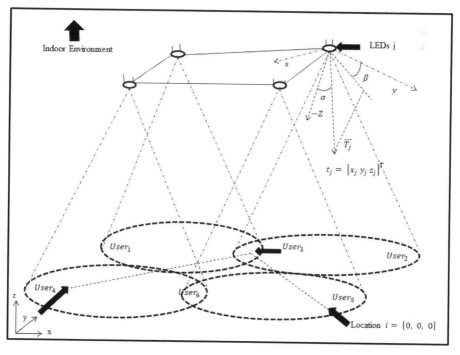

**FIGURE 4.7** Joint PA technique with LED coordinates of normal vector in VL-NOMA communication systems.

et al., 2012), defining all the parameters and properties of the VL-NOMA communication system. We have considered LOS VLC in Figure 4.7. We define the channel gain equation of the location of $i$th to the $j$th LED such as discussed in Komine and Nakagawa (2004):

$$h_{ij} = \frac{(y+1)A_p g}{2\pi d_y^2} \cos^y \varnothing_{ij} \cos\theta_{ij}, \qquad (4.43)$$

where $A$ denotes the set of K light received in the plane receiver PD, $N$ denotes the number of LED, $y$ denotes the Lambertian order, $\varnothing_{ij}$ denotes the irradiance angle of the $j$th LED in line with $i$th PD, $\theta_{ij}$ denotes the incident angle of the $j$th LED in line with $i$th PD, $d_{ij}$ denotes the distance within $j$th LED and $i$th PD. From Figure 1.7, the Lambertian coefficient is considered where $y = 1$ (Sewaiwar et al., 2015; Varma et al., 2018; Wang et al., 2012, 2014); therefore the equation for channel coefficient is given as (Tran and Kim, 2019):

$$h_{ij} = \frac{A_p g}{\pi d_y^2} \cos\varnothing_{ij} \cos\theta_{ij}. \qquad (4.44)$$

Relating the equation, it is given as:

$$\cos\varnothing_{ij} = \frac{\vec{T}_j \cdot \vec{V}_{ij}}{\|\vec{T}_j\| \|\vec{V}_{ij}\|}, \qquad (4.45)$$

and

$$\cos\theta_{ij} = \frac{-\vec{V}_{ij} \cdot \vec{U}_i}{\|\vec{V}_{ij}\| \|\vec{U}_i\|}, \qquad (4.46)$$

where $\vec{T}_j$ denotes the irradiance direction of the normal vector of the LED $j$–$th$, $\vec{V}_{ij}$ represents the PD $i$–$th$ within the LED $j$–$th$ vectors, and $\vec{U}_i$ represents the incident direction of the PD $i$–$th$ normal vector, normally $\vec{U}_i = [0\,0\,1]^T$. Equations (4.45) and (4.46) can be given as:

$$\cos\varnothing_{ij} = \frac{t_j^T v_{ij}}{\|t_j\| \|v_{ij}\|}, \qquad (4.47)$$

and

$$\cos\theta_{ij} = \frac{-v_{ij}^T U_i}{\|v_{ij}\| \|U_i\|}, \qquad (4.48)$$

where $t_j = \vec{T}_j$ vectors, $v_{ij} = \vec{V}_{ij}$ vectors, and $U_i = \vec{U}_j$ vectors. We considered that the PDs and the LEDs location are stationed in one place. Then the LED $j$th parameter with normal vector is given as:

$$\vec{T}_j(\alpha_j \beta_j), \ 0 \le \alpha_j \le \pi, 0 \le \beta_j \le 2\pi \qquad (4.49)$$

# Power Allocation Techniques for Visible Light

where the azimuth angle is denoted by $\alpha_j$ and the elevation angle is denoted by $\beta_j$, both parameters can be use to adjust the LED illumination quality for better performance of the system orientation. Hence, we have provided the equation for the $j$th vector of the LED (Nuwanpriya et al., 2015), which is given as:

$$t_j \equiv (\alpha_j, \beta_j) \underline{\underline{\Delta}} \begin{cases} x_j = \|\vec{x_j}\| \sin\alpha_j \cos\beta_j \\ y_j = \|\vec{y_j}\| \sin\alpha_j \sin\beta_j \\ z_j = \|\vec{z_j}\| \cos\alpha_j \end{cases}, \tag{4.50}$$

From Figure 4.7 to optimize the $j$th LED vectors $t_j$, we can consider converting $t_j$ back to spherical coordinates $(\alpha_j \beta_j)$. The equations for channel coefficient are given when combined with Equations (4.47) and (4.48) as:

$$h_{ij} = \frac{A_p g}{\pi d_y^2} \frac{-V_{ij}^T U_i}{\|v_{ij}\| \|U_i\|} \frac{v_{ij}^T t_j}{\|V_{ij}\| \|t_j\|}, \tag{4.51}$$

where $h_{ij}$ denotes normal vectors of the LED, which can be given as:

$$h_{ij} = q_{ij} \left( \frac{v_{ij}^T t_j}{\|V_{ij}\| \|t_j\|} \right), \tag{4.52}$$

where $q_{ij}$ is a constant for LED normal vectors for channel coefficient given as:

$$q_{ij} = \frac{A_p g}{\pi d_{ij}^2} \frac{-v_{ij}^T u_i}{\|v_{ij}\| \|u_i\|}, \tag{4.53}$$

and

$$\hat{v}_j^T = \frac{v_j^T}{\|v_j\|} \text{ and } \hat{t}_j = \left[ \frac{x_j}{\|t_j\|} \frac{y_j}{\|t_j\|} \frac{z_j}{\|t_j\|} \right]^T. \tag{4.54}$$

Equation (4.52) can also be given as:

$$h_{ij} = q_{ij} \hat{v}_{ij}^T \hat{t}_j, \tag{4.55}$$

where $h_i = [h_{i1} \ldots h_{iN}]^T$ denotes the entire LED channel coefficient vector to the $i$th location from Equation (4.55), which can be given as:

$$h_i = q_i o \, p_i, \tag{4.56}$$

where the element-wise product is denoted by $o$, $q_i$ denotes the vector $N \times 1$, where one element $q_{ij}$ relates within LED $j$th and PD $i$th, and $p_i$ represents the vector $N \times 1$, where one element $p_{ij}$ can be given as:

$$p_{ij} = \hat{v}_{ij}^T \hat{t}_j. \tag{4.57}$$

Moreover, $p_i$ can also be expressed as:

$$p_i = \hat{V}_i \hat{t}, \qquad (4.58)$$

where

$\hat{V}_i$ represents blkdiag $\left(\hat{v}_{i1}^T, \ldots, \hat{v}_{iN}^T\right)$, which is equal to a matrix of

$N \times 3N \begin{bmatrix} \hat{v}_{i1}^T & \cdots & 0 \\ \vdots & \ddots & \vdots \\ 0 & \cdots & \hat{v}_{iN}^T \end{bmatrix}$, $\hat{t}$, denotes $3N \times 1$ column vector that involves the entire

normal vectors for the LEDs, which is given as:

$$\hat{t} = \left[\hat{t}_1, \; \hat{t}_2, \ldots \hat{t}_N \right], \qquad (4.59)$$

where blkdiag denotes the function of the matrix block diagonalization, hence substituting Equations (4.58) and (4.56) it is given as:

$$h_i = q_i o \, p_i = q_i o \left(\hat{v}_i \hat{t}\right) = b_i \hat{t}, \qquad (4.60)$$

where the constant $b_i = \begin{bmatrix} q_{i1}\hat{v}_{i1}^T & \cdots & 0 \\ \vdots & \ddots & \vdots \\ 0 & \cdots & q_{iN}\hat{v}_{iN}^T \end{bmatrix}$.

From Figure 4.7, we consider PA for expression with LED for PA vector, which is given as:

$$w = \left[w_1 \ldots w_N\right]^T, \qquad (4.61)$$

of the size $N \times 1$.

Therefore, the received power at the PD $i$th can be given as:

$$r_i = w \, o \, h_i = w \, o \left(b_i \hat{t}\right), \qquad (4.62)$$

where $r_i$ is the total power received at the PD $i$th sum elements, which can also be given as:

$$R_i = w^T b_i \hat{t}, \qquad (4.63)$$

Therefore, we present the illuminance performance quality factor of the LED arrays evaluated from Equation (4.55), and it can be written as:

$$F_\Lambda = \frac{\overline{\Lambda}}{\sqrt[2]{\text{var}(\Lambda)}}, \qquad (4.64)$$

where $\Lambda_i = \frac{R_i^2}{\sigma_i^2}$ denotes the electrical SINR received of the PD$i$th, $\overline{\Lambda}$ denotes the mean of $\Lambda_i$, and *var* denotes the variance of $\Lambda_i$, $i = 1, ..., k$, correspondingly.

## 4.12 OPTIMAL PA TECHNIQUE IN VL-NOMA COMMUNICATION SYSTEM

The optimal allocation of resources such as power and channel to the users' VL-NOMA communication system is a potentially challenging task. In order to solve this optimization challenge, the researchers have approached it through maximizing minimal rate (MMR) and maximizing sum rate (MSR) (Shang et al., 2019). This objective is to advance the entire data rate of telecommunication network (Hanif et al., 2016), proposing an efficient PA and precoding architecture for single-carrier NOMA networks to tackle the maximization of its sum rate. This solves the problem of resource allocation using two ways, namely, the situation where the sum rate maximization challenge is transformed first into its corresponding form and, second, the complex precoding vectors and optimal PA are calculated by normally using algorithm of maximization and minimization (Hunter and Lange, 2004; Hunter and Li, 2005; Smola et al., 2005; Stoica and Selén, 2004). The MMR and MSR performance metrics are illustrated using the conduct of PA for channel assessment. In (Zhu et al., 2017), PA scheme is studied but in different dimension. In this equation, the authors have discussed how to solve the problem of optimal PA. The PA problem in MSR can be given as (Shang et al., 2019):

$$\max_{p} \sum_{k=1}^{k} \left[ R_1^k \left( P_1^k, P_2^k \right) + R_2^k \left( P_1^k, P_2^k \right) \right], \qquad (4.65)$$

$$s.t. \ R_n^k \geq \left( R_n^k \right) \min, \quad n = 1, 2, \quad \forall \mathcal{K} = 1, ..., K, \qquad (4.66)$$

$$\sum_{k=1}^{K} \left( P_1^k + P_2^k \right) \leq P_T, \qquad (4.67)$$

$$O \leq P_1^k \leq P_2^k, \quad \forall k = 1, ..., K, \qquad (4.68)$$

where $A_n^k = 2^{\frac{\left( R_n^k \right) \min}{B_c}}$ and assume $A_2^k \geq 2$, the solution from (4.65–4.68) is given as the following:

$$P_1^k = \frac{\Gamma_2^k q_k - A_2^k + 1}{A_2^k \Gamma_2^k}, \quad P_2^k = q^k - P_1^k. \qquad (4.69)$$

Let $q^k$ and $\gamma_k$ be written as:

$$q^k = \left[\frac{B_c}{\lambda} - \frac{A_2^k}{\Gamma_1^k} + \frac{A_2^k}{\Gamma_2^k} - \frac{1}{\Gamma_2^k}\right]_{\gamma_k}^{\infty}, \qquad (4.70)$$

where $\lambda$ satisfying $\sum_{k=1}^{k} q_k = P_T$. The optimization of MMR for power allocation is expressed as:

$$\max_{p} \min_{k=1,\ldots k} \left\{R_1^k\left(P_1^k, P_2^k\right), R_2^k\left(P_1^k, P_2^k\right)\right\}, \qquad (4.71)$$

$$\sum_{k=1}^{K}\left(P_1^k + P_2^k\right) \leq P_T, \quad 0 \leq P_1^k \leq P_2^k, \quad \forall k = 1,\ldots,K. \qquad (4.72)$$

The solution in (4.70) is given as:

$$P_1^k = \frac{-\left(\Gamma_1^k + \Gamma_2^k\right) + \sqrt{\left(\Gamma_1^k + \Gamma_2^k\right)^2 + 4\Gamma_1^k\left(\Gamma_2^k\right)^2 q_k}}{2\Gamma_1^k \Gamma_2^k}, \quad P_2^k = q^k - P_1^k. \qquad (4.73)$$

Let

$$q^k = \frac{\left(Z(\lambda)\Gamma_2^k + \Gamma_1^k\right)\left(Z(\lambda) - 1\right)}{\Gamma_1^k \Gamma_2^k}, \qquad (4.74)$$

$$Z(\lambda) = X + \sqrt{X^2 + \frac{B_c}{2\lambda \sum_{k=1}^{K} 1/\Gamma_1^k}}. \qquad (4.75)$$

## 4.13 COMPARISON OF POWER ALLOCATION TECHNIQUES

We compare different PA techniques used in this chapter for VL-NOMA systems in this section. We directly point out different PA techniques function with regard to their best way to enhance the performance in the VL-NOMA system. These comparisons cover the total achievable data rate, user fairness, user sum rate, decoding order, algorithm and power level in the channel, and optimally power and channel to user in a VL-NOMA system as shown in Table 4.1. We notice that these techniques enhance the performance of the communication system in different ways. We present one LED as an access lightning sending signals to two users for our demonstration. These users are stationed differently – one user near the LED and the other user far from the LED, both receiving the signal sent with different power accessibility, which might delay the speed the signals reach the far users. This concept is proven, as we notice that the farther the user, the lesser the intensity of the LED that it detects. The PA techniques are deployed in this scenario to enhance the performance of the system between the two users in terms of power that carries the signals to their destinations

# TABLE 4.1
## Comparison of PA techniques in VL-NOMA network.

| Comparisons | Conventional PA | Inverse PA | Adaptive PA | GRPA | Joint PA | Optimal PA |
|---|---|---|---|---|---|---|
| Users fairness (Yang et al., 2017; Tao et al., 2019) | Normalize fairness between two users | Serious unfairness between users and little improvement. | Better the balance between two users, which the user fairness varies smoothly. | Improvement in user's fairness. | ............ | ............ |
| Total achievable data rate (Yang et al., 2017; Tao et al., 2019) | The entire achievable data rate is high. | The entire achievable data rate is higher when users with a worse channel condition are allocated. | It balances between entire achievable data rate. | We find that the total achievable data rate increases significantly at a small cost of user fairness. | ............ | ............ |
| Power level in the channel (Yang et al., 2017; Tao et al., 2019) | Less power is allocated to users with worse channel condition at the transmitter side. | Users with a worse channel condition allocated less power from the transmitter | Users with a worse channel condition allocated less power from the transmitter. | Lower power levels will be sufficient for users with the right channel conditions. | ............ | ............ |
| Algorithm (Yang et al., 2017; Tao et al., 2019; Obeed et al., 2018; Guo et al., 2009) | Selects its command according to mathematical conditions | Selects its command according to mathematical conditions. | MADM function is required to improve the conditions. | Mathematical expressions are used for better signal. | Combine two techniques and solve them jointly with the assistance of an algorithm | ............ |
| Decoding order (Yang et al., 2017; Tao et al., 2019) | A higher decoding order is required in a signal message at the receiver | A higher decoding order is required in a signal message at the receiver. | A higher decoding order is required especially at the receiver. | The right channel conditions decode after subtracting the signals of the user or multiple users with lesser decoding order. | ............ | ............ |
| Optimally Power and channel to users (Hanif et al., 2016; He et al., 2019) | ............ | | | | Optimally assign resources | MMR and MSR |

(end users) successfully. These PA techniques are conventional, which focuses mostly on providing high power to users with a worse channel condition to enhance the data rate of the signals and uses decoding order to better its performance. The inverse PA technique is performed from the transmitter side, less power assigned to users with low-quality channel condition, and at the receiver side, the decoding order is deployed to users with a low-quality channel condition. An adaptive PA technique performs based on multiattribute decision making (MADM), making the decision parameters to apply in the entire user fairness and achievable data rate. An adaptive PA is considered to be used for the combination of conventional and inverse PA. Hence, the GRPA considers using the improvement of the user sum rate and mathematical approach to solve the problems of distributing power allocation signals power domain control evenly to the weaker and stronger users. Therefore, the joint PA techniques combine two techniques and solve them jointly with the assistance of an algorithm. We conclude with optimal PA technique, which in turn optimally allocates resources to NOMA network in the area of power and channel to the users. In order to solve this optimization challenge, the researchers have approached it through maximizing minimal rate (MMR) and maximizing sum rate (MSR). This PA technique discussed enhances the performance of the system throughput in their various capacity. These are conventional, inverse, adaptive, GRPA, joint, and optimal power allocation techniques. We have only compared this PA technique in terms of their user fairness, total achievable data rate, power level in the channel, algorithm, decoding order, and optimal power and channel to user.

## 4.14 SUMMARY

In this chapter, we have discussed different power allocation techniques and challenges of multielement VL-NOMA communication systems. We have discussed the LED-user assignment problem in a downlink VLC scenario where multiple LEDs serve for multiple users. As a case study for the presented communication system, the LED is used as a source of carriers signal, and the power allocation techniques for two users (near-user and far-user) are applied over its intensity. Further, the comparison of conventional, inverse, adaptive, gain-ratio, joint, and optimal power allocation techniques are presented systematically with context to the performance parameters of the presented communication systems. The power control techniques are shown to provide substantial gains in sum rate and fairness, especially for a larger number of users. Different frameworks are designed for better illustrations. We observe that the entire PA techniques demonstrated in these enhance the performance of the system using different approaches. Further, the potential challenges like an improvement in the total achievable data rate and user fairness have to be further investigated for VL-NOMA communication system for next generation wireless systems. As each receiver has to perform SIC to decode the signal that is intended for the required SIC process in the NOMA-based systems, which imposes intensive computational requirements and its performance is susceptible to small SIC errors, corresponding receiver complexity is a challenging open research issue. Further, in-depth investigation has to be carried out to address the efficiency and sensitivity of SIC receivers. In addition to this, more practical investigations to improve multiuser power allocation algorithms are required along with extensive tests on the corresponding performance particularly, in the case of large numbers of users. Moreover, the relationship between FOV and both the decoding order and the distance of user from the transmitter should be characterized and quantified. To ensure accurate and reliable results that will be useful in the successful deployment and operation of VLC systems, the detailed analysis of realistic mobility scenarios is an important issue. The performance of VL-NOMA communication systems depends on the quality of channel estimation such as imperfect or outdated channel state information (CSI), which lead to increased error rate performance. Further, the associated issue of LED nonlinearities also needs to be addressed thoroughly in the context of NOMA, particularly in the presence of imperfect SIC and CSI. Therefore, novel compensation and mitigation techniques must be proposed that provide an increased LED linearity and immunity to slight SIC and CSI imperfections such as those incurred in practical communications. The VL-NOMA communication systems can coexist efficiently with conventional multiple access schemes. It is important to quantify the benefits of any new methods in the context of realistic indoor scenarios. This is anticipated to provide important solutions that maximize the potential of VL-NOMA communication systems.

## REFERENCES

3rd Generation Partnership Project (3GPP). March 2015. Study on downlink multiuser superposition transmission for LTE.

5G radio access: Requirements, concepts and technologies. July 2014. NTT DOCOMO, Inc., Tokyo, Japan, 5G White Paper.

Al-Imari, Mohammed, Xiao, Pei, Imran, Muhammad Ali, and Tafazolli, Rahim. 2014. Uplink non-orthogonal multiple access for 5G wireless networks. *Proceedings of the 11th International Symposium on Wireless Communications Systems (ISWCS)*, Barcelona, Spain, 781–785.

Andrews, J. G., and Meng, T. H. 2003. Optimum power control for successive interference cancellation with imperfect channel estimation. *IEEE Transaction on Wireless Communications*, 2: 375–383.

Bawazir, Sarah S., Sofotasios, Paschalis C., Muhaidat, Sami, Al-Hammadi Yousof, and Karagiannidis, George K. 2018. Multiple access for visible light communications: Research challenges and future trends. *IEEE Access*, 6: 26167–26174.

Chen, Chen, Zhong, Wen-De, and Wu, Dehao. 2017. On the coverage of multiple-input multiple-output visible light communications [Invited]. *Journal of Optical Communications and Networking*, 9(9): D31–D41.

Cole, Marty, and Driscoll, Tim. 2012. The lighting revolution: If we were experts before, we're novices now. *Proceedings of the 59th Annual IEEE Conference on Petroleum and Chemical Industry Technical Conference (PCIC)*, Chicago, USA, 1–12

Dai, Linglong, Wang, Bichai, Yuan, Yifei, Han, Shuangfeng, Chih, Lin I., and Wang, Zhaocheng. 2015. Non-orthogonal multiple access for 5G: solutions, challenges, opportunities, and future research trends. *IEEE Communications Magazine*, 53(9): 74–81.

Di, Boya, Bayat, Siavash, Song, Lingyang, and Li, Yonghui. 2015. Radio resource allocation for downlink non-orthogonal multiple access (NOMA) networks using matching theory. *Proceedings of the Global Communications Conference (GLOBECOM)*, San Diego, CA, 1–6.

Dimitrov, Svilen, and Haas, Harald. 2015. *Principles of LED Light Communications: Towards Networked Li-Fi*. Cambridge, UK: Cambridge University Press.

Ding, Zhiguo, Lei, Xianfu, Karagiannidis, George K., Schober, Robert, Yuan, Jinhong, and Bhargava, Vijay K. 2017. A survey on non-orthogonal multiple access for 5G networks: Research challenges and future trends. *IEEE Journal on Selected Areas in Communications*, 35(10): 2181–2195.

Ding, Zhiguo, Yang, Zheng, Fan, Pingzhi, and Vincent, Poor H. 2014. On the performance of non-orthogonal multiple access in 5G systems with randomly deployed users. *IEEE Signal Processing Letters*, 21(12): 1501–1505.

Dong, Zanyang, Shang, Tao, Li, Qian, and Tang, Tang. 2019.Adaptive power allocation scheme for mobile NOMA visible light communication system. *Electronics*, 8(4): 381/1–20.

Feng, Zhen, Guo, Caili, Ghassemlooy, Zabih, and Yang, Yang. 2018. The spatial dimming scheme for the MU-MIMO-OFDM VLC system. *IEEE Photonics Journal*, 10(5): 7907013/1–140.

Gong, Chen, Li, Shangbin, Gao, Qian, and Xu, Zhengyuan. 2015. Power and rate optimization for visible light communication system with lighting constraints. *IEEE Transactions on Signal Processing*, 63(16): 4245–4256.

Guo, Wenxuan, Huang, Xinming, and Zhang, Kai. 2009. Joint optimization of antenna orientation and spectrum allocation for cognitive radio networks. *2009 Conference Record of the Forty-Third Asilomar Conference on Signals, Systems and Computers*, Pacific Grove, CA, 419–423.

Guo, Zi-Quan, Liu, Kai, Zheng, Li-Li, et al. 2019. Investigation on three-hump phosphor-coated white light-emitting diodes for healthy lighting by genetic algorithm, *IEEE Photonics Journal*, 11(1): 8200110/1–10.

Hanif, Muhammad, Ding, Zhiguo, Ratnarajah, Tharmalingam, and Karagiannidis, George K. 2016. A minorization-maximization method for optimizing sum rate in the downlink of non-orthogonal multiple access systems. *IEEE Transactions on Signal Processing*, 64(1): 76–88.

He, Chanfan, Hu, Yang, Chen, Yan, and Zeng, Bing. 2019. Joint power allocation and channel assignment for NOMA with deep reinforcement learning. *IEEE Journal on Selected Areas in Communications*, 37(10): 2200–2210.

Higuchi, Kenichi, and Benjebbour, Anass. 2015. Non-orthogonal multiple access (NOMA) with successive interference cancellation for future radio access. *IEICE Transactions on Communications*, E98(B3): 403–414.

Hunter, David R., and Lange, Kenneth. 2004. A tutorial on MM algorithms. *American Statistician*, 58(1): 30–37.

Hunter, David R., and Li, Runze. 2005. Variable selection using MM algorithms. *Annals of Statistics*, 33(4): 1617–1642.

Islam, S. M. R., Avazov, Nurilla, Dobre Octavia A., and Kwak, Kyung-Sup. 2017. Power domain non-orthogonal multiple access (NOMA) in 5G systems: Potential and challenges. *IEEE Communications Survey and Tutorials*, 19(2): 721–742.

Jiang, Rui, Wang, Qi, Haas, Harald, and Wang, Zhaocheng. 2018. Joint user association and power allocation for cell-free visible light communication networks. *IEEE Journal on Selected Areas in Communications*, 36(1): 136–148.

Kahn, J. M.. and Barry, J. R. 1997. Wireless infrared communications. *Proceedings of the IEEE*, 85(2): 265–298.

Kashef, Mohamed, Abdallah, Mohamed, Qaraqe, Khalid, Haas, Harald, and Uysal, Murat. 2014. On the benefits of cooperation via power control in OFDM-based visible light communication systems. *Proceedings of the IEEE 25th Annual International Symposium on Personal, Indoor, and Mobile Radio Communication (PIMRC)*, Washington, DC, 856–860.

Kazmi, S. M. A., Tran Nguyen, H., Ho Tai, Manh, Manzoor, Aunas, Niyato, Dusit, and Hong, Choong Seon. 2018. Coordinated device-to-device communication with non-orthogonal multiple access in future wireless cellular networks. *IEEE Access*, 6: 39860–39875.

Kim, Kyuntak, Lee, Kyesan, and Lee, Kyujin. 2016. Appropriate RLL coding scheme for effective dimming control in VLC. *Electronics Letters*, 52(19): 1622–1624.

Kizilirmak, Refik Caglar, Rowell, Corbett, and Uysal, Murat. 2015. Non-orthogonal multiple access (NOMA) for indoor visible light communications. *2015 4th International Workshop on Optical Wireless Communications (IWOW)*, Istanbul, 98–101.

Komine, Toshihiko, and Nakagawa, Masao. 2004. Fundamental analysis for visible-light communication system using LED lights. *IEEE Transactions on Consumer Electronics*, 50(1): 100–107.

Lahby, Mohamed, Cherkaoui, Leghris, and Adib, Abdellah. 2014. Performance analysis of normalization techniques for network selection access in heterogeneous wireless networks. *Proceedings of the 9th International Conference on Intelligent Systems: Theories and Applications (SITA-14)*, Rabat, 1–5.

Lapidoth Amos, Moser Stefan M. and Wigger Michele. 2009. On the capacity of free-space optical intensity channels. *IEEE Transactions on Information Theory*, 55(10): 4449–4461.

Li, Juan, Bao, Xu, Zhang, Wance, and Bao, Nan. 2020. QoE probability coverage model of indoor visible light communication network. *IEEE Access*, 8: 45390–45399.

Li Xuan, Zhang Rong and Hanzo Lajos. 2015. Cooperative load balancing in hybrid visible light communications and WiFi. *IEEE Transactions on Communications*, 63(4): 1319–1329.

Lin Bangjiang, Tang Xuan and Ghassemlooy Zabih. 2019. Optical power domain NOMA for visible light communications. *IEEE Wireless Communications Letters*, 8(4): 1260–1263.

Marshoud, Hanaa, Kapinas Vasileios, M., Karagiannidis, George K., and Muhaidat, Sami. 2016. Non-orthogonal multiple access for visible light communications. *IEEE Photonics Technology Letters*, 28(1): 51–54.

Marshoud, Hanaa, Sofotasios Paschalis, C., Muhaidat, S., Karagiannidis, George K., and Sharif, Bayan. 2017. On the performance of visible light communication systems with non-orthogonal multiple access. *IEEE Transaction on Wireless Communication*, 16(10): 6350–6364.

Nuwanpriya, Asanka, Ho Siu, Wai, and Chen, Chung Shue. 2015. Indoor MIMO visible light communications: Novel angle diversity receivers for mobile users. *IEEE Journal on Selected Areas in Communications*, 33(9): 1780–1792.

Obeed, Mohanad, Salhab Anas, M., Zummo Salam, A., and Alouini, M.-S. 2018. Joint power allocation and cell formation for energy-efficient VLC networks. *Proceedings of the IEEE International Conference on Communications (ICC)*, Kansas City, MO, 1–6.

Pathak, Parth, Feng, Xiaotao, Hu, Pengfei, and Mohapatra, Prasant. 2015. Visible light communication, networking, and sensing: A survey, potential and challenges. *IEEE Communications Surveys & Tutorials*, 17(4): 2047–2077.

Proposed solutions for new radio access. 2015. Mobile and Wireless Communications Enablers for the Twenty-twenty Information Society (METIS), Deliverable D.2.4, February.

Rodoplu, Volkan, Hocaoğlu, Kemal, Adar, Anil, Çikmazel Rifat, Orhan, and Saylam, Alper. 2020. Characterization of line-of-sight link availability in indoor visible light communication networks based on the behavior of human users. *IEEE Access*, 8: 39336–39348.

Selvam, K., and Kumar, Krishan. 2019. Energy and spectrum efficiency trade-off of non-orthogonal multiple access (NOMA) over OFDMA for machine-to-machine communication. *Proceedings of the 5th International Conference on Science Technology Engineering and Mathematics (ICONSTEM)*, Chennai, India, 523–528.

Sewaiwar, Atul, Tiwari, Samrat Vikramaditya, and Chung, Yeon Ho. 2015. Smart LED allocation scheme for efficient multiuser visible light communication networks. *Optics Express*, 23(10): 13015–13024.

Shahjalal, M., Islam, M. Mainul, Hasan, M. Khalid, Chowdhury, Mostafa Zaman, and Jang, Yeong Min. 2019. Multiple access schemes for visible light communication. *Proceedings* of the Eleventh International Conference on Ubiquitous and Future Networks (ICUFN), Zagreb, Croatia, 115–117.

Shang, Qian Li, Tao, Tang, and Dong, Zanyang. 2019. Optimal power allocation scheme based on multi-factor control in indoor NOMA-VLC systems. *IEEE Access*, 7: 82878–82887.

Shen, Cong, Lou, Shun, Gong, Chen, and Xu, Zhengyuan. 2016a. User association with lighting constraints in visible light communication systems. *2016 Annual Conference on Information Science and Systems (CISS)*, Princeton, NJ, 222–227.

Shen, Hong, Deng, Yuqin, Xu, Wei, and Zhao, Chunming. 2016b. Rate maximization for downlink multiuser visible light communications. *IEEE Access*, 4: 6567–6573.

Shen, Hong, Wu, Yanfei, Xu, Wei, and Zhao, Chunming. 2017a. Optimal power allocation for downlink two-user non-orthogonal multiple access in visible light communication. *Journal of Communications and Information Networks*, 2(4): 57–64.

Shen, Hong, Deng, Yuqin, Xu, Wei, and Zhao, Chunming. 2017b. Rate maximization for downlink multiuser visible light communications. *IEEE Access*, 4(99): 6567–6573.

Smola Alex, J., Vishwanathan, S. V. N., and Hofmann, Thomas. March 2005. Kernel methods for missing variables. *Proceedings of the 10th International Workshop on Artificial Intelligence and Statistics*, Barbados, 325–332.

Sohail, Muhammad Farhan, Leow, Chee Yen, and Won Seung, Hwan. 2018. Non-orthogonal multiple access for unmanned aerial vehicle assisted communication. *IEEE Access*, 6: 22716–22727.

Stoica, Petre, and Selén, Yngve. 2004. Cyclic minimizers, majorization techniques, and the expectation-maximization algorithm: a refresher. *IEEE Signal Processing Magazine*, 21(1): 112–114.

Tao Si-yu, Yu Hongyi, Li Qing and Yanqun Tang. 2018. Performance analysis of gain ratio power allocation strategies for non-orthogonal multiple access in indoor visible light communication networks. *EURASIP Journal of Wireless Communication Network*, 1: 154.

Tao, Siyu, Yu, Hongyi, Li, Qing, and Tang, Yanqun. 2019. Strategy-based gain ratio power allocation in non-orthogonal multiple access for indoor visible light communication networks. *IEEE Access*, 7: 15250–15261.

Thakur, Prabhat, and Singh, G. November 2019. Sum-rate analysis of MIMO based CR-NOMA communication systems. *Proceedings of the 4th IEEE International Conference on Image Information Processing (ICIIP -2019)*, Waknaghat, India, 1–6.

Thakur, Prabhat, and Singh, G. April 2020. Performance analysis of MIMO based CR-NOMA communication systems. *IET Communication*, 14(6), 2676–2687, 2020.

Thakur, Prabhat, Kumar, Alok, Pandit, S., Singh, G., and Satashia, S. N. 2019. Frameworks of non-orthogonal multiple access techniques in cognitive radio communication systems. *China Communication*, 16(6): 129–149.

Timotheou, Stelios, and Krikidis, Ioannis. 2015. Fairness for non-orthogonal multiple access in 5G systems. *IEEE Signal Processing Letters*, 22(10): 1647–1651.

Tran, Manh Le, and Kim, Sunghwan. 2019. Joint power allocation and orientation for uniform illuminance in indoor visible light communication. *Optics Express*, 27(20): 28575–28587.

Varma Praneeth, G. V. S. S. 2018. Optimum power allocation for uniform illuminance in indoor visible light communication. *Optics Express*, 26(7): 8679–8689.

Varma Praneeth, G. V. S. S., Kumar, Abhinav, and Sharma, Govind. 2018. Resource allocation for visible light communication using stochastic geometry. *2018 11th International Symposium on Communication Systems, Networks & Digital Signal Processing (CSNDSP)*, Budapest, 1–6.

Vega, Maria Torres, Koonen, Antonius Marcellus Jozef, Liotta, Antonio, and Famaey, Jeroen. 2018. Fast millimeter wave assisted beam-steering for passive indoor optical wireless networks. *IEEE Wireless Communications Letters*, 7(2): 278–281.

Wang, Ying Ming. 2003. A method based on standard and mean deviations for determining the weight coefficients of multiple attributes and its applications. *Application of Statics and Management*, 22: 22–26.

Wang, Zhaocheng, Wang, Qi, Chen, Sheng, and Hanzo, Lajos. 2014. An adaptive scaling and biasing scheme for OFDM-based visible light communication systems. *Optics Express*, 22(10): 12707.

Wang, Zixiong, Yu, Changyuan, Zhong, Wen-De, Chen, Jian, and Chen, Wei. 2012. Performance of a novel LED lamp arrangement to reduce SNR fluctuation for multi-user visible light communication systems. *Optics Express*, 20(4): 4564–4573.

Wei, Zhiqiang, Yuan, Jinhong, Ng, Derrick Wing Kwan, Elkashlan, Maged, and Ding, Zhiguo. 2016. A survey of downlink non-orthogonal multiple access for 5G wireless communication networks. *ZTE Communications*, 14(4):17–25

Xiao, Kaiyi, and Li, Changgeng. 2018. Vertical handoff decision algorithm for heterogeneous wireless networks based on entropy and improved TOPSIS. *Proceedings of the IEEE 18th International Conference on Communication Technology (ICCT)*, Chongqing, 706–710.

Xu, Fangcheng, Yu, Xiangbin, Li, Minglu, and Wen, Benben. 2019. Energy-efficient power allocation scheme for hybrid precoding mmWave-NOMA system with multi-user pairing. *2019 IEEE International Conference on Communications Workshops (ICC Workshops)*, Shanghai, China, 1–5.

Yang, Kai, Yang, Nan, Ye, Neng, Jia, Min, Gao, Zhen, and Fan, Rongfei. 2019. Non-orthogonal multiple access: Achieving sustainable future radio access. *IEEE Communications Magazine*, 57(2): 116–121.

Yang, Zhaohui, Xu, Wei, and Li, Yiran. 2017. Fair non-orthogonal multiple access for visible light communication downlinks. *IEEE Wireless Communications Letters*, 6(1): 66–69.

Yapici, Yavuz, and Guvenc, Ismail. 2019. Non-orthogonal multiple access for mobile VLC networks with random receiver orientation. *2019 IEEE Global Communications Conference (GLOBECOM)*, Waikoloa, HI, USA, 1–6.

Yin, Liang, Popoola, Wasiu, Wu, Xiping, and Haas, Harald. 2016. Performance evaluation of non-orthogonal multiple access in visible light communication. *IEEE Transactions on Communications*, 64(12): 5162–5175.

Yuichi, Tanaka, Toshihiko, Komine, Haruyama S., and Masao, Nakagawa. October/September 2001. Indoor visible communication utilizing plural white LEDs as lighting. *12th IEEE International Symposium on Personal, Indoor and Mobile Radio Communications. PIMRC 2001. Proceedings (Cat. No.01TH8598)*, San Diego, CA, USA, F-81–F-85.

Zhang, Xiaoke, Gao, Qian, Gong, Chen, and Xu, Zhengyuan. 2017a. User grouping and power allocation for NOMA visible light communication multi-cell networks. *IEEE Communications Letters*, 21(4): 777–780.

Zhang, Yi, Wang, Hui-Ming, Zheng, Tong-Xing, and Yang, Qian. 2017b. Energy-efficient transmission design in non-orthogonal multiple access. *IEEE Transactions on Vehicular Technology*, 66(3): 2852–2857.

Zhang, Yi, Yang, Qian, Zheng, Tong-Xing, Wang, Hui-Ming, Ju, Ying, and Meng, Yue. 2016. Energy efficiency optimization in cognitive radio inspired non-orthogonal multiple access. *Proceedings of the 27th Annual International Symposium on Personal, Indoor, and Mobile Radio Communications (PIMRC)*, Valencia, Spain, 1–6.

Zhao, Xiang, Chen, Hongbin, and Sun, Jinyong. 2018. On physical-layer security in multiuser visible light communication systems with non-orthogonal multiple access. *IEEE Access*, 6: 34004–34017.

Zhou, Jing, and Zhang, Wenyi. 2017. On the capacity of bandlimited optical intensity channels with Gaussian noise. *IEEE Transactions on Communications*, 65(6): 2481–2493.

Zhu, Jianyue, Wang, Jiaheng, Huang, Yongming, He, Shiwen, You, Xiaohu, and Yang, Luxi. 2017. On optimal power allocation for downlink non-orthogonal multiple access systems. *IEEE Journal on Selected Areas in Communications*, 35(12): 2744–2757.

Zhu, Lipeng, Xiao, Zhenyu, Xia, Xiang-Gen, and Wu, Dapeng Oliver. 2019. Millimeter-wave communications with non-orthogonal multiple access for B5G/6G. *IEEE Access*, 7: 116123–116132.

# 5 Multiantenna Systems

## Large-Scale MIMO and Massive MIMO

*Bhasker Gupta*
CCET

## CONTENTS

5.1 Introduction ................................................................................................ 79
5.2 Massive MIMO Uplink and Downlink ................................................... 82
5.3 Spectral Efficiency (SE) .............................................................................. 82
5.4 Area Throughput ........................................................................................ 86
5.5 Precoding .................................................................................................... 86
  5.5.1 Single-Cell (SC) Precoding Techniques ..................................... 87
    5.5.1.1 Matched Filter (MF) ...................................................... 87
    5.5.1.2 Zero Forcing Precoding ................................................ 87
    5.5.1.3 Regularized Zero Forcing ............................................ 88
    5.5.1.4 Truncated Polynomial Expansion (TPE) .................... 88
    5.5.1.5 Phased Zero Forcing .................................................... 88
  5.5.2 Precoding Techniques for the Downlink of Multicell Scenario ........ 89
    5.5.2.1 Max SINR Precoding .................................................... 89
    5.5.2.2 Multilayer Precoding .................................................... 89
    5.5.2.3 Quantized Precoders for Massive MIMO .................. 90
    5.5.2.4 Nonlinear Quantized Precoding for Massive MIMO ......... 90
    5.5.2.5 Multiuser Massive MIMO with One-Bit Quantized Precoding ........ 91
5.6 Hybrid Precoding ...................................................................................... 92
5.7 Massive MIMO with Linear Precoding and Detection ........................ 92
5.8 Energy Efficiency (EE) .............................................................................. 96
References ............................................................................................................ 97

## 5.1 INTRODUCTION

The wireless communication technology has witnessed significant improvements, especially in the past decade. These improvements were triggered by the foreseeable growth in network traffic (Cudak et al. 2013) and technology up-gradations like MIMO systems. Various broadband standards like WLAN, WiMAX, and Long Term

Evolution (LTE) have incorporated the MIMO technology. The MIMO technology helps to overcome most of the challenges in wireless channels along with various constraints. The basic idea behind the MIMO system is introduction of spatial domain along with time and frequency domains. The MIMO system explores the spatial dimension (provided by the multiple antennas at the transmitter and the receiver) as shown in Figure 5.1.

MIMO benefits offer a significant increase in data throughput, data rate, and spectral efficiency without an additional frequency spectrum or power. The MIMO system shows improved throughput even when the channel is heavily faded or when there is too much interference. This improvement is credited to the fact that MIMO systems make use of a diversity techniques to estimate channel. Due to the above reasons, MIMO technology emerged as 4G standard/interface for current/future wireless systems. Mathematically, the theoretical upper bound of channel capacity is given by the Shannon–Hartley theorem (Shannon 1948)

$$C = B\log_2(1+SNR), \tag{5.1}$$

where

- $C$ : capacity in bps (bits / second)
- $B$ : bandwidth
- $SNR$ : the signal-to-noise power ratio.

Equation (5.1) shows that the capacity can be enhanced by allocating more bandwidth or by improving signal-to-noise ratio or both. However, in MIMO systms (Rappaport et al. 2013) channel capacity can be expressed as

$$C = N \times B\log_2(1+SNR), \tag{5.2}$$

where N is the number of spatial streams. Thus, in addition to the rather costly bandwidth, the capacity improvement can be achieved by escalating the count of spatial

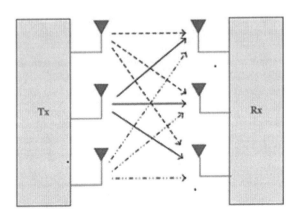

**FIGURE 5.1** Generic MIMO System.

# Multiantenna Systems

streams (Gutierrez et al. 2009). The greater the number of streams, the greater is the improvement in capacity. MIMO technology undoubtedly improved the quality of service (QoS) performance of existing wireless networks. But due to continuously improving technology and increase in customer base, the overall network capacity requires even more expansion. As a result, the next phase of wireless generation, i.e., "beyond 4G" or "5G" generation is explored.

Massive MIMO (Rappaport, Murdock, and Gutierrez 2011; Rappaport et al. 2012; Marzetta 2015) is a key technology for 5G system, which breaks scalability barriers of conventional MIMO systems. In general, any wireless system can be referred to as a "Massive MIMO" system (Hoydis, Ten Brink, and Debbah 2013) if it contains very large antenna numbers either at base station or at both ends, as shown in Figure 5.2.

However, a quantitative answer to count of antennas may vary from system to system (Hoydis, Ten Brink, and Debbah 2013). In practice, a few hundred antennas may be installed at the base station, and the user equipment may have tens of antennas. This upscaling of antennas will significantly improve the parameters like energy efficiency, reliability, coverage, and spectral efficiency.

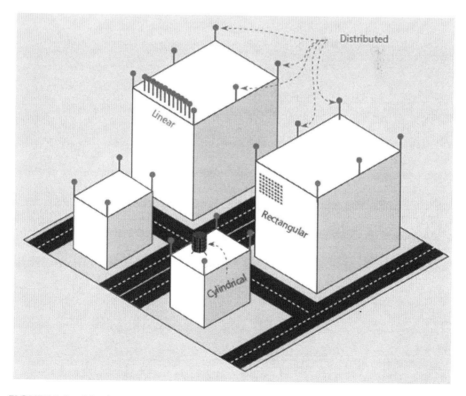

**FIGURE 5.2** Massive MIMO Antenna Configurations.

## 5.2 MASSIVE MIMO UPLINK AND DOWNLINK

Figure 5.3 illustrates the downlink scenario of massive MIMO (Marzetta 2015) communication system for $K$ active users and $M$ base station antennas.

Usually, M base station antennas are in the form of a small antenna array, and K users would have a single antenna each. The basic idea is that each user transmits an individual data stream and receives only the intended data stream. The above task is achieved by data streams in different time frames and frequencies. The data streams are transmitted in different spatial domain but in the same time-frequency bins as referred in the massive MIMO system. Data streams occupy the same time-frequency resources but differ in spatial dimension. In line-of sight (LOS) propagation conditions, active terminals will receive focused data streams. However, in non-LOS (NLOS) environments, the data streams can be combined constructively to get high SNR and destructively to mitigate intersymbol interference (ISI). The precoded block in transmitter section is used to obtain channel state information (CSI), which helps to determine channel frequency response between each of the antenna elements and users. Energy efficiency can also be improved by narrowing the beam toward an intended user with the help of antennas at the base station.

During uplink transmission, data streams are time and frequency multiplexed as shown in Figure 5.4.

The antenna array at the base station receives the modified data streams and passes them to the decoder. The decoder utilizes CSI of individual users to produce their data stream. The benefits of having a larger M/K ratio offers improved spectral efficiency, higher data throughput, less radiated power, effective power control, and simplicity in signal processing.

## 5.3 SPECTRAL EFFICIENCY (SE)

Spectral efficiency (SE) (Björnson, Hoydis, and Sanguinetti 2017) is expressed as bit/s/Hz. The maximum SE is defined by (5.3) (Shannon 1948).

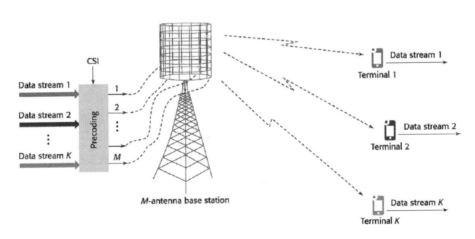

**FIGURE 5.3** Massive MIMO Downlink Operation.

# Multiantenna Systems

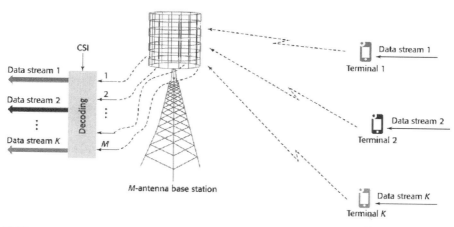

**FIGURE 5.4** Massive MIMO Uplink Operation.

$$C = \sup_{f(x)} \left( H(y) - H(y|x) \right), \qquad (5.3)$$

where $H(y)$ is entropy at output of system, and $H(y|x)$ is conditional entropy of output with respect to input, The supremum is calculated over $f(x)$, and channel capacity for deterministic channels ($h$) can be defined as

$$C = \log_2 \left( 1 + \frac{\rho(h)^2}{\sigma^2} \right), \qquad (5.4)$$

where $\rho(h)^2$ is received signal power over noise variance or power $\sigma^2$. The cellular system are corrupted by interference from same and other nearby cells. The lower bounds of such channels can be obtained as expectation of random realizations as shown below.

$$C \geq \mathbb{E}\left\{ \log_2 \left( 1 + \frac{\rho(h)^2}{\rho_I + \sigma^2} \right) \right\}, \qquad (5.5)$$

where $\rho_I$ is interference power, and interference I has zero mean and known variance but independent of input x, i.e., $\mathbb{E}(x * I) = 0$. The cellular networks are designed such that these interfering signal are not very strong, thus they can be treated as addition noise. In case of strong interferes, they can be decoded first and then subtracted from received signal before detecting desired signal. In low-interference regime, SE given by (5.5) will be optimal in nature. Equation (5.5) can be modified as shown below.

$$C \geq \mathbb{E}\left\{ \log_2 \left( 1 + SINR \right) \right\}, \qquad (5.6)$$

where signal-to-interference-plus-noise ratio $(SINR) = \dfrac{\rho(h)^2}{\rho_I + \sigma^2}$. In MIMO systems, both transmitters and receivers are equipped with multiple antennas. This leads to increases in the data rates and spectral efficiencies of MIMO (Marzetta 2015) systems as compared to SISO system. In MIMO, multistream transmissions from the multiple antennas would further lead to an increase in system capacity as given below.

$$C = \min(M,K)\log_2(SNR). \tag{5.7}$$

The above equation shows that throughput can be enhanced by adding more antennas rather than increasing spectral bandwidth or radiated power. Equation (5.7) is valid for sufficiently high SNR values, and elements of channel matrix should be independent, identically distributed. Equation (5.7) can be further generalized as follows.

$$C = \log_2 \det\left(I_K + \frac{SNR}{M} H^H H\right), \tag{5.8}$$

where $H^H$ is M×K frequency response of channel between base station antennas and user antennas and $I_K$ is identity matrix of order K×K. The superscript "$H$" represents Hermitian transpose of a matrix. The assumptions for (5.8) include noise to be complex Gaussian and the receiver knows downlink CSI, but the transmitter doesn't. The operation will be simplified if both transmitter and receiver know CSI, but it will increase overheads. Equation (5.8) signifies downlink mode equipped with M antennas on base station and K antennas on the receiver side. Uplink operation on same link indicates K antennas on transmitter and M antennas on receiver end. Equation (5.8) can be rewritten into (5.9) with terms $\dfrac{SNR}{M}$ replaced by $\dfrac{SNR}{K}$ and $I_K$ replaced by $I_M$, as shown below.

$$C = \log_2 \det\left(I_M + \frac{SNR}{K} HH^H\right). \tag{5.9}$$

The uplink and downlink channels differ for FDD, but they are theoretically the same for TDD systems. For channel estimation, the transmitter would send known pilots through the channel under power constraints. For optimum performance, these pilots should be mutually orthogonal. Downlink data transmission requires the sample duration of the pilot sequences should be higher than M and similarly uplink data transmission requires uplink pilot duration to be higher than K. Thus, total pilots burden is greater than the sum of M and K in either the FDD or TDD system.

Multi-user MIMO is equivalent to replacing the single K antenna user with k users having a single antenna. Multi-user MIMO shows lesser achievable throughput as compared to point-to-point MIMO because users are not able to communicate with each other. The capacity formula of (5.9) is also valid for multi-user MIMO in uplink

# Multiantenna Systems

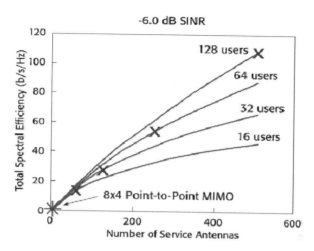

**FIGURE 5.5** Total Spectrum Efficiency with Variation in Number of Base Station Antennas for Particular K.

mode provided only base station knows the channel. The downlink capacity in multi-user MIMO is the solution of the following equation.

$$C = \sup_a \left\{ \log_2 \det \left( I_M + \frac{SNR}{K} H diag H^H \right) \right\} \qquad a \geq 0, 1^T a = 1, \qquad (5.10)$$

where *diag* is M×1 diagonal matrix, and $a, 1$ are M×1 vector of one's. The downlink multi-user capacity exceeds that of (5.9) due to additional CSI involved in it. Further, the precoding techniques can be employed to enhance capacity of MIMO systems. To get optimum performance using dirty paper coding, very accurate CSI is essential. The biggest advantage of using multi-user MIMO is that it is less vulnerable to the environment as compared to point-to-point MIMO.

In massive MIMO systems, the lower bound of capacity is computed using (5.11) under perfect CSI and conjugate beamforming (Yang and Marzetta 2013b).

$$C > K \log_2 \left( 1 + \frac{MSNR}{k(1+SNR)} \right). \qquad (5.11)$$

Massive MIMO performance in terms of total spectrum efficiency with variation in number of base station antennas for particular K is plotted in Figure 5.5.

Figure 5.5 shows massive MIMO can work in a region that is not available for conventional MIMO systems. The results of Figure 5.5 show four plots correspond to different users with variation in M base station antennas at fixed SINR of −6.0 dB. The SE of 13.6 bits/s/Hz is achieved at point (64, 16), whereas 8×4 MIMO shows SE of 1.3 bits/s/Hz with full CSI.

## 5.4 AREA THROUGHPUT

The area throughput (Björnson, Hoydis, and Sanguinetti 2017) is another important performance metric of future wireless communication systems. Mathematically, it can be represented using (5.12).

$$\text{Area throughput}\left[\text{bit/s/km}^2\right] = B\left[\text{Hz}\right] \cdot D\left[\text{cells/km}^2\right] \cdot \text{SE}\left[\text{bit/s/Hz/cell}\right], \quad (5.12)$$

where $B$ is bandwidth, $D$ is average cell density, and SE is spectral efficiency for each cell. It is evident in (5.12) that area throughput depends upon $B$, $D$, and SE parameters.

Area throughput can be improved by

a. Allocating more frequency spectrum
b. Increasing network density
c. Employing techniques that increase SE in each cell.

To solve "the 1000× data challenge" posed by Qualcomm (Staff 2012), we require more than 1 THz of bandwidth. This requirement of bandwidth is impractical because frequency spectrum is shared globally for many services. This would limit the range of service and its reliability. Huge slots of bandwidth are available in 30–300 GHz band (mmWave band) (Qingling and Li 2006), but frequencies in these bands are valid for short range only. Thus, these frequencies have limited coverage and performance. The next option is to increase cellular density by increasing BSs in every predefined cell area. The intra–base station distances are in meters, and thus base stations are employed in such a way that they avoid shadowing in cells. It limits the base station scalability. Thus, we have no option other than moving base stations closer to user terminals. This leads to increased shadowing, which in turn reduces coverage of the system. mmWaves solves the above problem but within short range of cell. Thus, we can conclude that increasing bandwidth and densification of cells are not viable solutions, but using massive MIMO we can significantly improve SE as already discussed in Section 5.3.

## 5.5 PRECODING

Massive MIMO fundamentally contains a large number of base station antennas. The SE of such systems will improved further by employing precoding techniques. Let $P$ represent linear precoding matrix, $s$ represent the source symbols prior to precoding, and $\rho$ is base station power. The signal vector at base station is given below.

$$x = \sqrt{\rho}Ps. \quad (5.13)$$

The precoding matrix can be designed on the basis of availability of knowledge about the channel matrix $H$. Normalized transmit power is assumed, i.e., $\|s\|^2 = 1$. Also, $\text{tr}(PP^H) = 1$. The signals at user terminals are shown below.

$$y = H^T x + n \quad (5.14)$$

# Multiantenna Systems

$$y = \sqrt{\rho}H^T Ps + n, \qquad (5.15)$$

where, $n$ represents noise. Precoding techniques can be applied differently in single or multicell scenarios.

## 5.5.1 Single-Cell (SC) Precoding Techniques

The main SC techniques are discussed below.

### 5.5.1.1 Matched Filter (MF)

This filtering technique is comparatively simple to implement. This technique is based upon performing Hermitian transpose of channel matrix,

$$P_{MF} = \sqrt{\alpha}H^*. \qquad (5.16)$$

Power of all transmitted symbols is scaled by $\alpha$ to get normalized power. Equation (5.16) is rewritten in (5.17) to get matched filter response:

$$y_{MF} = \sqrt{\alpha\rho}H^T H^* s + n. \qquad (5.17)$$

This technique increases SNR of the intended user to the maximum extent, thus it can be termed as maximum ratio transmission (MRT) (Kammoun et al. 2014; Parfait, Kuang, and Jerry 2014). Selvan, Iqbal, and Al-Raweshidy (2014) exploits MRT in a downlink scenario and obtains its capacity and power expressions. Björnson et al. (2015b) calculated optimized values of M and K to get an energy-efficient system. However, the optimized values of M and K affect performance of massive MIMO systems. Thus, the MRT scheme shows optimal behavior when M >> K.

### 5.5.1.2 Zero Forcing Precoding

Matched filtering precoding is inefficient, when interferer power is considerably higher than noise power. In such a scenario, we preferred to use the zero forcing (ZF) precoding (Parfait, Kuang, and Jerry 2014) technique. The basic idea of ZF precoding is to mitigate ISI caused by unintended users. The ZF precoder matrix is given by

$$P_{ZF} = \sqrt{\alpha}H^*\left(H^T H^*\right)^{-1}. \qquad (5.18)$$

Equation (5.17) is rewritten using ZF precoder matrix as shown below:

$$y = \sqrt{\rho\alpha}H^T H^*\left(H^T H^*\right)^{-1} s + n. \qquad (5.19)$$

The correlation among users can be depicted from off-diagonal elements of $H^T H*$. The correlation between channels reduces capacity (Gao et al. 2011). The biggest drawback of ZF precoding is noise enhancement, but still it behaves ideally in a noise-free system.

### 5.5.1.3 Regularized Zero Forcing

Regularized ZF (RZF) (Hoydis, Ten Brink, and Debbah 2013) precoder optimally combines MRT and ZF precoders. The RZF precoder matrix is given below (Hoydis, Ten Brink, and Debbah 2013):

$$P_{RZF} = \sqrt{\alpha} H^* \left( H^T H^* + X + \lambda I_K \right)^{-1}. \tag{5.20}$$

The received signal in (5.17) can be rewritten as

$$y = \sqrt{\rho \alpha} H^T H^* \left( H^T H^* + X + \lambda I_K \right)^{-1} s + n. \tag{5.21}$$

It is evident that (5.21) becomes ZF precoder by setting X and $\lambda$ as zero, whereas it behaves as matched filter when X, $\lambda$ are 0 and $\infty$, respectively.

### 5.5.1.4 Truncated Polynomial Expansion (TPE)

This scheme implements RZF with lesser computational complexity. The corresponding precoding matrix (Kammoun et al. 2014) is

$$P_{TPE} = \sum_{j=0}^{J-1} w_j \left( H^T H^* \right)^j H^T, \tag{5.22}$$

where, $w_j$ is the set of precoder weight coefficients.

### 5.5.1.5 Phased Zero Forcing

This scheme also referred as the hybrid precoding technique (Liang, Xu, and Dong 2014) in which the precoder coefficients are a combination of its RF and baseband equivalents. Only analog precoding is not beneficial because of beamforming limitations, and only digital precoding is also not recommended because of its associated heavy complexity. Thus, a new hybrid combination of analog/digital precoding is proposed.

Initially, only phase transitions are allowed, which are then derived from Hermitian response of its corresponding channel. These phase alignments result in higher channel gains for massive MIMO systems. Analog precoder is given as

$$F_{i,j} = \frac{1}{\sqrt{N_t}} e^{-j\theta_{i,j}}, \tag{5.23}$$

where $\theta_{i,j}$ represents the phase of channel element. The channel H can be modified as

$$H_{eq} = H \, x \, F. \tag{5.24}$$

The PZF is then implemented as

$$P_{PZF} = H_{eq}^H \left( H_{eq} H_{eq}^H \right)^{-1} C, \tag{5.25}$$

# Multiantenna Systems

where diagonal matrix C is used for normalizing power.

### 5.5.2 Precoding Techniques for the Downlink of Multicell Scenario

Multicell scenario is more practical because user equipment will receive a signal from multiple cells simultaneously. The following key advantages may be viable:

- Improved mobility of user
- Enhanced coverage
- Diversity gains
- Capacity improvement
- Improved spectral efficiency

The traditional single-cell precoding techniques can be used in a multicell scenario as well. However, some of the best precoding techniques developed for a multicell regime are discussed below.

#### 5.5.2.1 Max SINR Precoding

Jing and Zheng (2014) propose an algorithm that improves the SINR by increasing the ratio of signal power to combined noise and intercell interference. The authors suggest that the power utilization rate must be as high as possible, and at the same time total power consumption should be minimum. Let $r$ represent power utilization rate and $\rho$ represent total power available. The objective function can be expressed as $\min_{r,w_{j,k}} r\rho$. The objective function can be modified as in (5.26),

$$\sum_{k=1}^{K}|w_{j,k}|^2 \leq r\rho \quad \& \, SINR_{j,k} \geq \gamma_{j,k}, \tag{5.26}$$

where $\gamma_{j,k}$ represents the threshold SINR required by a user $k$ in cell $j$ for successful communication.

#### 5.5.2.2 Multilayer Precoding

A novel multilayered solution to precoding is proposed by Alkhateeb, Leus, and Heath (2014), where each layer is derived from the previous layers and allows for reduction in complexity, as shown in Figure 5.6.

The system model assumes a scenario where intercellular interference is observed by user equipment along with downlink precoding. The received signal of user "k" in cell "c" is given by

$$y_{ck} = \sum_{b=1}^{B} h_{bck}^{*} F_b s_b + n_{ck}, \tag{5.27}$$

where

- $h_{bck}^{*}$ represents channel between base station "b" and user "k" for cell c,

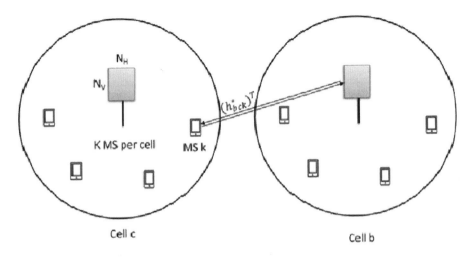

**FIGURE 5.6** Multilayer Precoding System Model.

- $s_b$ represents transmit symbol vector
- $n_{ck}$ represents the Gaussian noise.

Equation (5.27) can be expanded as

$$y_{ck} = h^*_{cck}\left[F_c\right]_{:,k} s_{c,k} + \sum_{m \neq k} h^*_{ccm}\left[F_c\right]_{:,m} s_{c,m} + \sum_{b \neq c} h^*_{bck} F_b s_b + n_{ck}. \qquad (5.28)$$

The precoding matrix $F_b$ of cell b is proposed to be designed as

$$F_b = F_b^{(1)} F_b^{(2)} F_b^{(3)}. \qquad (5.29)$$

Each one of these precoding matrices aims at a particular desired outcome. The first precoding matrix $F_b^{(1)}$ nulls the intercellular interference. The second precoding matrix $F_b^{(2)}$ is aimed at maximizing the signal power received by the user. The third precoding matrix aims at nullifying the intracellular interference.

### 5.5.2.3 Quantized Precoders for Massive MIMO

Quantized precoders (Jacobsson et al. 2017) are applied based on the fact that the lower the number of bits required for DAC, the lower is the complexity, and hence easier implementation may be achieved. While some of the researchers suggest usage of 1-bit resolution DACs, others support trade-off between number of bits and the efficiency achieved.

### 5.5.2.4 Nonlinear Quantized Precoding for Massive MIMO

Although nonlinear precoders are quite complex as compared to linear precoders, they have better performance. Jacobsson et al. (2017) proves that only 3 dB penalty is incurred for nonlinear precoding for the infinite resolution case as compared to linear precoders that incur a penalty of 8 dB, against 1 bit DACs. In order

# Multiantenna Systems

to deal with the challenges posed because of the finite-cardinality of DACs, two approaches have been explored – linear quantized precoding and nonlinear precoding. The system model assumes that there are 16 single antenna users and 128 antennas at the base station. The linear quantization problem can be mathematically expressed as

$$\begin{cases} \underset{P \in \mathbb{C}^{B \times U}, \beta \in \mathbb{R}}{\text{minimize}} & \mathbb{E}\left[\left\|s - \beta H_{eq} FWs\right\|_2^2\right] + \beta^2 U N_0 \\ \text{subject to} & \mathbb{E}\left[\left\|X\right\|_2^2\right] \leq P \quad \& \quad \beta > 0. \end{cases} \quad (5.30)$$

The research work proved that even if a 3–4 bit DAC is used, bit error rate (BER) and achievable rate performance is approximately the same as with infinite resolution DACs. Additionally, an asymptotic estimate of the effective signal-to-interference noise ratio + distortion ratio (SINDR) has been derived (Jacobsson et al. 2017). This estimate may be used to predict precise system performance.

### 5.5.2.5 Multiuser Massive MIMO with One-Bit Quantized Precoding

Saxena, Fijalkow, and Swindlehurst (2017) explores the benefits and supports usage of linear precoding with 1-bit DACs. Off course, the complexity of proposed techniques is much less, and in addition the BER performance is also quite appreciable. The author makes use of Bussgang theorem to analyze the ZF precoder. The author condemns the usage of relatively popular hybrid precoding, citing the suboptimal usage of same beamforming network for the entire band and additional complexity in the form of phase shifters. Also, the author supports usage of 1-bit DACs, as it requires no change in the RF design, which is required otherwise.

Figure 5.7 shows that massive MIMO network performance depends upon the ratio of antennas at the BSs to the number of single antenna users instead of individual values. The proposed precoding algorithm outperformed the NLS precoder up to moderate values of SNR with a bit of additional complexity.

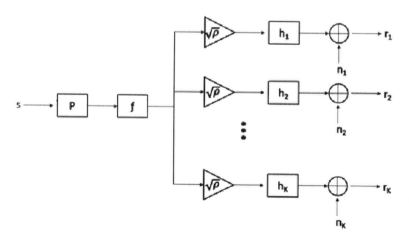

**FIGURE 5.7** One-Bit Quantized Precoded System Model.

## 5.6 HYBRID PRECODING

One of the recently proposed techniques is hybrid precoding (Sohrabi and Yu 2016). A hybrid precoder involves precoding at both baseband and RF. The baseband precoder is also known as digital precoder, and RF precoder is also known as analog precoder owing to the domain of signal processing. The hybrid precoders involve a low-dimensional baseband precoder followed by RF precoder, which is usually implemented by an array of phase shifters. Sohrabi and Yu (2016) reported the optimal hybrid precoder by making use of the following objective functions and along with some power constraints:

$$\max_{P_{BB} C_D P_{RF} C_{RF}} \sum_{k=1}^{K} \beta_k R_k. \tag{5.31}$$

Table 5.1 shows a summary of research work in this field.

## 5.7 MASSIVE MIMO WITH LINEAR PRECODING AND DETECTION

In Section 5.2, massive MIMO uplink and downlink operation was already discussed. In this section, we extend the discussion of Section 5.2 with linear precoding and detection method (Chien and Björnson 2017). Let us consider massive MIMO system with L cells, each cell contains (M, K) antennas. The channel across particular BS to user $k$ for $i$th cell is given by

$$\boldsymbol{h}_{i,k} = \{h_{i,k,1} \ldots \ldots h_{i,k,M}\}^T. \tag{5.32}$$

The average and random realizations of $\boldsymbol{h}_{i,k}$ can be expressed as

$$\vec{\boldsymbol{h}}_{i,k} = \mathbb{E}\{\boldsymbol{h}_{i,k}\} = \{\vec{h}_{i,k,1} \ldots \ldots \vec{h}_{i,k,M}\}^T. \tag{5.33}$$

Each uplink symbol yields corresponding received signal as shown below:

$$\boldsymbol{y} = \sum_{i=1}^{L} \sum_{k=1}^{K} \boldsymbol{h}_{i,k} \sqrt{\rho_{i,k}} x_{i,k} + \boldsymbol{n}, \tag{5.34}$$

where $x_{i,k}$ is normalized transmitted symbol, $\rho_{i,k}$ is transmitted power of user $k$ in cell $l$, and $\boldsymbol{n}$ is AWGN noise with zero mean and variance $\sigma^2$. Channel estimation is performed when every user transmits $\tau_{UL}$ pilots, as shown in Figure 5.8.

The system design should be in such a way that independent pilot sequence should be allocated to each user. The uplink received signal during pilot transmission is given as

$$\boldsymbol{y}_{pilot} = \sum_{i=1}^{L} \boldsymbol{H}_i \boldsymbol{P}_i^{\frac{1}{2}} \varphi_i^H + \boldsymbol{n}_{pilot}, \tag{5.35}$$

where $\boldsymbol{y}_{pilot} \in \mathbb{C}^{M \times \tau_{UL}}$, $\boldsymbol{H}_i = \{\boldsymbol{h}_{i,1} \ldots \ldots \boldsymbol{h}_{i,K}\}^T$, $\boldsymbol{P}_i = diag(\rho_{i,1} \ldots \ldots \rho_{i,K}) \in \mathbb{C}^{K \times K}$, and pilot matrix $\varphi_i^H = \{\varphi_{i,1} \ldots \ldots \varphi_{i,K}\} \in \mathbb{C}^{\tau_{UL} \times K}$. Channel mean and variances along

# TABLE 5.1
## Latest Work in Hybrid Precoding

| S. No. | Author & Year | Title | Findings |
|---|---|---|---|
| 1 | Ribeiro et al. (2018) | "Energy Efficiency of mmWave Massive MIMO Precoding with Low-Resolution DACs" | Research work proves improved energy efficiency of partially connected hybrid precoders as compared to digital precoders. Poor energy efficiency is demonstrated by fully connected hybrid precoders in general. |
| 2 | Ratnam et al. (2018) | "Hybrid Beamforming with Selection for Multiuser Massive MIMO Systems" | Authors suggest a generic architecture for HBwS that reduces complexity and also the overall cost of massive MIMO systems |
| 3 | Zhao et al. (2018) | "Multi-cell Hybrid Millimeter Wave Systems: Pilot Contamination and Interference Mitigation" | Authors approximate average achievable rate per user in a closed form expression. Authors also prove effective mitigation of interference (intercell and intracell) by increased count of antennas at base station. |
| 4 | Du et al. (2018) | "Hybrid Precoding Architecture for Massive Multiuser MIMO with Dissipation: Sub Connected or Fully-Connected Structures?" | Research work concludes that the sub-connected structure has better system spectral efficiency as compared to complex, fully connected structures in the systems with low SNR. A fully connected structure performs better in other cases. |
| 5 | Xie et al. (2018) | "Geometric Mean Decomposition Based HybridPrecoding for Millimeter-Wave Massive MIMO" | The proposed technique avoids complicated bit allocation and also yields improved performance. |
| 6 | Zhu, Zhang, and Yang (2018) | "Low-Complexity Hybrid Precoding with Dynamic BeamAssignment in mmWave OFDM Systems" | Authors achieve improved performance along with reduction in complexity by assignment of different (multiple, >1) beams from multiple subarrays to one user. |
| 7 | Buzzi et al. (2018) | "Single-Carrier Modulation versus OFDM forMillimeter-Wave Wireless MIMO" | Research work establishes performance improvement by use of single-carrier modulation along with time-domain equalization. Poor performance observed for lower transmit powers and increased distances (>90 m) |

with linear minimum mean square error (LMMSE) estimator (Kammoun et al. 2014) is used to estimate channel coefficients ($\hat{h}_{i,K}$) from received pilot signal defined in (5.35). The channel estimator ($\hat{h}_{i,K}$) for kth user in ith cell using LMSSE is

$$\hat{h}_{j,K} = \vec{h}_{j,k} + \frac{\sqrt{\rho_{j,k}}\alpha_{j,k}}{\sum_{i=1}^{L}\sum_{k=1}^{K}\rho_{i,k}\tau_{UL}\alpha_{i,k} + \sigma^2}\left(y_{pilot}\varphi_{j,K} - \sum_{i=1}^{L}\sum_{k=1}^{K}\sqrt{\rho_{i,k}}\tau_{UL}\vec{h}_{i,k}\right). \quad (5.36)$$

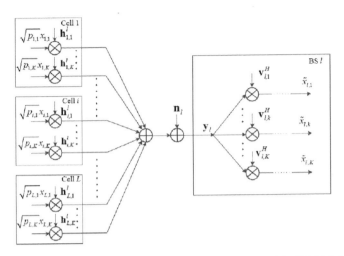

**FIGURE 5.8** Uplink Transmission with Linear Detection in Multicell Multi-User MIMO System.

The estimation error $e_{j,k} = h_{j,K} - \hat{h}_{j,K}$ has zero mean and variance $(n_{j,k}^v)$:

$$n_{j,k}^v = \alpha_{j,k}\left(1 - \frac{\rho_{j,k}\tau_{UL}\alpha_{j,k}}{\sum_{i=1}^{L}\sum_{k=1}^{K}\rho_{i,k}\tau_{UL}\alpha_{i,k} + \sigma^2}\right). \quad (5.37)$$

An equation (5.37) shows that $n_{j,k}^v$ is not altered by increasing M at the base stations, but it depends upon noise as well as interference, which comes from same cell. Thus, we can conclude that users in same cell should have different pilot sequences. The performance of massive MIMO can be improved if BS belongs to particular cell, treats its own received signal as a desired signal, and signals from nearby users are called intercell interference. To extract a desired signal, (5.34) is multiplied with $\beta_{i,k} \in C^M$:

$$\beta_{i,k}^H y = \sum_{i=1}^{L}\sum_{l=1}^{K}\beta_{i,k}^H h_{i,l}\sqrt{\rho_{i,l}}\,x_{i,l} + \beta_{i,k}^H n. \quad (5.38)$$

Equation (5.38) can be rewritten as

$$\beta_{i,k}^H y = \beta_{i,k}^H h_{i,k}\sqrt{\rho_{i,k}}\,x_{i,k} + \sum_{\substack{i=1\\l\neq k}}^{L}\sum_{l=1}^{K}\beta_{i,k}^H h_{i,l}\sqrt{\rho_{i,l}}\,x_{i,l} + \sum_{\substack{i=1\\i\neq j}}^{L}\sum_{l=1}^{K}\beta_{i,k}^H h_{i,l}\sqrt{\rho_{i,l}}\,x_{i,l} + \beta_{i,k}^H n. \quad (5.39)$$

The received signal in (5.39) comprises four parts, which represents intended signal, interference within same cell, interference from nearby cells, and remedial noise. The main aim is to amplify the first term and suppress all other remaining terms. This task can be achieved by employing MR or ZF detection techniques on linear detection vector $\beta_{UL} = \{\beta_1 \ldots \ldots \beta_k\} \in C^{M \times K}$. MR detection generally works on maximizing the ratio of average signal gain to norm of detection vector as given here:

$$\mathbb{E}\left\{\frac{\beta_{i,k}^H h_{i,k}}{\beta_{i,k}}\right\} = \frac{\beta_{i,k}^H \hat{h}_{i,k}}{\beta_{i,k}} \leq \hat{h}_{i,k}. \tag{5.40}$$

On other side, ZF detection minimizes interference within same cell and amplifies the intended signal as shown in (5.41):

$$\mathbb{E}\left\{\beta_{UL}^H H P^{\frac{1}{2}} X\right\} = \beta_{UL}^H \hat{H} P^{\frac{1}{2}} X = \left(\left(\hat{H}\right)^H H\right)^{-1}\left(\left(\hat{H}\right)^H H\right) P^{\frac{1}{2}} X. \tag{5.41}$$

During downlink operation of massive MIMO system, the transmitted signal vectors are linearly precoded as done in (5.42),

$$X^n = \sum_{l=1}^{K} \sqrt{\rho_l^n}\, \vartheta_l^n x_l^n, \tag{5.42}$$

where $X^n$ is a transmitted signal vector from the $n$th base station, $x_l^n$ is symbol intended for the $l^{th}$ user in the $n$th cell with unit transmitted power, and $\vartheta_l^n \in \mathbb{C}^M$ is a linear precoding vector. The received signal for user $k$ in the $n$th cell is as follows:

$$y_k^n = \sum_{i=1}^{L} \left(h_{i,k}^n\right)^H X_i + n_k^n, \tag{5.43}$$

where $h_{i,k}^n$ is channel frequency response, which is analogous to uplink channel because of TDD mode of operation. As there are no downlink pilots, we don't have instantaneous CSI, which will affect the performance of MIMO system. Due to the large value of M antennas, the precoded channels will quickly approach to their mean value. Thus, even the lack of CSI won't affect massive MIMO systems. As uplink and downlink channels are in coherence interval, it would connect uplink and downlink performance. Let us consider downlink precoding vectors as

$$\vartheta_k^n = \frac{\beta_k^n}{\sqrt{\mathbb{E}\{\beta_k^{n2}\}}}. \tag{5.44}$$

Due to uplink-downlink duality, total powers in both directions would be the same but it would be different among users. Thus, MR and ZF precoding can be changed as follows in (5.45).

$$\vartheta_k^n = \begin{cases} \dfrac{\hat{\boldsymbol{h}}_k^n}{\sqrt{\mathbb{E}\left\{\hat{\boldsymbol{h}}_k^{n2}\right\}}} & \text{for MR} \\[2ex] \dfrac{\hat{\boldsymbol{H}}_k^n \left(\left(\hat{\boldsymbol{H}}\right)^H \hat{\boldsymbol{H}}\right)^{-1}_{kth-column}}{\sqrt{\mathbb{E}\left\{\left|\hat{\boldsymbol{H}}_k^n \left(\left(\hat{\boldsymbol{H}}\right)^H \hat{\boldsymbol{H}}\right)^{-1}_{kth-column}\right|^2\right\}}} & \text{for ZF} \end{cases} \qquad (5.45)$$

## 5.8 ENERGY EFFICIENCY (EE)

Energy efficiency (Björnson et al. 2014) is referred to as energy spent to do certain work. This definition is applicable to all fields of science but particularly for wireless communication. It can be derived from the definition of SE as shown below.

The energy efficiency (EE) of a wireless network is total bits that can be transmitted reliably per unit of energy

Mathematically EE can be expressed as

$$\text{Energy Efficiency}(\text{EE}) = \frac{\text{Throughput}(\text{bit / s / cell})}{\text{Power Consumption}(\text{W / cell})}. \qquad (5.46)$$

In this section, we will discuss power scaling law (Björnson, Matthaiou, and Debbah 2015a) and then SE/EE trade-off. An uplink scenario of a single-cell system consists of K users with single-antenna and base station with M antennas such that M>>K. We further assume uncorrelated channels and users. In such scenario, detectors like MF, ZF, and MMSE will show better results. The ergodic uplink data rate (Ngo, Larsson, and Marzetta 2013) for MF detector at $k$th user with $M \to \infty$ under perfect channel estimation is given by (5.47).

$$\mathcal{R}_k^{UL} \approx \log_2\left(1 + M\gamma_k \rho_{UL}\right), \qquad (5.47)$$

where $\mathcal{R}_k^{UL}$ is uplink data rate for $k$th user, $\gamma_k$ is large-scale fading coefficient for $k$th user, and $\rho_{UL}$ transmitted power. Data rate in (5.47) can be rewritten for SISO system:

$$\mathcal{R}_k^{UL} \approx \log_2\left(1 + \gamma_k \rho_{UL}\right). \qquad (5.48)$$

From (5.47) and (5.48), it can be observed that for large M, user performance with transmit power $\dfrac{\rho_{UL}}{M}$ in MU-MIMO system (Lu et al. 2014) will exhibit the same performance in SISO system with transmit power $\rho_{UL}$. Thus, in the case of perfect CSI, power is scaled down by M times for single user and spectral efficiency increased by K times. In case of imperfect CSI, the ergodic uplink data rate (Ngo, Larsson, and Marzetta 2013) for MF detection is given by

$$\mathcal{R}_{k,imperfect}^{UL} \approx \log_2\left(1 + \tau_{UL} M \gamma_k^2 \rho_{UL}^2\right). \tag{5.49}$$

From (5.48) and (5.49), it can be observed that for large M, user performance with transmit power $\frac{\rho_{UL}}{\sqrt{M}}$ in MU-MIMO system will asymptotically exhibit the same performance in SISO system with modified power as $\tau_{UL} \gamma_k \rho_{UL}^2$. Thus, to achieve equal performance, we have to scale down power by factor $\frac{1}{\sqrt{M}}$. In multicell scenario, one user can scale down its transmit power by 1/M or $\frac{1}{\sqrt{M}}$ depending upon perfect or imperfect CSI.

Because of SE/EE trade-off (Yang and Marzetta 2013a; Björnson et al. 2014) under perfect CSI, the decrease in EE corresponds to increase in SE. Whereas with imperfect CSI, the increase in EE corresponds to increase in SE under lower power region, and it decreases with increase in SE under higher power regime.

Initially, circuit power consumption (Lu et al. 2014) is not considered as a factor that affects EE. But in Pei, Pham and Liang (2012), circuit power consumption plays an important rule to improve EE. Circuit power consumption generally depends upon how to perform antenna selection, which further improves EE. Pei, Pham, and Liang (2012), explores how SE is maximized under perfect/imperfect CSI with proper choice of RF chains. It was shown in the case of MISO system that the optimal RF chains are half of maximum RF chains (Shashank, Wajid, and Mandavalli 2012).

## REFERENCES

Alkhateeb, Ahmed, Geert Leus, and Robert W. Heath. 2014. "Multi-Layer Precoding for Full-Dimensional Massive MIMO Systems." In *2014 48th Asilomar Conference on Signals, Systems and Computers*, Pacific Grove, CA, 815–819.

Björnson, Emil, Jakob Hoydis, Marios Kountouris, and Merouane Debbah. 2014. "Massive MIMO Systems with Non-Ideal Hardware: Energy Efficiency, Estimation, and Capacity Limits." *IEEE Transactions on Information Theory* 60 (11): 7112–7139.

Björnson, Emil, Jakob Hoydis, and Luca Sanguinetti. 2017. "Massive MIMO Networks: Spectral, Energy, and Hardware Efficiency." *Foundations and Trends in Signal Processing* 11 (3–4): 154–655.

Björnson, Emil, Michail Matthaiou, and Mérouane Debbah. 2015a. "Massive MIMO with Non-Ideal Arbitrary Arrays: Hardware Scaling Laws and Circuit-Aware Design." *IEEE Transactions on Wireless Communications* 14 (8): 4353–4368.

Björnson, Emil, Luca Sanguinetti, Jakob Hoydis, and Mérouane Debbah. 2015b. "Optimal Design of Energy-Efficient Multi-User MIMO Systems: Is Massive MIMO the Answer?" *IEEE Transactions on Wireless Communications* 14 (6): 3059–3075.

Buzzi, S., C. D'Andrea, T. Foggi, A. Ugolini, and G. Colavolpe. 2018. "Single-Carrier Modulation versus OFDM for Millimeter-Wave Wireless MIMO." *IEEE Transactions on Communications* 66 (3): 1335–1348. doi:10.1109/TCOMM.2017.2771334.

Chien, van Trinh, and Emil Björnson. 2017. "Massive MIMO Communications." In *5G Mobile Communications*, 77–116. Springer.

Cudak, Mark, Amitava Ghosh, Thomas Kovarik, Rapeepat Ratasuk, Timothy A. Thomas, Frederick W. Vook, and Prakash Moorut. 2013. "Moving towards Mmwave-Based beyond-4G (B-4G) Technology." In *2013 IEEE 77th Vehicular Technology Conference (VTC Spring)*, Dresden, Germany, 1–5.

Du, J., W. Xu, H. Shen, X. Dong, and C. Zhao. 2018. "Hybrid Precoding Architecture for Massive Multiuser MIMO with Dissipation: Sub-Connected or Fully Connected Structures?" *IEEE Transactions on Wireless Communications* 17 (8): 5465–5479. doi:10.1109/TWC.2018.2844207.

Gao, Xiang, Ove Edfors, Fredrik Rusek, and Fredrik Tufvesson. 2011. "Linear Pre-Coding Performance in Measured Very-Large MIMO Channels." In *2011 IEEE Vehicular Technology Conference (VTC Fall)*, San Francisco, CA, 1–5.

Gutierrez, Felix, Shatam Agarwal, Kristen Parrish, and Theodore S. Rappaport. 2009. "On-Chip Integrated Antenna Structures in CMOS for 60 GHz WPAN Systems." *IEEE Journal on Selected Areas in Communications* 27 (8): 1367–1378.

Hoydis, Jakob, Stephan Ten Brink, and Mérouane Debbah. 2013. "Massive MIMO in the UL/DL of Cellular Networks: How Many Antennas Do We Need?" *IEEE Journal on Selected Areas in Communications* 31 (2): 160–171.

Jacobsson, Sven, Giuseppe Durisi, Mikael Coldrey, Tom Goldstein, and Christoph Studer. 2017. "Quantized Precoding for Massive MU-MIMO." *IEEE Transactions on Communications* 65 (11): 4670–4684.

Jing, Jiang, and Xu Zheng. 2014. "A Downlink Max-SINR Precoding for Massive MIMO System." *International Journal of Future Generation Communication and Networking* 7 (3): 107–116.

Kammoun, Abla, Axel Müller, Emil Björnson, and Mérouane Debbah. 2014. "Linear Precoding Based on Polynomial Expansion: Large-Scale Multi-Cell MIMO Systems." *IEEE Journal of Selected Topics in Signal Processing* 8 (5): 861–875.

Liang, Le, Wei Xu, and Xiaodai Dong. 2014. "Low-Complexity Hybrid Precoding in Massive Multiuser MIMO Systems." *IEEE Wireless Communications Letters* 3 (6): 653–656.

Lu, Lu, Geoffrey Ye Li, A. Lee Swindlehurst, Alexei Ashikhmin, and Rui Zhang. 2014. "An Overview of Massive MIMO: Benefits and Challenges." *IEEE Journal of Selected Topics in Signal Processing* 8 (5): 742–758.

Marzetta, Thomas L. 2015. "Massive MIMO: An Introduction." *Bell Labs Technical Journal* 20: 11–22.

Ngo, Hien Quoc, Erik G. Larsson, and Thomas L. Marzetta. 2013. "Energy and Spectral Efficiency of Very Large Multiuser MIMO Systems." *IEEE Transactions on Communications* 61 (4): 1436–1449.

Parfait, Tebe, Yujun Kuang, and Kponyo Jerry. 2014. "Performance Analysis and Comparison of ZF and MRT Based Downlink Massive MIMO Systems." In *2014 Sixth International Conference on Ubiquitous and Future Networks (ICUFN)*, Shanghai, China, 383–388.

Pei, Yiyang, The-Hanh Pham, and Y. Liang. 2012. "How Many RF Chains Are Optimal for Large-Scale MIMO Systems When Circuit Power Is Considered?" In *2012 IEEE Global Communications Conference (GLOBECOM)*, Anaheim, CA, 3868–3873, doi 10.1109/GLOCOM.2012.6503720.

Qingling, Zhao, and Jin Li. 2006. "Rain Attenuation in Millimeter Wave Ranges." In *2006 7th International Symposium on Antennas, Propagation & EM Theory*, Guilin, China, 1–4.

Rappaport, Theodore S., Eshar Ben-Dor, James N. Murdock, and Yijun Qiao. 2012. "38 GHz and 60 GHz Angle-Dependent Propagation for Cellular & Peer-to-Peer Wireless Communications." In *2012 IEEE International Conference on Communications (ICC)*, Ottawa, ON, Canada, 4568–4573.

Rappaport, Theodore S., James N. Murdock, and Felix Gutierrez. 2011. "State of the Art in 60-GHz Integrated Circuits and Systems for Wireless Communications." *Proceedings of the IEEE* 99 (8): 1390–1436.

Rappaport, Theodore S., Shu Sun, Rimma Mayzus, Hang Zhao, Yaniv Azar, Kevin Wang, George N Wong, Jocelyn K. Schulz, Mathew Samimi, and Felix Gutierrez. 2013. "Millimeter Wave Mobile Communications for 5G Cellular: It Will Work!" *IEEE Access* 1: 335–349.

Ratnam, V. V., A. F. Molisch, O. Y. Bursalioglu, and H. C. Papadopoulos. 2018. "Hybrid Beamforming with Selection for Multiuser Massive MIMO Systems." *IEEE Transactions on Signal Processing* 66 (15): 4105–4120. doi:10.1109/TSP.2018.2838557.

Ribeiro, L. N., S. Schwarz, M. Rupp, and A. L. F. de Almeida. 2018. "Energy Efficiency of mmWave Massive MIMO Precoding with Low-Resolution DACs." *IEEE Journal of Selected Topics in Signal Processing* 12 (2): 298–312. doi:10.1109/JSTSP.2018.2824762.

Saxena, Amodh Kant, Inbar Fijalkow, and A. Lee Swindlehurst. 2017. "Analysis of One-Bit Quantized Precoding for the Multiuser Massive MIMO Downlink." *IEEE Transactions on Signal Processing* 65 (17): 4624–4634.

Selvan, V. P., M. S. Iqbal, and H. S. Al-Raweshidy. 2014. "Performance Analysis of Linear Precoding Schemes for Very Large Multi-User MIMO Downlink System." In *Fourth Edition of the International Conference on the Innovative Computing Technology (INTECH 2014)*, Luton, UK, 219–224.

Shannon, Claude E. 1948. "A Mathematical Theory of Communication." *The Bell System Technical Journal* 27 (3): 379–423.

Shashank, S. B., M. Wajid, and S. Mandavalli. 2012. "Fault Detection in Resistive Ladder Network with Minimal Measurements." *Microelectronics Reliability* 52 (8). doi:10.1016/j.microrel.2011.12.012.

Sohrabi, Foad, and Wei Yu. 2016. "Hybrid Digital and Analog Beamforming Design for Large-Scale Antenna Arrays." *IEEE Journal of Selected Topics in Signal Processing* 10 (3): 501–513.

Staff, Qualcomm. 2012. "Rising to Meet the 1000x Mobile Data Challenge." QUALCOMM Incorporated.

Xie, T., L. Dai, X. Gao, M. Z. Shakir, and J. Li. 2018. "Geometric Mean Decomposition Based Hybrid Precoding for Millimeter-Wave Massive MIMO." *China Communications* 15 (5): 229–238. doi:10.1109/CC.2018.8388000.

Yang, Hong, and Thomas L. Marzetta. 2013a. "Performance of Conjugate and Zero-Forcing Beamforming in Large-Scale Antenna Systems." *IEEE Journal on Selected Areas in Communications* 31 (2): 172–179.

Yang, Hong, and Thomas L. Marzetta. 2013b. "Total Energy Efficiency of Cellular Large Scale Antenna System Multiple Access Mobile Networks." In *2013 IEEE Online Conference on Green Communications (OnlineGreenComm)*, Piscataway, NJ, 27–32.

Zhao, L., Z. Wei, D. W. K. Ng, J. Yuan, and M. C. Reed. 2018. "Multi-Cell Hybrid Millimeter Wave Systems: Pilot Contamination and Interference Mitigation." *IEEE Transactions on Communications* 66 (11): 5740–5755. doi:10.1109/TCOMM.2018.2846255.

# 6 Channel Estimation Techniques in the MIMO-OFDM System

*Asif Alam Joy, Mohammed Nasim Faruq, and Mohammad Abdul Matin*
North South University

## CONTENTS

6.1 Introduction ........................................................................................................ 101
6.2 Conventional MIMO ......................................................................................... 102
6.3 Massive MIMO ................................................................................................... 103
6.4 Channel Estimation Techniques ...................................................................... 106
    6.4.1 Least Square (LS) Estimation ............................................................... 106
    6.4.2 Maximum Likelihood Estimation ........................................................ 108
    6.4.3 MMSE Channel Estimation .................................................................. 109
    6.4.4 Pilot- or Training-Based Channel Estimation ................................... 112
    6.4.5 Blind Channel Estimation ..................................................................... 112
    6.4.6 Semi-Blind Channel Estimation .......................................................... 113
6.5 Strengths and Limitations of Existing Estimation Techniques ................... 113
6.6 Conclusion .......................................................................................................... 114
References .................................................................................................................... 115

## 6.1 INTRODUCTION

The recent existing wireless technologies are utilizing MIMO along with OFDM to improve data rate and higher spectral efficiency. The main objective behind MIMO is to attain enhanced throughput, and the objective behind OFDM is to transform frequency selective channels into flat fading parallel sets of channels. Hence, both the MIMO and OFDM are coupled to achieve high throughput as well as simplified processing at the receiver, i.e., the conversion of the frequency selective fading channels into a set of parallel flat fading channels. Moreover, coupling both the MIMO and OFDM enhance overall system capacity and reliability (Cho et al., 2010; Ahmed and Matin, 2015; Alizadeh et al., 2016). In case of flat fading MIMO channel, one needs to eliminate MIMO ISI, which implies that a MIMO equalizer is needed. So, MIMO-OFDM removes the need of the equalizer and simplifies the receiver processing by converting this into a set of flat fading MIMO channel. The application of inverse fast Fourier transform (IFFT) and fast Fourier transform (FFT) at the transmitter and receiver of the OFDM helps to have an ISI free transmission with higher reliability.

To attain maximum throughput and accomplish maximum diversity gain, the channel state information (CSI) of the MIMO-OFDM framework should be precisely evaluated and tracked. In addition, the amazing execution of the symbol detection at the receiver is based on the known state of CSI (Suraweera and Armstrong, 2005). Yet, the received signal suffers from difficulty while estimating the CSI. Thus the MIMO-OFDM receivers have pulled up a considerable amount of research interest due to the influence of channel estimation errors (Li, Seshadri, and Ariyavisitakul, 1999; Liu, Ma, and Giannakis, 2002; Stuber et al., 2004; Balakumar, Shahbazpanahi, and Kirubarajan, 2007; Ho and Chen, 2007). In general, a two-step procedure has to be followed to obtain the CSI. All the users need to receive the training data stream from the base station (BS) and send it back to the BS after estimating the channel. Therefore, the time needed for the channel state information depends on the number of antennas used at the transmitter (Pappa, Ramesh, and Kumar, 2017).

The goal of this chapter is to provide information to the researcher on how different estimation techniques in MIMO-OFDM systems address their weaknesses and strengths so that researchers can get a clear picture about channel estimation techniques at a glance. The rest of the chapter has been organized as follows: Section 6.2 describes conventional MIMO in brief. The fundamental idea of massive MIMO-OFDM is explained in Section 6.3. In Section 6.4, different channel estimation techniques are described. Section 6.5 demonstrates the strengths and weakness of different estimation schemes. Finally, we draw a conclusion in Section 6.6.

## 6.2 CONVENTIONAL MIMO

MIMO is an innovative antenna-based technology that has created a huge impact in the world of wireless communication. Here in MIMO, at either ends of the transmitter and receiver, numerous antennas are being installed. The installment of these numerous antennas at the either ends make the appropriate execution of the performance of the system in terms of data rate and throughput. Basically, the communication framework that happens in a MIMO is by means of a BS with different portable terminals (MS), so a numerous multi-user terminal can communicate to the BS (Wei et al., 2016).

The numerous antennas installed at either end in this communication structure are shared to reduce errors and optimize data speed. MIMO can be considered as one of the many types of smart antenna technologies, where SISO, MISO, and SIMO are the special cases. When both of the transmit ($N_T$) and receive ($N_R$) antennas are 1, then it is SISO. When $N_T = 1$ and $N_R = 2$, such a system is referred to SIMO. When $N_T = 2$ and $N_R = 1$, such a system is referred to MISO (Goldsmith, 2005). The application of MIMO enhances the spectral efficiency keeping the transmission power and bandwidth fixed. Basically, two fundamental gains are achieved by implementing MIMO: rate gain and diversity gain.

In common wireless communications, each end of the transmitter and receiver is equipped with a single antenna only. This creates the unwanted effect of multipath

# Channel Estimation Techniques in the MIMO-OFDM System

fading in a few cases of the communication framework. The utilization of two or more antennas at both the ends of source and destination can mitigate the unwanted multipath fading. The digital signal processing is applied in the communication from the antennas, which enhances the spectral efficiency as well (Larsson and Van der Perre, 2017).

## 6.3 MASSIVE MIMO

Massive MIMO is the extension of MIMO, which essentially groups together a huge number of antennas at the transmitter and receiver to offer better throughput and higher spectral efficiency. Moreover, implementation of a massive number of antennas can improve the execution of wireless communication systems by enhancing the data rate (Foschini and Gans, 1998; Amihood et al., 2007), as these large numbers of antennas emphasize energy into ever smaller regions of space to achieve massive improvement in throughput. So, a massive MIMO system is a combination of a large number of transmit antennas at the transmitter in which multiple inputs are provided to the wireless channel and multiple receive antennas at the receiver in which multiple elements or multiple measurements are expected as the output of the wireless communication channel (Stuber et al., 2004; Pun, Koivunen, and Poor, 2010),. The block diagram of *NxN* massive MIMO is depicted in Figure 6.1.

Between each transmit and receive antenna is a channel coefficient for fading channel. So, in a MIMO system, it is a collection of a large number of coefficients, one between each transmit and receive antenna pair. Due to this, the MIMO system is able to transmit several information streams in parallel. This property of MIMO

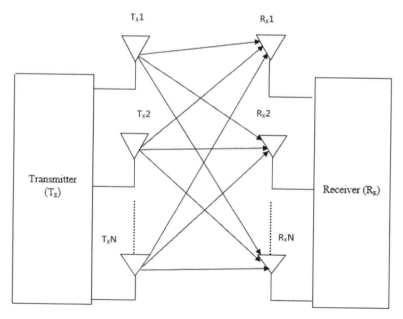

**FIGURE 6.1** NxN massive MIMO.

distinguishes it from conventional MIMO with increased reliability. The transmission of several data information in parallel between the transmitter and receiver is termed as spatial multiplexing. Moreover, every transmit antenna is connected to every received antenna, and hence it enhances the reliability of data transfer. This is possible through the techniques of diversity principle of gain (Stuber et al., 2004). The conventional MIMO system lacks this, and hence massive MIMO prevails. This is the attractive feature of massive MIMO, which generates a lot of attention to the researchers.

The channels that connect the transmitter and receiver are all fading and hence are frequency selective. The channels are all prone to undergo deep fading. So, the coupling of OFDM with MIMO converts the frequency selectivity of the channels into a parallel flat fading. For a MIMO-OFDM system, it is needed to perform inverse discrete Fourier transform (IDFT) or IFFT at each of the transmit antennas (Larsson, Danev, et al., 2017; Nahar et al., 2017). The symbols that are to be transmitted now are not only one symbol, rather there is one symbol transmitted from each of the transmitted antennas. Let illustrate the idea of this transmitter schematic using an example. Let's say there are 256 subcarriers, i.e., $N = 256$ and transmit antenna $t = 4$, and a block is given that consists of 1,024 incoming symbols. Now these incoming symbols have to be distributed among the 4 transmit antennas the way it is illustrated in Figure 6.2. This results in 256 symbols along each antenna.

Here 256 symbols are transmitted among the 4 transmit antennas. So, it results in $N \times t$ symbols, and then IFFT is performed on them. Here the process begins with a serial-to-parallel operation. Here 256 streams of symbols are to be converted into a set of parallel symbols by the DEMUX operation to load onto the subcarrier along each of the 4 antennas. Then the IFFT is performed on these symbols to generate the corresponding samples (Nahar et al., 2017). These samples are now again converted back to a serial stream by MUX operation. Now some specified samples are

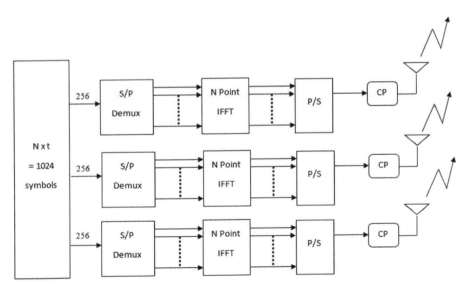

**FIGURE 6.2** The schematic diagram of the MIMO-OFDM transmitter.

# Channel Estimation Techniques in the MIMO-OFDM System

taken from the tail prefixed at the head of the sequence, and this is known is cyclic prefix (CP).

Let's say that the symbols that were loaded are $[X(0), X(1), X(3) \ldots\ldots\ldots\ldots X(N)]$. The corresponding samples, or the IFFT outputs, are $[x(0), x(1), x(3) \ldots\ldots\ldots\ldots x(n)]$. The sample becomes $[x(n), x(n-1), x(n-2) \ldots\ldots (x(0), x(1), x(3) \ldots\ldots\ldots\ldots x(n)]$ after adding CP. Now these CP-added samples are transmitted across the faded or ISI channel by the transmitting antennas. After this transmission, the output becomes a circular convolution of the channel in time domain with the transmitted samples plus the noise, i.e.,

$$y = h \otimes x + v. \qquad (6.1)$$

The receiver antennas in Figure 6.3 receive the sample sequences. Then the first thing the receiver does is the removal of the outputs that belong to the cyclic prefix. Because these samples are meant for the transmission only, they do not carry information of their own and are only meant for transmission.

Let the corresponding received transmitted samples be $y(0)$, $y(1)$, $y(3)$, $y(4)\ldots\ldots\ldots y(n)$. Now a serial-to-parallel operation is carried out, i.e., DEMUX at the output of the Demultiplexer. Then the FFT operation at the receiver is performed. The outputs of the FFT block are $Y(0)$, $Y(1)$, $Y(2)$, $Y(3)\ldots\ldots Y(N)$. After taking the FFT operation of the circular convolution, it becomes the multiplication in the frequency domain.

$$FFT(y) = FFT(h).FFT(x) + FFT(v) \Rightarrow Y(k) = H(k)X(k) + V(k) \qquad (6.2)$$

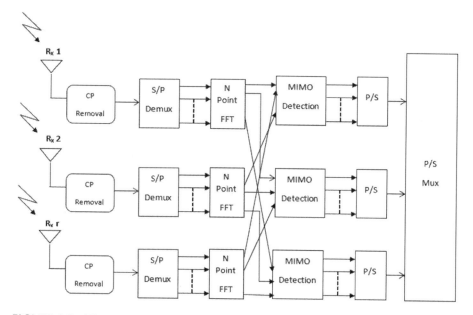

**FIGURE 6.3** The schematic diagram of the MIMO-OFDM receiver.

It is clear that, the subcarriers become ISI free after the FFT operation at the receiver. Now these symbols contain errors, degradation from the channels. So, using a better detection technique the symbols are to be extracted. Here ZERO FORCING is considered to detect the symbols.

$$\hat{X} = \left[H^H H\right]^{-1} H^H Y \qquad (6.3)$$

After the detection is done, the symbols are then multiplexed into one stream.

## 6.4 CHANNEL ESTIMATION TECHNIQUES

The method of estimating the channel coefficients of a system is known as channel estimation. The information transmitted by the channel can be faded before the receiver can have it to detect.

For the proper communication to occur, an essential knowledge about the channel is much needed, as it is considered one of the fundamental features (Farzamnia et al., 2018). To estimate the channel coefficients, various schemes have been proposed in open literature, which are delineated in the subsequent section.

### 6.4.1 Least Square (LS) Estimation

The LS estimation is mainly utilized when the channel and noise distribution are not known (Kaur, Khosla, and Sarin 2018).

The entire model of the system in the frequency domain is written in the vector-matrix form as:

$$\bar{Y} = X\bar{H} + \bar{V}, \qquad (6.4)$$

where, $\bar{Y}$ = Pilot output, $X$ = Pilot matrix in the subcarrier, $\bar{H}$ = Channel coefficient matrix, and $\bar{V}$ = noise vector

Now the channel estimation of the coefficient matrix can be done using the LS technique (Chow et al., 2013).

The LS expression of the system is $\|\bar{Y} - X\bar{H}\|^2$. Now the LS can be differentiated. Then taking it equal to zero, the value of $\bar{H}$ can be found, which minimizes the LS function. According to Hussein, Alias, and Abdulkafi (2016), the solution of the LS is written as

$$\hat{H} = (X^H X)^{-1} X^H Y. \qquad (6.5)$$

Y is the FFT of the outputs received of the sample received in the time domain. Basically, $\bar{Y}$ is the vector of outputs across the various subcarriers, and X is the diagonal matrix consisting of the pilot symbols, which are loaded onto the subcarriers, and something interesting can be observed: the diagonal matrix, X, is invertible. Because X is invertible,

$$\left(X^H X\right)^{-1} = X^{-1} \left(X^H\right)^{-1}. \qquad (6.6)$$

# Channel Estimation Techniques in the MIMO-OFDM System

So,

$$\hat{H} = X^{-1}(X^H)^{-1} X^H Y \qquad (6.7)$$

$$\hat{H} = X^{-1}\bar{Y}. \qquad (6.8)$$

Because X is diagonal, $X^{-1}$ is easy to compute,

$$\therefore X^{-1} = \begin{bmatrix} \frac{1}{X(0)} & 0 & 0 & 0 \\ 0 & \frac{1}{X(1)} & 0 & 0 \\ 0 & 0 & \frac{1}{X(2)} & 0 \\ 0 & 0 & 0 & \frac{1}{X(N)} \end{bmatrix} \qquad (6.9)$$

$$\therefore \begin{bmatrix} \hat{H}(0) \\ \hat{H}(1) \\ \hat{H}(2) \\ \vdots \\ \hat{H}(l) \end{bmatrix} = \begin{bmatrix} \frac{1}{X(0)} & 0 & 0 & 0 \\ 0 & \frac{1}{X(1)} & 0 & 0 \\ 0 & 0 & \frac{1}{X(2)} & 0 \\ \vdots & \vdots & \vdots & \vdots \\ 0 & 0 & 0 & \frac{1}{X(l)} \end{bmatrix} \begin{bmatrix} Y(0) \\ Y(1) \\ Y(2) \\ \vdots \\ Y(l) \end{bmatrix} \qquad (6.10)$$

$$\therefore \hat{H}(0) = \frac{Y(0)}{X(0)} \qquad (6.11)$$

$$\hat{H}(1) = \frac{Y(1)}{X(1)} \qquad (6.12)$$

$$\hat{H}(2) = \frac{Y(2)}{X(2)} \qquad (6.13)$$

$$\hat{H}(l) = \frac{Y(l)}{X(l)} \qquad (6.14)$$

i.e., the estimate of the channel coefficient across each subcarrier is the pilot output across the subcarrier divided by the pilot input across the subcarrier.

## 6.4.2 MAXIMUM LIKELIHOOD ESTIMATION

The channel estimation for MIMO-OFDM is analyzed based on the ML estimate of channel coefficients. The MIMO-OFDM system's model can be depicted as follows:

$$y(k) = hx(k) + v(k). \tag{6.15}$$

Now based on this, the joint PDF can be modified as $\left(\dfrac{1}{2\pi\sigma^2}\right)^{\frac{N}{2}} e^{-\dfrac{\sum_{k=1}^{N}(y(k)-hx(k))^2}{2\sigma^2}}$.

Now differentiating the logarithm of this and setting that equal to zero and finding out the ML estimate results in

$$\hat{h} = \dfrac{\sum_{k=1}^{N} x(k)y(k)}{\sum_{k=1}^{N} (x(k))^2}. \tag{6.16}$$

Now one can consider it in a much more compact form by using vector notation like

$$\bar{y} = \begin{pmatrix} y(1) \\ y(2) \\ \vdots \\ y(N) \end{pmatrix} \text{ and } \bar{x} = \begin{pmatrix} x(1) \\ x(2) \\ \vdots \\ x(N) \end{pmatrix}.$$

Now after doing some mathematical working, the channel estimate can be written as

$$\hat{h} = \dfrac{\sum x(k)y(k)}{\sum (x(k))^2}. \tag{6.17}$$

This is the maximum likelihood estimate of a real parametric channel coefficient.

Now Equation (6.17) can also be arranged as

$$\hat{h} = \dfrac{\bar{x}^T \bar{y}}{\bar{x}^T \bar{x}}. \tag{6.18}$$

Now, one can extend this to the complex parameter scenario. For the complex parameter, a simple trick is to replace the transpose by the Hermitian operator. So, it turns out like this:

$$\hat{h} = \dfrac{\bar{x}^H \bar{y}}{\bar{x}^H \bar{x}}. \tag{6.19}$$

# Channel Estimation Techniques in the MIMO-OFDM System

$$\Rightarrow \hat{h} = \frac{\sum x^*(k) y(k)}{\sum |(x(k))|^2}. \quad (6.20)$$

This is the ML estimate for a complex parameter channel coefficient (Yang and Kwak, 2006; Tiiro et al., 2009).

### 6.4.3 MMSE Channel Estimation

Minimum mean squared error (MMSE) estimation is a special form of Bayesian estimation (White, 1982). The MMSE channel estimator is capable of using the spatial relationship of the MIMO channel but entails earlier estimation of the channel correlation matrix. There are two important components in this process:

1. Probability of measurement P($\bar{y}$|h), $\bar{y} = \begin{bmatrix} y(1) \\ y(2) \\ \vdots \\ y(N) \end{bmatrix}$

2. P(h) → Prior PDF of parameter h

$\hat{h}$ → Estimate of h. $\hat{h}$ is a function of the measurement vector, i.e., $\hat{h}(\bar{y})$. The estimate, $\hat{h}$, can only be a function of $\bar{y}$ because h itself is unknown, and this is the reason of the estimation of h. Now the task is to come up with an estimate that minimizes the estimation error. The estimation error is the difference between the estimate and the parameter, h $\hat{h}(\bar{y}) - h$ → estimation error (Ahmed and Matin, 2015).

After squaring the error, it becomes ($\hat{h}(\bar{y}) - h$)². Here the main interest is on minimizing the average estimation error; that way it is termed as the mean squared error (MSE). The mean of the ($\hat{h}(\bar{y}) - h$)² is denoted by the expectation operator E(($\hat{h}(\bar{y}) - h$)²). Now the attention is to find an estimate that minimizes the MSE. The MSE can be expanded as

$$E\left\{\left(\hat{h}(\bar{y}) - h\right)^2\right\} = \int_{-\infty}^{\infty}\int_{-\infty}^{\infty} \left(\left(\hat{h}(\bar{y}) - h\right)^2\right) P(\bar{y}, h) \, dh \, d\bar{y}. \quad (6.21)$$

One of the most important properties of the correlation coefficient is that its magnitude is always less than 1 and lies between −1 & 1 (Saha et al., 2009). It indicates the extent of the correlation between h and Y. For Gaussian random variables, if the $\rho = 0$ then h and y are uncorrelated. However, for Gaussian random variables if $\rho = 0$, then h and y are independent. At this point, we want to come up with an MMSE estimate of h and y. So first, we want to construct a joint distribution of h and y. so let's construct the vector

$$\begin{bmatrix} h \\ y \end{bmatrix}.$$

Let us express the covariance matrix of this vector as R

$$R = E\left(\begin{bmatrix}h\\y\end{bmatrix}\begin{bmatrix}h & y\end{bmatrix}\right) = E\left(\begin{bmatrix}h^2 & hy\\yh & y^2\end{bmatrix}\right). \quad (6.22)$$

Now let's take the expected value of each of the terms. After taking the expected value of each of the terms, now we have

$$R = \begin{bmatrix}\sigma_h^2 & \rho\sigma_h\sigma_y\\ \rho\sigma_h\sigma_y & \sigma_y^2\end{bmatrix}. \quad (6.23)$$

Taking the inverse of the matrix R, it turns out to be

$$R^{-1} = \begin{bmatrix}\sigma_h^2 & \rho\sigma_h\sigma_y\\ \rho\sigma_h\sigma_y & \sigma_y^2\end{bmatrix}^{-1} = \frac{1}{\det(R)}\begin{bmatrix}\sigma_y^2 & -\rho\sigma_h\sigma_y\\ \rho\sigma_h\sigma_y & \sigma_y^2\end{bmatrix}. \quad (6.24)$$

Now we need to calculate MMSE estimate, i.e., E(h|y). For the calculation of MMSE estimate, we need conditional PDF, which can be written as

$$F_{H|Y}(h|y) = \frac{F_{H,Y}(h,y)}{F_Y(y)}. \quad (6.25)$$

Now we have to compute each of them:

$$F_Y(y) = \frac{e^{-\frac{1}{2}\frac{y^2}{\sigma_y^2}}}{\sqrt{2\pi\sigma_y^2}} \quad (6.26)$$

Now we have to find the joint PDF of $F_{H,Y}(h,y)$, and the expression to find it is as follows (Alam, Kaddoum, and Agba, 2018)

$$F_{(H,Y)}(h,y) = \frac{\exp\left(-\frac{1}{2}\begin{bmatrix}h & y\end{bmatrix}R^{-1}\begin{bmatrix}h\\y\end{bmatrix}\right)}{\sqrt{(2\pi)^2\det(R)}}. \quad (6.27)$$

The power of the exponential can be more simplified. and then the joint PDF becomes

$$F_{(H,Y)}(h,y) = \frac{1}{\sqrt{(2\pi)^2(1-\rho^2)\sigma_y^2\sigma_h^2}} \times \exp\left(-\frac{1}{2}\frac{h^2\sigma_y^2 + y^2\sigma_h^2 - 2\rho\,\sigma_h\,\sigma_y\,h_y}{\sigma_y^2\sigma_h^2(1-\rho^2)}\right) \quad (6.28)$$

# Channel Estimation Techniques in the MIMO-OFDM System

Now we have two aspects:

1. Marginal PDF of y
2. Joint Distribution (Yang and Kwak, 2006)

Now we have both of the quantities. Plugging both the values, we can find the expression of the conditional PDF.

$$F_{HY}(h|y) = \frac{F_{H,Y}(h,y)}{F_Y(y)} \quad F_{HY}(h|y) = \frac{1}{\sqrt{2\pi(1-\rho)^2 \sigma_h^2}}$$
$$\times \exp\left\{-\frac{1}{2}\left(\frac{h^2\sigma_y^2 + y^2\sigma_h^2 - 2\rho\sigma_h\sigma_y hy}{(1-\rho)^2 \sigma_y^2 \sigma_h^2} - \frac{y^2}{\sigma_y^2}\right)\right\}. \tag{6.29}$$

Now the simplified form of the exponent can be written as $\dfrac{\left(h - \rho\dfrac{\sigma_h}{\sigma_y} y\right)^2}{(1-\rho^2)\sigma_h^2}$ after the mathematical manipulation. Equation (6.29) can more simplified by taking out the factors as $\tilde{\sigma}^2 = (1-\rho^2)\sigma_h^2$; $\tilde{\mu} = \rho\dfrac{\sigma_h}{\sigma_y} y$. Now putting it again in Equation (6.29), the expression becomes more convenient, and this is the required joint PDF

$$F_{H|Y}(h|y) = \frac{1}{\sqrt{2\pi\tilde{\sigma}^2}} \exp\left(-\frac{1}{2}\left(\frac{(h-\tilde{\mu})^2}{\tilde{\sigma}^2}\right)\right). \tag{6.30}$$

The parameter H given Y is a Gaussian distribution with mean $= \tilde{\mu}$ and variance $= \tilde{\sigma}^2$, i.e.,
($H | Y \sim \mathbb{N}(\tilde{\mu}, \tilde{\sigma}^2)$, and the expected value of h|y equals to the $\tilde{\mu}$

$$E(h|y) = \tilde{\mu} = \rho\frac{\sigma_h}{\sigma_y} y. \tag{6.31}$$

The expected value of the h given y is basically the channel estimate for MMSE, i.e.,

$$\hat{h} = E(h|y) = \rho\frac{\sigma_h}{\sigma_y} y. \tag{6.32}$$

Now let's simplify this MMSE estimate and the variance to give alternate expression, which is much more convenient

$$E(h^2) = \sigma_h^2 = r_{hh} \tag{6.33}$$

$$E(y^2) = \sigma_y^2 = r_{yy} \qquad (6.34)$$

$$E(hy) = \rho\sigma_h\sigma_y = r_{hy} \qquad (6.35)$$

Equation (6.32) can be further simplified by multiplying the numerator and denominator by $\sigma_y$

With the substitution from Equations (6.34) and (6.35), the MMSE channel estimation simplifies to

$$\hat{h} = r_{hy}r_{yy}^{-1}y. \qquad (6.36)$$

This is the simplified and convenient MMSE estimate that can be used for channel estimation.

### 6.4.4 Pilot- or Training-Based Channel Estimation

The transmitter transmits known data symbols or standard set of symbols, which are called as pilot or training symbols. At the receiver, corresponding output or pilot output has been observed with respect to the training sequences. The receiver mitigates the channel effects by the help of known pilot symbols and determines the channel. MMSE estimator is implemented at the receiver to eliminate the noise interference during the transmission (Hayter, 2012). As the pilots are already known, the underlying unknown channel behavior can be extracted. In Pappa, Ramesh, and Kumar (2017), the author addresses a solution for pilot sequences by designing a set of short orthogonal sequences, equi-powered and the placement of the pilot subcarriers that helps to reduce the interference among pilot sequences of transmit antenna.

### 6.4.5 Blind Channel Estimation

Based on the received data signal, it is possible to evaluate the channel, which is Blind channel estimation technique. This method is focused on natural constraints instead of using pilot sequences. In other words, it eliminates the use of trained or pilot sequences and has the advantage of being spectrally efficient. Moreover, adding a few pilot sequences improves the channel estimation performances. The channel model is presumed to be a Rayleigh fading channel. Apart from using the preestimated channel parameters, the data streams can be estimated blindly while traveling through the noisy channels. It is observed that it is bandwidth efficient but with large computational complexity as compared to the pilot techniques.

Blind channel estimation has been explored in several articles (e.g., Zhang et al., 2007; Sarmadi, Shahbazpanahi, and Gershman, 2009; Chen and Wu, 2012; Ercan and Kurnaz, 2015). In Zhang et al. (2007), the author determines the channel product matrices from a series of decoupled linear equations. A blind channel estimation for Zero padding (ZP) MIMO OFDM system has been proposed based on the statistical covariance matrix of the received data. In addition, the proposed method can trace out more input channels under a quite mild condition. Another blind channel estimation process is quoted from (Chen and Wu, 2012) that uses semidefinite relaxation (SDR) method, which helps the channel estimation problem to transfer to a convex form and then be solved efficiently using modern convex optimization tools. In Sarmadi, Shahbazpanahi, and Gershman (2009), a blind algorithm based on Independent component analysis (ICA) and comb type pilot algorithm is proposed by using two different frequency selective channels. It is found that the channel estimation performance varies with respect to pilot-based algorithm and channel frequency selectivity.

### 6.4.6 SEMI-BLIND CHANNEL ESTIMATION

Semi-blind methods create a compromise between the spectral efficiency and computational complexity. Both small numbers of pilot bits and statistical channels are used in this process. This algorithm is also used for the estimation of a frequency selective MIMO channel matrix. The major improvement is that it is more effective in transmission capacity than pilot-based channel estimation because it is of less pilot utilization (Ercan and Kurnaz, 2015).

The channel matrix estimation is updated by the author in Zhang, Gao, and Yin (2015) as shown here:

$$\hat{G}_{l+1} = \left( Y_p S_P^H + Y_d \mathrm{E}\left( S_d |, \hat{G}_l, Y \right)^H \right) \times \left( S_p S_P^H + E\left( S_p S_d^H |, \hat{G}_l, Y \right) \right)^{-1} \quad (6.37)$$

The optimized orthogonal pilot is used instead of conventional or STBC pilots for estimating a channel as only one training sequence is transmitted from each antenna to each subcarrier, which is not suitable for next generation mobile systems. This reduces the error of estimation while maintaining high spectrum efficiency.

## 6.5 STRENGTHS AND LIMITATIONS OF EXISTING ESTIMATION TECHNIQUES

Channel estimation for the MIMO-OFDM system has been studied comprehensively (Enescu, Roman, and Koivunen, 2003; Miao and Juntti, 2004; Zhang et al., 2005; Rana, 2011; Sagar and Palanisamy, 2015; Trimeche, Sakly, and Mtibaa, 2015). The most common estimation techniques include maximum likelihood (ML), least squares (LS), and minimum mean squared error (MMSE). In the MIMO-OFDM framework, the MMSE estimator appears better in performance than the LS estimator, and additional change on LS and MMSE estimators using the DFT-based estimation technique is obtained incorporating a large number of pilots (Sagar and Palanisamy, 2015) (Table 6.1).

It is revealed that ML outperforms MMSE significantly in overall system performance while eliminating the intersymbol interference (ISI). As it is stated in Trimeche, Sakly, and Mtibaa (2015), the LS estimator is presented to decrease the squared differences between the source and the receiver. An iterative channel estimation method based on Wiener filtering is inspected to upgrade the accuracy of the estimator. But the major drawback that limits it is the idea about the channel correlation.

**TABLE 6.1**
**Table of Mathematical Equations for Different Channel Coefficients**

| | |
|---|---|
| LS estimate | $\therefore \hat{H}(0) = \dfrac{Y(0)}{X(0)}, \hat{H}(1) = \dfrac{Y(1)}{X(1)}, \hat{H}(l) = \dfrac{Y(l)}{X(l)}$ |
| Maximum likelihood estimate | $\hat{h} = \dfrac{\sum x^*(k) y(k)}{\sum \lvert (x(k)) \rvert^2}$ (when h is a complex parameter) |
| | $\hat{h} = \dfrac{\sum x(k) y(k)}{\sum (x(k))^2}$ (when h is a real parameter) |
| MMSE estimate | $\hat{h} = r_{hy} r_{yy}^{-1} y$ |

LS-based and LMMSE method had been at first considered as a pilot-based channel estimation procedure within the cluster space channel. As LMMSE based procedure furthermore utilizes earlier information of the channel relationship. It give way to better execution compared to LS estimator (Enescu, Roman, and Koivunen, 2003; Miao and Juntti, 2004; Rana, 2011). A new blind channel estimation method is presented in Zhang et al. (2005) to lower the computational complexity as well as achieve better performance by estimating the channel from a limited number of received blocks even when the symbols are drawn from the low-order constellation.

Semi-blind methods have been preferred by many researchers to overcome the drawbacks of pilot- and blind-based methods (Aldana, de Carvalho, and Cioffi, 2003a; Wautelet et al., 2007; Zhang, Gao, and Yin, 2015). In this case, both training and data symbols are focused to estimate the channel. This technique is way more efficient as compared to the blind channel estimation method, as the channel coefficients can be fully detected under certain conditions (De Carvalho and Slock, 1997). On the other hand, the channel can only be identified within some ambiguities. This technique can be processed even if the data symbols are unavailable, whereas the first technique uses the pilot sequences to estimate the channel.

An iterative semi-blind method is expectation-maximization (EM), which works in those areas where there is lack of sufficient data. The only disadvantage is that its computational complexity significantly increases with respect to the constellation size, and its implementation is inadequate to time invariant channels (Aldana, de Carvalho, and Cioffi, 2003b; Abuthinien, Chen, and Hanzo, 2008).

## 6.6 CONCLUSION

Massive MIMO is one of the most encouraging techniques for current wireless technologies that has acquired a lot of interest recently from the research community. The effective channel estimation imposes the key challenge in addressing pilot contamination. This chapter has offered a comprehensive survey on various channel estimation methods for MIMO-OFDM wireless communication systems. It is observed that the blind channel estimation technique provides a better solution to the pilot contamination problem than other channel estimation techniques. On the other hand, the semi-blind provides less computation complexities than other estimation techniques.

# REFERENCES

Abuthinien, M., S. Chen, and L. Hanzo. 2008. "Semi-blind Joint Maximum Likelihood Channel Estimation and Data Detection for MIMO Systems." *IEEE Signal Processing Letters* 15: 202–205.

Ahmed, B., and M. A. Matin. 2015. *Coding for MIMO-OFDM in Future Wireless Systems.* Cham: Springer.

Alam, M. S., G. Kaddoum, and B. L. Agba. 2018. "Bayesian MMSE Estimation of a Gaussian Source in the Presence of Bursty Impulsive Noise." *IEEE Communications Letters* 22 (9): 1846–1849.

Aldana, C. H., E. de Carvalho, and J. M. Cioffi. 2003a. "Channel Estimation for Multicarrier Multiple Input Single Output Systems Using the EM Algorithm." *IEEE Transactions on Signal Processing* 51 (12): 3280–3292.

Aldana, C. H., E. de Carvalho, and J. M. Cioffi. 2003b. "Channel Estimation for Multicarrier Multiple Input Single Output Systems Using the EM Algorithm." *IEEE Transactions on Signal Processing* 51 (12): 3280–3292.

Alizadeh, M. R., G. Baghersalimi, M. Rahimi, M. Najafi, and X. Tang. 2016. "Performance Improvement of a MIMO-OFDM Based Radio-over-Fiber System Using Alamouti Coding." *2016 10th International Symposium on Communication Systems, Networks and Digital Signal Processing (CSNDSP)*, Prague, 1–5.

Amihood, P., E. Masry, L. B. Milstein, and J. G. Proakis. 2007. "Performance Analysis of High Data Rate MIMO Systems in Frequency-Selective Fading Channels." *IEEE Transactions on Information Theory* 53 (12): 4615–4627.

Balakumar, B., S. Shahbazpanahi, and T. Kirubarajan. 2007. "Joint MIMO Channel Tracking and Symbol Decoding Using Kalman Filtering." *IEEE Transactions on Signal Processing* 55 (12): 5873–5879.

Chen, Y. S., and J. Y. Wu. 2012. "Statistical Covariance-matching Based Blind Channel Estimation for Zero-padding MIMO–OFDM Systems." *EURASIP Journal on Advances in Signal Processing* 2012 (1): 139.

Cho, Y. S., J. Kim, W. Y. Yang, and C. G. Kang. 2010. *MIMO-OFDM Wireless Communications with MATLAB.* Singapore: John Wiley & Sons.

Chow, C. W., C. H. Yeh, Y. F. Liu, and P. Y. Huang. 2013. "Background Optical Noises Circumvention in LED Optical Wireless Systems Using OFDM." *IEEE Photonics Journal* 5 (2): 7900709–7900709.

De Carvalho, E., and D. T. Slock. 1997. "Cramer-Rao Bounds for Semi-blind, Blind and Training Sequence Based Channel Estimation." *First IEEE Signal Processing Workshop on Signal Processing Advances in Wireless Communications*, Paris, 129–132.

Enescu, M., T. Roman, and V. Koivunen. 2003. "Channel Estimation and Tracking in Spatially Correlated MIMO OFDM Systems." *IEEE Workshop on Statistical Signal Processing*, St. Louis, 347–350.

Ercan, S. Ü., and Ç. Kurnaz. 2015. "Investigation of Blind and Pilot Based Channel Estimation Performances in MIMO-OFDM System." *2015 23nd Signal Processing and Communications Applications Conference (SIU)*, Malatya, 1869–1872.

Farzamnia, A., E. Moung, N. W. Hlaing, L. E. Kong, M. K. Haldar, and L. C. Fan. 2018. "Analysis of MIMO System through Zero Forcing and Minimum Mean Square Error Detection Scheme." *2018 9th IEEE Control and System Graduate Research Colloquium (ICSGRC)*, Shah Alam, 172–176.

Foschini, G., and M. Gans. 1998. "On Limits of Wireless Communications in a Fading Environment When Using Multiple Antennas." *Wireless Personal Communications* 6 (3): 311–335.

Goldsmith, A. 2005. *Wireless Communications.* Cambridge University Press.

Hayter, A. 2012. *Probability and Statistics for Engineers and Scientists*, 4th edition. Boston, MA: Brooks/Cole.

Ho, T., and B. Chen. 2007. "Tracking of Dispersive DS-CDMA Channels: An AR-embedded Modified Interacting Multiple-model Approach." *IEEE Transactions on Wireless Communications* 6 (1): 166–174.

Hussein, Y. S., M. Y. Alias, and A. A. Abdulkafi. 2016. "On performance analysis of LS and MMSE for channel estimation in VLC systems." *2016 IEEE 12th International Colloquium on Signal Processing & Its Applications (CSPA)*, Malacca City, 204–209.

Kaur, H., M. Khosla, and R. K. Sarin. 2018. "Channel Estimation in MIMO-OFDM System: A Review." *2018 Second International Conference on Electronics, Communication and Aerospace Technology (ICECA)*, Coimbatore, 974–980.

Larsson, E. G., and L. Van der Perre. 2017. "Massive MIMO for 5G." *IEEE 5G Tech Focus* 1 (1).

Larsson, E. G., D. Danev, M. Olofsson, and S. Sorman. 2017. "Teaching the Principles of Massive MIMO: Exploring Reciprocity-based Multiuser MIMO Beamforming Using Acoustic Waves." *IEEE Signal Processing Magazine* 34 (1): 40–47.

Li, Y., N. Seshadri, and S. Ariyavisitakul. 1999. "Channel Estimation for OFDM Systems with Transmitter Diversity in Mobile Wireless Channels." *IEEE Journal on Selected Areas in Communications* 17 (3): 461–471.

Liu, Z., X. Ma, and G. B. Giannakis. 2002. "Space-time Coding and Kalman Filtering for Time-selective Fading Channels." *IEEE Transactions on Communications* 50 (2): 183–186.

Miao, H., and M. J. Juntti. 2004. "Spatial Signature and Channel Estimation for Wireless MIMO-OFDM Systems with Spatial Correlation." *IEEE 5th Workshop on Signal Processing Advances in Wireless Communications, 2004*, Lisbon, 522–526.

Nahar, A. K., S. A. Gitaffa, M. M. Ezzaldean, and H. K. Khleaf. 2017. "FPGA Implementation of MC-CDMA Wireless Communication System Based on SDR—a Review." *Review of Information Engineering and Applications* 4 (1): 1–19.

Pappa, M., C. Ramesh, and M. N. Kumar. 2017. "Performance Comparison of Massive MIMO and Conventional MIMO Using Channel Parameters." *2017 International Conference on Wireless Communications, Signal Processing and Networking (WiSPNET)*, Chennai, 1808–1812.

Pun, M. O., V. Koivunen, and H. V. Poor. 2010. "Performance Analysis of Joint Opportunistic Scheduling and Receiver Design for MIMO-SDMA Downlink Systems." *IEEE Transactions on Communications* 59 (1): 268–280.

Rana, M. M. 2011. "Performance Comparison of LMS and RLS Channel Estimation Algorithms for 4G MIMO OFDM Systems." *14th International Conference on Computer and Information Technology (ICCIT 2011)*, Dhaka, 635–639.

Sagar, K., and P. Palanisamy. 2015. "Optimal Pilot-aided Semi blind Channel Estimation for MIMO-OFDM System." *2015 Global Conference on Communication Technologies (GCCT)*, Thuckalay, 290–293.

Saha, S., Y. Boers, H. Driessen, P. K. Mandal, and A. Bagchi. 2009. "Particle Based MAP State Estimation: A Comparison." *2009 12th International Conference on Information Fusion*, Seattle, WA, 278–283.

Sarmadi, N., S. Shahbazpanahi, and A. B. Gershman. 2009. "Blind Channel Estimation in Orthogonally Coded MIMO-OFDM Systems: A Semidefinite Relaxation Approach." *EEE Transactions on Signal Processing* 57 (6): 2354–2364.

Stuber, G. L., J. R. Barry, S. W. McLaughlin, Y. Li, M. A. Ingram, and T. G. Pratt. 2004. "Broadband MIMO-OFDM Wireless Communications." *Proceedings of the IEEE* 92 (2): 271–294.

Suraweera, H. A., and J. Armstrong. 2005. "Alamouti Coded OFDM in Rayleigh Fast Fading Channels—Receiver Performance Analysis." *TENCON 2005—2005 IEEE Region 10 Conference*, Melbourne, 1-5.

Tiiro, S., J. Ylioinas, M. Myllyla, and M. Juntti. 2009. "Implementation of the Least Squares Channel Estimation Algorithm for MIMO-OFDM Systems." *Proceedings of the International ITG Workshop on Smart Antennas (WSA 2009)*, Berlin, 16–18.

Trimeche, A., A. Sakly, and A. Mtibaa. 2015. "FPGA Implementation of ML, ZF and MMSE Equalizers for MIMO Systems." *Procedia Computer Science* 73: 226–233.

Wautelet, X., C. Herzet, A. Dejonghe, J. Louveaux, and L. Vandendorpe. 2007. "Comparison of EM-Based Algorithms for MIMO Channel Estimation." *IEEE Transactions on Communications* 55 (1): 216–226.

Wei, H., D. Wang, H. Zhu, J. Wang, S. Sun, and X. You. 2016. "Mutual Coupling Calibration for Multiuser Massive MIMO Systems." *IEEE Transactions on Wireless Communications* 15 (1): 606–619.

White, H. 1982. "Maximum Likelihood Estimation of Misspecified Models." *Econometrica: Journal of the Econometric Society* 50 (1): 1–25.

Yang, Q., and K. S. Kwak. 2006. "Superimposed Pilot Aided Multiuser Channel Estimation for MIMO-OFDM Uplinks." *ETRI Journal* 28 (5): 688–691.

Zhang, H., Y. Li, A. Reid, and J. Terry. 2005. "Channel Estimation for MIMO OFDM in Correlated Fading Channels." *IEEE International Conference on Communications, 2005. ICC 2005*, Seoul, 2626–2630.

Zhang, J., W. Zhou, H. Sun, and G. Liu. 2007. "A Novel Pilot Sequences Design for MIMO OFDM Systems with Virtual Subcarriers." *2007 Asia-Pacific Conference on Communications,* Bangkok, 501–504.

Zhang, W., F. Gao, and Q. Yin. 2015. "Blind Channel Estimation for MIMO-OFDM Systems with Low Order Signal Constellation." *IEEE Communications Letters* 19 (3): 499–502.

# 7 Localization Protocols for Wireless Sensor Networks

*Ash Mohammad Abbas and
Hamzah Ali Abdul Rahman Qasem*
Aligarh Muslim University

## CONTENTS

7.1 Introduction ........................................................................................................ 119
7.2 Classification of Localization Schemes .......................................................... 120
7.3 Range-Based Localization ................................................................................ 121
7.4 Range-Free Localization ................................................................................... 124
7.5 Anchor-Based Localization .............................................................................. 127
7.6 Anchor-Free Localization ................................................................................. 129
7.7 Directional Localization ................................................................................... 131
7.8 Conclusion .......................................................................................................... 133
References .................................................................................................................. 133

## 7.1 INTRODUCTION

A *wireless sensor network* (WSN) is a cooperative engagement of a collection of sensors that communicate using wireless links. Sensors in a WSN can be used to gather information about specific parameters needed for a designated purpose or task. Many characteristics of a WSN differentiate it from a wired or a wireless network. Sensor nodes (also called *motes*) have a very small transmission range; therefore, their communication capability is limited. The storage and processing capabilities of sensor nodes are much less as compared to their wired or wireless network counterparts. Further, sensor nodes are often operated through batteries, whose power depletion may cause node and associated link and/or path failures. These characteristics pose different kinds of issues and challenges in designing a protocol or a scheme for a WSN as compared to other wired or wireless networks.

Many applications of a WSN require information about the location(s) of an object or sensor node themselves. For example, if the temperature in the vicinity of a particular sensor is suddenly increasing, it is required to know its location so as to take corrective measures. Otherwise, it may result in hazardous situations such as forest fires. Movement and positions of kids in a kindergarten classroom can be monitored by locating the sensors (Srivastava, Muntz, and Potkonjak, 2001). The use of WSNs together with a review of research carried out in the area of healthcare is presented in Alemdar and Ersoy (2010). Specifically, patients suffering from Alzheimer disease can be located using a WSN. Also, it is needed in vehicular networks for traffic monitoring (Li et al. 2008) and road surface monitoring (Eriksson et al., 2008). In a forest, wild animals can be located using a WSN. A WSN called GreenOrbs for localization in a forest is presented in J. Zhao et al. (2012).

In addition to civilian applications, localization is also required in military applications. For example, detection of infiltration across the line of control (LoC) requires locating the sensors sending the data about infiltration. Similarly, tracking an object is one such application that requires the locations of the object to be estimated that are varying with time. Audio-based surveillance can be used for tracking and monitoring an object in a wireless acoustic sensor network (WASN) (Cobos et al., 2014). Further, in some applications of WSNs, the knowledge about the geometric positions is assumed to be an inherent part of the data recorded by sensor nodes because it is required by network protocols for their operation. The examples of such protocols include topology control, clustering, and geographic routing. Computing the locations of nodes in a WSN is thus a major problem and is often called the *localization problem*.

One may argue about why all nodes in WSN are not equipped with a Global Positioning System (GPS) (Hofmann-Wellenhof, Lichtenegger, and Collins, 2012) receiver. The reason is that providing each node a GPS receiver is not a cost-effective solution. As a result, only a small fraction of nodes in a WSN is equipped with a GPS receiver, and these special nodes are called *anchor* or *beacon* nodes. The location of other ordinary nodes can be computed using the location information of anchor nodes.

In this chapter, we provide an extensive survey of the protocols or schemes for localization in WSNs. We identify issues and challenges faced by a localization protocol. We classify the protocols in different categories and describe the key points of each protocol. We point out relative merits and demerits of each protocol or a group of protocols with a similar strategy.

The remaining part of this chapter is organized as follows. Section 7.2 contains a classification of the protocols or schemes of localization in WSNs. In Section 7.3, we describe research works related to range-based localization. Section 7.4 contains a description of range-free localization techniques. In Section 7.5, a description of anchor-based localization is provided. Section 7.6 contains a description of anchor-free localization techniques. In Section 7.7, a description of research related to directional localization is provided. The last section is for conclusion.

## 7.2 CLASSIFICATION OF LOCALIZATION SCHEMES

Localization schemes or protocols for WSNs can be classified in several ways, depending on the basis that one takes into account for classification. For example, depending on whether the scheme uses a technique for measuring the distances (or

Localization Protocols for Wireless Sensor Networks

ranges) or not, a classification can be distance-based (or *range-based*) and distance-free (*range-free*), which is as follows.

- *Range-Based Localization*: In order to estimate the position of a node, a scheme that measures the distance of an unlocalized node from a beacon (or an anchor) node is often called a distance-based (or range-based) scheme.
- *Range-Free Localization*: A scheme that is independent of the information that a node receives from its neighbors in the form of either position information or hop count is called a range-free scheme. To compute the location of an unlocalized sensor, range-free techniques often employ either mathematical or geometric methods.

Another classification can be on the basis of whether the network contains special nodes that are aware of their own locations or not. Accordingly, the classification can be *anchor-based* and *anchor-free*.

- *Anchor-Based Localization*: Localizations schemes can be distinguished on the basis of the use of anchor or beacon nodes, which are special nodes whose coordinates are known. Anchors are deployed to either help other nodes in the network to determine their locations or to introduce static coordinates in the network.
- *Anchor-Free Localization*: In a network where no anchors are used, nodes need to create their own coordinate system to establish their relative positions.

Also, the localization schemes can be classified depending on the type of the networks to which they are applied, e.g., whether the scheme is applied to a static network or mobile network, and whether the network is isotropic or anisotropic. Some of the schemes can be applied to specific WSNs, e.g., underwater sensor networks (UWSNs) and some of the schemes may require the sensors to use a directional antenna. Figure 7.1 provides a list of different categorizations of localization schemes for WSNs on the basis of a specific consideration. One should keep in mind that these categories may not be disjoint. In other words, there can be a scheme that belongs to more than one category. For example, a localization scheme can be anchor-based, range-based, and it can be applied to mobile WSNs. In other words, there can be a scheme that may belong to more than one category.

## 7.3 RANGE-BASED LOCALIZATION

In order to compute the location of a sensor, a scheme that measures the distance of the sensor from an anchor node is often called a distance-based (or range-based) scheme. To estimate the distance, these schemes use different techniques such as radio signal strength indicator (RSSI) (Bahl and Padmanabhan, 2000; Liu et al., 2007), angle of arrival (AoA) (Niculescu and Nath, 2003; Boushaba, Hafid, and Benslimane, 2009), or time-difference of arrival (TDoA) (Priyantha, Chakraborty, and Balakrishnan, 2000). The distances (also called ranges) measured using these techniques are sensitive to range errors and often require additional hardware.

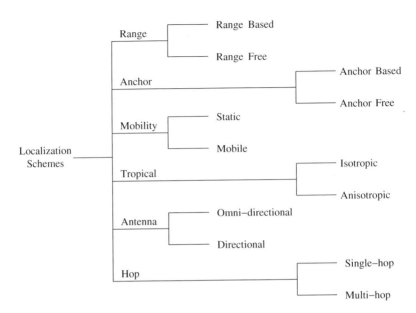

**FIGURE 7.1** A list of classifications of localization schemes for WSNs on the basis of a specific consideration.

A localization scheme called *time-based positioning scheme* (TPS) is proposed in Cheng et al. (2004) for outdoor WSNs. An improved version of TPS called *i*TPS is proposed in Thaeler, Ding, and Cheng (2005). As in the case of TPS, *i*TPS is also based on *time-difference of arrival* (TDoA) method. The difference between TPS and *i*TPS is that TPS uses range differences from three sensors, while *i*TPS uses range differences from four sensors. The range differences are then used to compute the location of a sensor using trilateration. It has been argued in Thaeler et al. (2005) that the number of ambiguous estimates about the position of a sensor are substantially reduced in *i*TPS as compared to TPS.

The major features of TPS and *i*TPS are as follows. There is no need of synchronization among base stations and sensors due to the use of local clocks for measuring TDoA of radio signals. As opposed to other schemes, there is no need for directional antennas in the case of TPS for measuring AoA. Also, there is no need of a rigorous refinement of position estimates in TPS because the time difference is measured by computing an average over a number of beacon intervals. Consequently, the effects of instantaneous interference and fading are reduced. In TPS, sensors are not needed to transmit any beacon signal, as these are transmitted by base stations. It helps to conserve energy and also reduces RF channel use.

A framework for localization based on estimated ranges called transformed least-squares (TLS) is described in Yan et al. (2010). Therein, the problem of localization is solved by transforming a higher-dimensional problem to a lower-dimensional problem, which is then solved iteratively. In TLS, the problem of localization is transformed to a single dimension (1-D) and then iterated. For the proposed method called 1-D iterative (1DI), the amount of computational overhead in each iteration is

# Localization Protocols for Wireless Sensor Networks

significantly reduced as compared to the existing methods such as nonlinear least squares (NLS). In Yan et al. (2010), a method for how to choose 1-D parameter is described, and an analysis of the rate of convergence and error (in terms of *root-mean squared error* [RMSE]) is carried out. The framework is validated in office environment with ultra-wide band (UWB) measurements with signal bandwidths 0.5 GHz–7.5 GHz. It is observed that 1DI outperforms the linear least-squares (LLS) method in terms of RMSE.

An analysis of the impact of various factors such as geometry of the deployment region, node distribution, distance-based error distribution, and anchor distribution on the parameters such as *mean square error* (MSE) and *Cramér–Rao lower bound* (CRLB) is presented in Zhou and Shi (2011). The major findings of the work are as follows. The localization error is minimized when anchors are uniformly distributed. A high density of anchors improves the accuracy of localization; however, the improvement for the number of anchors (as the neighbors of a node) beyond 8 is not significant. Further, the discrepancy between expected position values obtained using CRLB and MSE decreases with an increase in the number of anchors.

A localization algorithm called *d*istributed *w*eighted *M*ulti-*D*imensional *S*caling (dwMDS) is proposed in Costa, Patwari, and Hero III (2006), where each node in the network selects its neighborhood in an adaptive manner. The node, then, updates the estimate of its location such that the function representing the cost incurred in carrying out the update is optimized. Consequently, it passes the update to its neighbors. It has been pointed out that distance-based methods are prone to localization errors, specifically, in a noisy environment. A major contribution of the work presented in Costaet al. (2006) is that the issue of *biasing effect* is addressed. A scenario where nodes, whose measured distances are less than the actual distances even then those that are selected as neighbors, is called the biasing effect. In Costa et al. (2006), the biasing effect is alleviated using a two-stage process for selection of neighboring nodes.

A *range-based* version of Monte Carlo Localization (rMCL) is proposed in Dil, Dulman, and Havinga (2006) that improves the accuracy of localization by using the range information obtained from anchors that are lying at a distance of not more than two hops from the sensor to be localized. Therein, only the well-connected nodes are considered, i.e., nodes that have heard location information from three or more anchors. However, the improvement in accuracy is at the cost of spending more energy in communication with one another for forwarding positions of anchors.

A localization scheme called ratiometric vector iteration (RVI) is proposed in Lee et al. (2006), which uses an estimate of the ratio of distances as opposed to the estimate of the absolute distance. It iteratively updates estimated location using ratio of distances obtained with the help of three sensors. It tracks the location of the target by reporting its location to the subscriber. For a reduction in the number of report messages, it dynamically adjusts the reporting frequency. Further, RVI employs a report scheduling algorithm that adapts according to the movements of the target to reduce the error of localization and the number of transmitted messages. A range-based localization scheme for three-dimensional WSNs is proposed in X. Liu et al. (2018). We abbreviate it as component based localization in 3D for sparse networks (CBL3DS). The conditions for merging two or more subnetworks in

### TABLE 7.1
### A Comparison of Range-based Localization Techniques for Sensor Networks

| Scheme | Basis | Features | Remarks |
| --- | --- | --- | --- |
| iTPS (Thaeler, Ding, and Cheng, 2005) | Relies on TDoA | Trilateration | Asynchronous |
| TLS (Yan et al., 2010) | 1-D iterative | Validated on UWB | MDLP transformed to LDLP |
| dwMDS (Costa, Patwari, and Hero III, 2006) | Adaptively selects neighborhood of sensors | Optimization | Alleviates biasing effect |
| rMCL (Dil, Dulman, and Havinga, 2006) | Range information | 2-hop anchors | Trade-off between accuracy and energy |
| RVI (Lee et al., 2006) | Iterative | Ratio of distances | Uses adaptive report scheduling algorithm |
| CBL3DS (X. Liu et al., 2018) | Component based | Merging of subnetworks | Sparse networks |

three-dimensional space are derived using nodes that are common and edges that connect adjacent subnetworks. A comparison of range-based localization techniques for WSNs is presented in Table 7.1.

## 7.4 RANGE-FREE LOCALIZATION

Range-free techniques do not require measuring the distance between sensors and anchor nodes. These schemes depend on the information that a node receives from its neighbors in the form of either position information or hop count. A merit of range-free schemes is that these are insensitive to inaccuracies in distance measurements and do not require additional hardware. To compute the position of a sensor node, range-free techniques often use mathematical methods (Bulusu, Heidemann, and Estrin, 2000; He et al., 2003; Nagpal, Shrobe, and Bachrach, 2003; Niculescu and Nath, 2003; Moore et al., 2004). For example, in Bulusu et al. (2000), a range-free localization scheme is presented that uses a mathematical method based on the centroid of the locations of the anchors.

A range-free approach based on relative distances between neighboring nodes that are one hop away, using their neighborhood orderings, is proposed in Zhong and He (2009). It can be used to provide unique location signatures for nodes in a sensor network. Therein, the idea of *regulatory signature distance* (RSD), a measure of proximity, is used. Given the received signal strengths of neighboring nodes, a node may compute a neighborhood ordering as follows. Arrange the neighbors according to their signal strengths in descending order. Add itself as the first element in the ordered list of nodes. The resulting ordered list of nodes, say $S_i$, is unique for each node $i$. It depends on the position of node $i$, and it is a range-free sensing observation.

The problem of localization in WSN in an untrusted environment is addressed in Lazos and Poovendran (2005). Therein, a range-free distributed localization scheme called *Se*cure and *R*obust *Loc*alization (SeRLoc) is proposed. SeRLoc uses a two-level network architecture, and locations of sensor nodes are determined passively without interacting with other sensor nodes. A feature of SeRLoc is its robustness against some of the known security attacks on WSNs including wormhole, Sybil, and a compromise of the components of the network. Each type of attack is analyzed, and the probability that an attack will be successful is computed. It has been shown that SeRLoc is capable of providing accurate measurements even in the presence of such types of attacks.

A range-free localization scheme called Monte Carlo Localization (MCL) is proposed in Hu and Evans (2004) for a sensor network where anchor and nonanchor nodes may move about randomly. The assumptions are that sensor as well as anchor nodes have common radio range, and each of them may move with a speed that is uniformly and randomly distributed between a minimum speed and a maximum speed. The nodes are assumed to be mobile with the following objectives: (i) improvement in the accuracy of localization, and (ii) reduction in the number of anchor nodes. MCL consists of three major phases: (i) initialization, (ii) prediction, and (iii) filtering. In the *initialization* phase, a node selects a set of samples randomly containing its possible random locations within the area of deployment. In the *prediction* phase, a sensor computes a new set of samples using the previous set of samples. The locations in the set are constrained by the maximum speed of the node and the previous location samples. In the *filtering* phase, those locations that are not possible are removed from the set. The prediction and filtering phases are repeated to obtain a required number of samples.

An improved localization algorithm based on MCL called Monte Carlo localization Boxed (MCB) is presented in Baggio and Langendoen (2006). As the name suggests, an anchor box around a node with unknown location is constructed in such a way that it covers the area where the transmission ranges of anchors heard by the node overlap. The major differences between MCL and MCB are in terms of (i) the way to use anchor information, and (ii) the method of generating new samples from the old ones. These differences can be understood as follows. In MCL (Hu and Evans 2004), a node applies the information obtained from neighbors that are not more than two hops away at the time of filtering so as to reject the samples that are irrelevant. In MCB (Baggio and Langendoen, 2006), the information from the anchors is used to confine the area that can be used to draw samples. A reduction in the sampling area enables a node to draw relevant samples more easily and faster as compared to MCL. Generation of *relevant* samples implies that the samples are rejected less frequently in the filtering phase, and therefore, the number of iterations needed by the algorithm to obtain the desired number of samples is reduced.

A range-free localization scheme called *Monte-Carlo Sensor Localization* (MSL*) is proposed in Rudafshani and Datta (2007). MSL* provides an improvement over MCL using information from one-hop and two-hop sensors and anchors. A sensor updates its location estimate if it receives a better location estimate from its neighbors. However, in high mobility scenarios, the accuracy of localization in the case of MSL* is reduced.

A distributed localization scheme for mobile WSNs called improved MCL (IMCL) is proposed in Sheu, Hu, and Lin (2009). The intuition behind IMCL is that there will be much less difference between the estimated location and the actual locations of a sensor if the locations contained in the valid samples are closer to the actual location of the sensor. In IMCL, three types of constraints, namely anchor, neighbor, and moving direction, are used to bound the region confined by the valid samples in the vicinity of the actual location of the sensor. The assumptions made in designing IMCL are the same as those in case of MCL. IMCL consists of the following three phases: (i) selection of samples, (ii) exchange of neighbor constraints, and (iii) refinement. In the first phase, each sensor chooses a set of samples from a number of samples obtained from anchors to form a set of possible locations. In the second phase, the sensor sends the set of its allowable locations to its neighbors. Last, the samples are refined using neighbor as well as moving direction constraints so as to improve the accuracy of localization. Thereafter, the average of the samples is taken over the set of refined locations to estimate the location of the sensor.

All these Monte Carlo localization schemes use the sequential Monte Carlo method, and most of them either require high density of anchors or suffer from low sampling efficiency. To address these issues, a scheme called Weighted Monte Carlo Localization (WMCL) is proposed in Zhang, Cao, Chen, and Chen (2008). Similar to existing Monte Carlo algorithms, WMCL is also based on the bounding box technique, which is used to improve the frequency of sampling by a reduction in the range of values from which the candidate samples are drawn. However, the factor of reduction in the scope is significantly large in case of WMCL as compared to other schemes, and as a result, there is a significant improvement in the sampling efficiency of WMCL, and consequently there is a significant reduction in the computational overhead. To address the localization in high mobility scenarios, an iterative version of WMCL called IWMCL is also presented in Zhang et al. (2008). A comparison of range-free localization techniques for WSNs is presented in Table 7.2.

**TABLE 7.2**
**A Comparison of Range-free Localization Techniques for Sensor Networks**

| Scheme | Basis | Features | Remarks |
|---|---|---|---|
| Centroid (Bulusu, Heidemann, and Estrin, 2000) | Range-free | Anchor-based | Geometric |
| RSD (Zhong and He, 2009) | Range-free | Relative distance | Neighborhood orderings |
| SeRLoc (Lazos and Poovendran, 2005) | Range-free | Secure and robust | Two-tier network architecture |
| MCL (Hu and Evans, 2004) | Random movement and filtering | Initialization, prediction, estimation | Not recommended in presence of obstacles |
| MSL* (Rudafshani and Datta, 2007) | Uses information from sensors and anchors | Improvement over MCL | Reduced accuracy in high-mobility scenarios |
| WMCL (Zhang et al., 2008) | Bounding box | High sampling efficiency | Low anchor density |
| MCB (Baggio and Langendoen, 2006) | Relevant samples and anchor box | Faster | Two-hop anchor can generate samples |
| IMCL (Sheu, Hu, and Lin, 2009) | Valid samples are near the actual position | Uses anchor, neighbor, & moving direction constraints | Sample selection, exchange of neighbor constraints |

On the other hand, combinations of range-based and range-free techniques are also possible. One such scheme, called Combined and Differentiated Localization (CDL), is presented in Zhao et al. (2012) as part of a WSN, called GreenOrbs, which is deployed in a forest to localize wild animals. Therein, it is observed that the quality of distance measurement has a significant impact on the quality (or accuracy) of localization.

## 7.5 ANCHOR-BASED LOCALIZATION

Anchors or beacons are distinct nodes whose locations are known either through GPS or by some other means. The algorithms that compute the locations of ordinary nodes in the network using the locations of anchor nodes are called anchor-based localization schemes. Anchors are deployed to either help other nodes in the network to determine their locations or to introduce static coordinates in the network. In a network where no anchors are used, nodes need to create their own coordinate system to establish their relative positions. In general, is the larger the number of anchors, the better is the accuracy of location estimation. However, increasing the number of anchors increases the cost of the system and is not advisable for enhancement of the accuracy of localization.

A localization scheme called Distributed Range-free Localization Scheme (DRLS) is presented in Sheu, Hu, and Lin (2009). DLRS enables a node to localize itself using anchors lying within a distance of two hops. There are two major phases of the DRLS: (i) grid scan or initialization, and (ii) refinement. The grid scan algorithm is used to compute the locations of unlocalized sensors. Thereafter, a refinement algorithm, which is based on vectors, is used to improve the accuracy of the estimated positions of sensors.

With a motivation of enhancement in the accuracy and the overhead incurred in localization, a scheme called anchor-based distributed localization (ADL) is proposed in Kim, Shon, Kim, Kim, and Choo (2012). ADL is based on grid scan algorithm that uses information from anchors that might be at a distance of two hops from the node to be localized. The accuracy of localization of ADL is observed to be larger than that of DRLS. However, in case of ADL, there are still chances of propagation of errors due to estimation, specifically, for networks with low anchor density. After estimating its position, the node examines whether it should correct its position. The correction might be needed to reduce inaccuracies caused due to insufficient number of anchors within its two-hop neighborhood. If a correction is needed, the initial position of the node is recomputed using a probabilistic localization information obtained from anchors that are beyond a distance of two hops from the node. The probabilistic information is called *hop progress* and is based on the probability of finding a given number of sensor nodes.

An anchor-based localization scheme for a mobile WSN is proposed in Mourad, Snoussi, Abdallah, and Richard (2009), which is based on an interval analysis framework. The errors in localization are estimated using a state space model. The problem of localization is transformed to the problem of satisfying the constraints. The problem is then solved using Waltz algorithm. The proposed technique guarantees an online estimation of the positions of mobile nodes. It takes into account observation errors and anchor node imperfections and is capable of estimating the locations when the anchors are multiple hops away from the node to be localized.

In some schemes, a mobile anchor node broadcasts its position while moving along a specified path so as to help ordinary sensors to localize themselves. Such a method reduces the cost of localization and has a potential of being relatively more accurate as compared to other methods. However, it introduces the following issue: How to plan the path along which the anchor node has to move? Ideally, the anchor should move along a path so that all nodes are localized and the localization error is minimized. A scheme called *Localization with a Mobile Anchor based on Trilateration* (LMAT) is proposed in Han et al. (2013). In LMAT, the mobile anchor moves along a trajectory and broadcasts its location so that the nodes are able to compute their positions using trilateration.

A distributed scheme for localization of nodes in a WSN using a moving beacon is proposed in Xiao, Chen, and Zhou (2008) where the estimated location is computed using a range-free technique. We call it distributed localization with a moving beacon (DLMB). It is argued in Xiao et al. (2008) that the accuracy of location estimation depends on the transmission range of the beacon and the number of times the beacon needs to broadcast its position information. More precisely, it is shown that the transmission range of beacon and how frequently it broadcasts the position information determines an upper bound on the error estimation when the route traversed by the beacon is a straight line. In Xiao et al. (2008), the position estimation is extended to incorporate the localization even when the traversed route of the beacon is randomly chosen and may not be a straight line. Further, it is pointed out that the mobility patterns of the beacon nodes play a major role in the process of localization. To minimize the effect of localization errors, sensors may use a variety of algorithms in accordance with mobility patterns of the beacons.

A scheme for locating the position of nodes in a delay-tolerant sensor network using *mobile robots* (MR) is proposed in Pathirana et al. (2005). Note that in a *delay-tolerant network* (DTN), the delay is not a stringent constraint. Therefore, in a delay-tolerant sensor network, it is not necessary that all nodes are localized in real time. In Pathirana et al. (2005), a model of the delay-tolerant sensor network is assumed where sensor devices are often organized into clusters that may be mutually disconnected, and mobile robots are used to collect data from clusters. As a robot passes, it can be used to estimate the location of sensors based on the strengths of the received signals. As opposed to other methods, this method is not affected by processing constraints of static sensor nodes, and it does not use static anchor nodes. However, a mobile robot can be regarded as a mobile anchor. As opposed to a standard Kalman Filter, the scheme in Pathirana et al. (2005) uses robust extended Kalman filter (REKF). Compared to the standard Kalman filter REKF is more robust and is computationally efficient.

An anchor-based localization scheme called path planning algorithm (PPA) for mobile WSNs is presented in Ou and He (2012), which attempts to minimize the trajectory of mobile anchors in such a fashion that localization error is minimized and provides a guarantee to localize all nodes that are part of the network. In order to alleviate the impact of obstacles, the trajectory is chosen in such a fashion so that there is no obstacle lying along it.

## TABLE 7.3
### A Comparison of Anchor-based Localization Rechniques for Sensor Networks

| Scheme | Basis | Features | Remarks |
| --- | --- | --- | --- |
| DLRS (Sheu, Hu, and Lin, 2009) | Anchor based & range-free | Refines estimates | Utilizes anchors up to two hops |
| ADL (Kim et al., 2012) | Anchor based | Enhanced accuracy & overhead reduction | Grid scan algorithm |
| DLMB (Xiao et al., 2008) | Moving beacon and filtering | Also applicable for non-straight-line path | Transmission range and frequency of beacon transmission decides error |
| LMAT (Han et al., 2013) | Trilateration | Mobile anchor | Tries to minimize localization error |
| MR (Pathirana et al., 2005) | Anchor-based | Mobile robots | Use of robust extended Kalman filter |
| PPA (Ou and He, 2012) | Path planning | Trajectory avoids obstacles | Tries to minimize trajectory |
| AGM (Chang et al., 2012) | Mobile anchor | Determines the locations & constructs a path | Benefit & distance-based location selection |

Note that in range-free algorithms, nodes estimate their locations using geometric constraints that are imposed on the locations of a mobile anchor node. However, the issue of how the anchor should move so that localization errors are minimized with the minimum movement of the anchor remains unaddressed. A scheme called anchor guiding mechanism (AGM) is proposed in Chang, Chang, and Lin (2012) that determines the locations of the moving anchor and constructs a path for its movement so that the above objective is fulfilled. AGM consists of the four major phases: (i) identifying promising regions, (ii) weighting, (iii) selection of locations of anchors, and (iv) construction of paths. Further, for selecting the locations of the mobile anchor, two policies are proposed that are based on either benefit or distance. A comparison of anchor-based localization techniques for WSN is given in Table 7.3.

## 7.6 ANCHOR-FREE LOCALIZATION

Although, the cost of an anchor-based location discovery scheme is significantly less than a global positioning scheme such as GPS, however, anchors are still expensive, and in order to achieve a robust and accurate location discovery, the number of anchor nodes has to be large enough, making the system more expensive. Therefore, it is desirable to devise a location discovery scheme that is free of anchors. However, in a positioning scheme that is free of anchors, the major issue is that one still needs to find some form of reference points so as to enable the sensors to compute their locations.

In Priyantha et al. (2003), a scheme called anchor-free localization (AFL) is proposed in which all nodes in the network concurrently compute and refine their location information. AFL is based on the concept of an acyclic embedded graph. There are two major steps in AFL. In the first step, an acyclic embedding is constructed from the original embedding of the graph. Thereafter, five reference nodes are selected on an ad hoc basis to approximate the coordinates of the node to be localized. In the second step, a mass-spring-based optimization is carried out to adjust the localization errors.

An anchor-free localization scheme called deployment *K*nowledge-based *P*ositioning *S*ystem (KPS) is proposed in Fang, Du, and Ning (2007). In KPS, there are two underlying assumptions. The first assumption is that the sensors are prearranged in smaller groups, and the coordinates of the deployment points of the sensor groups are predetermined. These deployment points are used as the reference points. To estimate the location of a sensor, a spatial relationship among the sensor and the reference points is established. The second assumption is that the locations of sensors might not follow a uniformly random distribution in the region of their deployment. Consequently, sensors belonging to different deployment points may have different sets of neighbors. The prior knowledge about how likely a sensor is dropped, at a certain distance away from its deployment point, is modeled using a probability distribution function. In KPS, each sensor identifies a set of neighbors from each group, which is called the *observation* of the sensor. The sensor, then, estimates its location using the principle of the *maximum likelihood estimation*.

A localization scheme called Connectivity-Based and Anchor Free *T*hree Dimensional *L*ocalization (CATL) is proposed in Tan et al. (2013) for large-scale WSNs with concave regions. CATL is based on the identification of special nodes called *notches*, where end-to-end shortest paths between the source and the destination bend. Also, at notches a significant difference between the Euclidean distance and the hop-count-based distance is started to be observed. CATL consists of an iterative algorithm using multilateration-based localization to avoid notches. The method of identifying notch nodes in the network is based on the intuition that in the global shortest path tree, such type of nodes often possess a subtree that is relatively fatter as compared to ordinary nodes. To identify notches, one needs to detect the fat-tree abnormality in the shortest path tree. A major feature of CATL is that it is independent of the network boundaries.

An anchor-free localization scheme called multi-scale dead-reckoning (MSDR) algorithm is proposed in Efrat et al. (2010). MSDR is distributed, scalable, and resilient to noise. MSDR is based on computing layouts of *force-directed graphs* (FDG). MSDR uses multiscale extension of FDG to handle the issue of scalability and an extension of dead-reckoning to alleviate the problems arising with relatively simple topologies. Note that dead-reckoning or *deduced reckoning* is a method of estimating the current position of a moving object using the direction and distance traveled by the object with reference to a previously determined position.

A localization algorithm called Anchor-Free Mobile Geographic Distributed Localization (MGDL) is proposed in Xu et al. (2007) for a sensor network that contains a subset of mobile nodes. The distance traveled by a mote is estimated using accelerometers that are present in each node. The procedure for localizing a node contains a series of steps. A localization scheme called anchor-free localization algorithm (AFLA) is proposed in Guo and Liu (2013) for UWSN. Instead of using anchors, AFLA utilizes the neighborhood of each node to determine its location.

There is a class of localization scheme called patch and stitch localization algorithms (PSLA). In PSLA, a local map is built by the node, which is called a *patch*. A patch, typically, contains a node and its neighbors and can be considered to be an

## TABLE 7.4
### A Comparison of Anchor-free Localization Techniques in Sensor Networks

| Scheme | Basis | Features | Remarks |
| --- | --- | --- | --- |
| AFL (Priyantha et al., 2003) | Fold free embedding of a graph | Mass-spring-based optimization | Concurrent computation of locations |
| KPS (Fang, Du, and Ning, 2007) | Prior knowledge about deployment points | Minimum likelihood estimation | Requires accurate locations of deployment points |
| CATL (Tan et al., 2013) | Notch nodes | Independent of network boundaries | Global shortest path tree |
| MSDR (Efrat et al., 2010) | Force-directed graphs | Scalable, distributed | Dead-reckoning, Multiscaling |
| MGDL (Xu et al., 2007) | Distance-based | Subset of mobile nodes | Use of accelerometer |
| PSLA (Kwon, Song, and Park, 2010) | Patch & Stitch | Local and global maps | Filtering schemes prevent flip-errors |
| AFLA (Guo and Liu, 2013) | Neighborhood based | Anchor-free | UWSN |

embedding of constituent sensors into a relative coordinate system. The phenomena of joining the patches or local maps to construct a global map is called *stitching* the patches. In PSLAs, a wrong reflection of the map to which a patch is stitched is called *flip-error*. For detection and prevention of such errors an anchor-free localization scheme is presented in Kwon, Song, and Park (2010). The flip errors are prevented using the following two filtering schemes: (i) ambiguity test, and (ii) detection of flip conflict. A comparison of anchor-free localization techniques in sensor networks is provided in Table 7.4.

## 7.7 DIRECTIONAL LOCALIZATION

Angle-of-arrival (AoA) is the angle between the direction of propagation between a reference direction and an incident wave. The reference direction is also called orientation, and it is the direction against which all AoAs are measured. A localization scheme based on AoA information between neighbors is proposed in Peng and Sichitiu (2006). The scheme utilizes the information from nodes that can be multiple hops away from it. We call the approach proposed in Peng and Sichitiu (2006) as Probabilistic Localization with Orientation Information (PLOI). The algorithm is probabilistic in the sense that the positions of nodes are estimated using probability distribution functions. It is assumed that the orientation of each sensor node is known a priori. Note that for the node whose orientation is known, there has to be two or more anchor nodes to estimate the position of the node. However, in a sparse WSN, the number of anchor nodes for each node might be less than two. Consequently, the location information obtained from anchors that are multiple hops away is required to be used so that each node is localized.

A Directionality based Location Discovery (DLD) for WSNs is proposed in Nasipuri and Li (2002). In DLD, a sensor node may determine its location using wireless transmissions from three or more fixed beacon nodes. It is based on the AoA technique. In DLD, the beam width of the directional beacon signals is the main source of error. It has been argued in Nasipuri and Li (2002) that beam width should be below 15° for relatively small errors. An advantage of DLD is that its performance does not depend on the absolute dimensions of the region of deployment of the network because it is based on angular estimates. However, the scheme requires the beacon nodes to be equipped with specialized antennas for transmitting a rotating directional beam.

An autonomous localization system called smart antenna based movable localization system (SAM LOST) is presented in Falletti, Presti, and Sellone (2006). SAM LOST uses direction-of-arrival (DoA) signals that are transmitted by mobile sensor nodes. A feature of the system is that it may move on-demand wherever its services are required. It is autonomous in the sense that localization and orientation of the localization station is performed automatically. The users may work with simple transmitters and receivers and they need not possess a specific localization instrument.

A localization scheme called dual wireless radio localization (DWRL) is proposed in Akcan and Evrendilek (2012), which is directional in nature. DWRL is based on measuring the distances and does not require the use of anchor nodes. The limitations of DWRL are that it can be used to localize sensors in a static WSN only, and it requires an additional radio for each node. A merit of DWRL is its robustness, i.e., it is capable of localizing a node in the presence of noise.

A group of two algorithms for localization in a WSN using directional information is proposed in Akcan and Evrendilek (2012) without a need of anchor nodes. The algorithms are distributed and allow movements of nodes in a collaborative manner. The algorithms are named as GPS-free Directed Localization (GDL), and GPS and Compass-free Directed Localization (GCDL). The difference between the two algorithms is that GDL assumes each node is equipped with a digital compass pointing toward north as a reference direction; however, GCDL does not require the availability of compass at each node. In GDL, localization is carried out in three stages, and each stage is called an *epoch*. There are two major phases of GDL: (i) localization of core, and (ii) verification. In the first phase, for each neighbor participating in the localization process, a set of two possible relative locations is generated. In the second phase, another neighbor is used to compute the final locations with the help of the set of two relative locations.

On the other hand, the requirement of compass is relaxed in GCDL because of additional hardware cost and unfavorable conditions that may change the magnetic field and thus reduce the chances of using a compass, especially in a hostile environment. GCDL controls the collaborative movement of sensor nodes. The nodes are divided into two groups, namely *red* and *blue*, and each group is allowed to move in a stepwise manner such that when one group moves, the other group remains stationary. Using geometric properties, neighbors are localized during the movement of each group in a stepwise manner. Instead of using a compass, nodes agree for a common *virtual* north that enables them to move as a cohort. As opposed to GDL, GCDL is free from compass-related failures and errors.

Localization utilizing collaboration in visual WSNs is presented in Karakaya and Qi (2014), where the issue of inaccurate information received from malfunctioning

**TABLE 7.5**
**A Comparison of Directional Localization Techniques in Sensor Networks**

| Scheme | Basis | Features | Remarks |
|---|---|---|---|
| DLD (Nasipuri and Li, 2002) | Anchor-free | Based on AoA | Rotating directional beam |
| DWRL (Akcan and Evrendilek, 2012) | Anchor-based | Robust | Dual wireless radio |
| SAMLOST (Falletti, Presti, and Sellone, 2006) | Direction of arrival (DoA) | On-demand movement | Autonomous |
| GDL (Akcan and Evrendilek, 2012) | Compass-based | Core localization & verification | North as reference |
| GCDL (Akcan and Evrendilek, 2012) | Anchor-free | Group mobility | Virtual north |

sensors is addressed. Therein, a distributed fault-tolerant algorithm based on a voting mechanism is developed to mitigate the effect of faulty sensors. Another focus of Karakaya and Qi (2014) is to possibly detect sensor faults and to take corrective measures pertaining to orientations of cameras. A comparison of directional localization techniques for WSNs is presented in Table 7.5.

A testbed for rapid prototyping of localization algorithms for *software-defined radio* (SDR) networks is presented in Goverdovsky et al. (2016). The testbed is cost effective, is wideband, and can be reconfigured. A *C*luster *B*ased localization scheme using *M*ulti-*D*imensional *S*caling (CB-MDS) for *cognitive radio networks* (CRNs) is proposed in Saeed and Nam (2016). A deep learning based device-free localization and activity recognition (DFLAR) scheme is described in Wang et al. (2017). Selection of sensors to minimize the localization using Semi Definite Programming is described in Y. Zhao et al. (2019).

## 7.8 CONCLUSION

The design of a localization technique for a WSN is a challenging task. In this chapter, we presented a survey of the localization techniques proposed in the literature for WSNs. We described different classes of localization techniques such as anchor-based, anchor free, range-based, range-free, and directional, for static and mobile sensor networks. We compared the basis and salient features of different techniques and discussed their merits and demerits.

## REFERENCES

Akcan, Hüseyin, and Cem Evrendilek. 2012. "GPS-Free Directional Localization via Dual Wireless Radios." *Computer Communications* 35 (9): 1151–1163.

Alemdar, Hande, and Cem Ersoy. 2010. "Wireless Sensor Networks for Healthcare: A Survey." *Computer Networks* 54 (15): 2688–2710.

Baggio, Aline, and Koen Langendoen. 2006. "Monte-Carlo Localization for Mobile Wireless Sensor Networks." *Ad Hoc Networks* 6 (5): 713–718.

Bahl, Paramvir, and Venkata N. Padmanabhan. 2000. "RADAR: An In-Building RF-Based User Location and Tracking System." In *Proceedings IEEE INFOCOM 2000. Conference on Computer Communications. Nineteenth Annual Joint Conference of the IEEE*

*Computer and Communications Societies (Cat. No. 00CH37064)*, Tel Aviv, Isreal, 2:775–784.

Boushaba, Mustapha, Abdelhakim Hafid, and Abderrahim Benslimane. 2009. "High Accuracy Localization Method Using AoA in Sensor Networks." *Computer Networks* 53 (18): 3076–3088.

Bulusu, Nirupama, John Heidemann, and Deborah Estrin. 2000. "GPS-Less Low-Cost Outdoor Localization for Very Small Devices." *IEEE Personal Communications* 7 (5): 28–34.

Chang, Chao-Tsun, Chieh Young Chang, and Chih-Yu Lin. 2012. "Anchor-Guiding Mechanism for Beacon-Assisted Localization in Wireless Sensor Networks." *IEEE Sensors Journal* 12 (5): 1098–1111.

Cheng, Xiuzhen, Andrew Thaeler, Guoliang Xue, and Dechang Chen. 2004. "TPS: A Time-Based Positioning Scheme for Outdoor Wireless Sensor Networks." In *IEEE INFOCOM 2004* 4:2685–2696.

Cobos, Maximo, Juan J. Perez-Solano, Santiago Felici-Castell, Jaume Segura, and Juan M. Navarro. 2014. "Cumulative-Sum-Based Localization of Sound Events in Low-Cost Wireless Acoustic Sensor Networks." *IEEE/ACM Transactions on Audio, Speech, and Language Processing* 22 (12): 1792–1802.

Costa, Jose A., Neal Patwari, and Alfred O. Hero III. 2006. "Distributed Weighted-Multidimensional Scaling for Node Localization in Sensor Networks." *ACM Transactions on Sensor Networks (TOSN)* 2 (1): 39–64.

Dil, Bram, Stefan Dulman, and Paul Havinga. 2006. "Range-Based Localization in Mobile Sensor Networks." In *European Workshop on Wireless Sensor Networks*, Zurich, Switzerland, 164–179.

Efrat, Alon, David Forrester, Anand Iyer, Stephen G. Kobourov, Cesim Erten, and Ozan Kilic. 2010. "Force-Directed Approaches to Sensor Localization." *ACM Transactions on Sensor Networks (TOSN)* 7 (3): 1–25.

Eriksson, Jakob, Lewis Girod, Bret Hull, Ryan Newton, Samuel Madden, and Hari Balakrishnan. 2008. "The Pothole Patrol: Using a Mobile Sensor Network for Road Surface Monitoring." In *Proceedings of the 6th International Conference on Mobile Systems, Applications, and Services*, Breckenridge, CO, 29–39.

Falletti, Emanuela, Letizia Lo Presti, and Fabrizio Sellone. 2006. "SAM LOST Smart Antennas-Based Movable Localization System." *IEEE Transactions on Vehicular Technology* 55 (1): 25–42.

Fang, Lei, Wenliang Du, and Peng Ning. 2007. "A Beacon-Less Location Discovery Scheme for Wireless Sensor Networks." In *Secure Localization and Time Synchronization for Wireless Sensor and Ad Hoc Networks*, Miami, FL, 33–55. Springer.

Goverdovsky, Valentin, David C. Yates, Marc Willerton, Christos Papavassiliou, and Eric Yeatman. 2016. "Modular Software-Defined Radio Testbed for Rapid Prototyping of Localization Algorithms." *IEEE Transactions on Instrumentation and Measurement* 65 (7): 1577–1584.

Guo, Ying, and Yutao Liu. 2013. "Localization for Anchor-Free Underwater Sensor Networks." *Computers and Electrical Engineering*, 39 (6): 1812–1821.

Han, Guangjie, Huihui Xu, Jinfanf Jiang, Lei Shu, Takahiro Hara, and Shojiro Nishio. 2013. "Path Planning Using a Mobile Anchor Node Based on Trilateration in Wireless Sensor Networks." *Wireless Communication and Mobile Computing* 13 (14): 1324–1336.

He, Tian, Chengdu Huang, Brian M. Blum, John A. Stankovic, and Tarek Abdelzaher. 2003. "Range-Free Localization Schemes for Large Scale Sensor Networks." In *Proceedings of the 9th Annual International Conference on Mobile Computing and Networking*, San Deigo, CA, 81–95.

Hofmann-Wellenhof, Bernhard, Herbert Lichtenegger, and James Collins. 2012. *Global Positioning System: Theory and Practice*. Wein, Austria: Springer-Verlag.

Hu, Lingxuan, and David Evans. 2004. "Localization for Mobile Sensor Networks." In *Proceedings of the 10th Annual International Conference on Mobile Computing and Networking*, Philadelphia, PA, 45–57.

Karakaya, Mahmut, and Hairong Qi. 2014. "Collaborative Localization in Visual Sensor Networks." *ACM Transactions on Sensor Networks (TOSN)* 10 (2): 1–24.

Kim, Taeyoung, Minhan Shon, Mihul Kim, Dongsoo S. Kim, and Hyunseung Choo. 2012. "Anchor-Node-Based Distributed Localization with Error Correction in Wireless Sensor Networks." *Hindawi International Journal of Distributed Sensor Networks* 2012:1-14. doi: 10.1155/2012/975147

Kwon, Oh-Heum, Ha-Joo Song, and Sangjoon Park. 2010. "Anchor-Free Localization through Flip-Error-Resistant Map Stitching in Wireless Sensor Network." *IEEE Transactions on Parallel and Distributed Systems* 21 (11): 1644–1657.

Lazos, Loukas, and Radha Poovendran. 2005. "SeRLoc: Robust Localization for Wireless Sensor Networks." *ACM Transactions on Sensor Networks (TOSN)* 1 (1): 73–100.

Lee, Jeongkeun, Kideok Cho, Seungjae Lee, Taekyoung Kwon, and Yanghee Choi. 2006. "Distributed and Energy-Efficient Target Localization and Tracking in Wireless Sensor Networks." *Computer Communications* 29 (13–14): 2494–2505.

Li, Xu, Wei Shu, Minglu Li, Hong-Yu Huang, Pei-En Luo, and Min-You Wu. 2008. "Performance Evaluation of Vehicle-Based Mobile Sensor Networks for Traffic Monitoring." *IEEE Transactions on Vehicular Technology* 58 (4): 1647–1653.

Liu, Chong, Tereus Scott, Kui Wu, and Daniel Hoffman. 2007. "Range-Free Sensor Localisation with Ring Overlapping Based on Comparison of Received Signal Strength Indicator." *International Journal of Sensor Networks* 2 (5–6): 399–413.

Liu, Xuan, Jiangjin Yin, Shigeng Zhang, Bo Ding, Song Guo, and Kun Wang. 2018. "Range-Based Localization for Sparse 3-D Sensor Networks." *IEEE Internet of Things Journal* 6 (1): 753–764.

Moore, David, John Leonard, Daniela Rus, and Seth Teller. 2004. "Robust Distributed Network Localization with Noisy Range Measurements." In *Proceedings of the 2nd International Conference on Embedded Networked Sensor Systems*, Baltimore, MD, 50–61.

Mourad, Farah, Hichem Snoussi, Fahed Abdallah, and Cedric Richard. 2009. "Anchor-Based Localization via Interval Analysis for Mobile Ad-Hoc Sensor Networks." *IEEE Transactions on Signal Processing* 57 (8): 3226–3229.

Nagpal, Radhika, Howard Shrobe, and Jonathan Bachrach. 2003. "Organizing a Global Coordinate System from Local Information on an Ad Hoc Sensor Network." In *Information Processing in Sensor Networks*, Palo Alto, CA. 333–348.

Nasipuri, Asis, and Kai Li. 2002. "A Directionality Based Location Discovery Scheme for Wireless Sensor Networks." In *Proceedings of the 1st ACM International Workshop on Wireless Sensor Networks and Applications*, Atlanta, GA, 105–111.

Niculescu, Dragocs, and Badri Nath. 2003. "DV Based Positioning in Ad Hoc Networks." *Telecommunication Systems* 22 (1–4): 267–280.

Ou, Chia-Ho, and Wei-Lun He. 2012. "Path Planning Algorithm for Mobile Anchor-Based Localization in Wireless Sensor Networks." *IEEE Sensors Journals* 13 (2): 466–475. doi: 10.1109/JSEN.2012.2218100

Pathirana, Pubudu N., Nirupama Bulusu, Andrey V. Savkin, and Sanjay Jha. 2005. "Node Localization Using Mobile Robots in Delay-Tolerant Sensor Networks." *IEEE Transactions on Mobile Computing* 4 (3): 285–296.

Peng, Rong, and Mihail L. Sichitiu. 2006. "Angle of Arrival Localization for Wireless Sensor Networks." In *2006 3rd Annual IEEE Communications Society on Sensor and Ad Hoc Communications and Networks*, Reston, VA, 1:374–382.

Priyantha, Nissanka B., Anit Chakraborty, and Hari Balakrishnan. 2000. "The Cricket Location-Support System." In *Proceedings of the 6th Annual International Conference on Mobile Computing and Networking*, Boston, MA, 32–43.

Priyantha, Nissanka B., Hari Balakrishnan, Erik Demaine, and Seth Teller. 2003. "Anchor-Free Distributed Localization in Sensor Networks." In *Proceedings of the 1st International Conference on Embedded Networked Sensor Systems*, Boston, MA, 340–341.

Rudafshani, Masoomeh, and Suprakash Datta. 2007. "Localization in Wireless Sensor Networks." In *Proceedings of International Conference on Information Processing in Sensor Networks (IPSN)*, Cambridge, MA, 51–60

Saeed, Nasir, and Haewoon Nam. 2016. "Cluster Based Multidimensional Scaling for Irregular Cognitive Radio Networks Localization." *IEEE Transactions on Signal Processing* 64 (10): 2649–2659.

Sheu, Jang-Ping, Wei-Kai Hu, and Jen-Chiao Lin. 2009. "Distributed Localization Scheme for Mobile Sensor Networks." *IEEE Transactions on Mobile Computing* 9 (4): 516–526.

Srivastava, Mani, Richard Muntz, and Miodrag Potkonjak. 2001. "Smart Kindergarten: Sensor-Based Wireless Networks for Smart Developmental Problem-Solving Environments." In *Proceedings of the 7th Annual International Conference on Mobile Computing and Networking*, Rome, Italy, 132–138.

Tan, Guang, Hongbo Jiang, Shengkai Zhang, Zhimeng Yin, and Anne-Marie Kermarrec. 2013. "Connectivity-Based and Anchor-Free Localization in Large-Scale 2D/3D Sensor Networks." *ACM Transactions on Sensor Networks (TOSN)* 10 (1): 1–21.

Thaeler, Andrew, Min Ding, and Xiuzhen Cheng. 2005. "ITPS: An Improved Location Discovery Scheme for Sensor Networks with Long-Range Beacons." *Journal of Parallel and Distributed Computing* 65 (2): 98–106.

Wang, Jie, Xiao Zhang, Qinhua Gao, Hao Yue, and Hongyu Wang. 2017. "Device-Free Wireless Localization and Activity Recognition: A Deep Learning Approach." *IEEE Transactions on Vehicular Technology* 66 (7): 6258–6267. doi:10.1109/TVT.2016.2635161

Xiao, Bin, Hekang Chen, and Shuigeng Zhou. 2008. "Distributed Localization Using a Moving Beacon in Wireless Sensor Networks." *IEEE Transactions on Parallel and Distributed Systems* 19 (5): 587–600. doi:10.1109/TPDS.2007.70773

Xu, Yurong, Yi Ouyang, Zhengyi Le, James Ford, and Fillia Makedon. 2007. "Mobile Anchor-Free Localization for Wireless Sensor Networks." In *International Conference on Distributed Computing in Sensor Systems*, Santa Fe, NM, 96–109.

Yan, Junlin, Christian C. J. M. Tiberius, Peter J. G. Teunissen, Giovanni Bellusci, and Gerard J. M. Janssen. 2010. "A Framework for Low Complexity Least-Squares Localization with High Accuracy." *IEEE Transactions on Signal Processing* 58 (9): 4836–4847.

Zhang, Shigeng, Jiannong Cao, Lijun Chen, and Daoxu Chen. 2008. "Locating Nodes in Mobile Sensor Networks More Accurately and Faster." In *2008 5th Annual IEEE Communications Society Conference on Sensor, Mesh and Ad Hoc Communications and Networks*, San Francisco, CA, 37–45.

Zhao, Jizhong, Wei Xi, Yuan He, Yunhao Liu, Xiang-Yang Li, Lufeng Mo, and Zheng Yang. 2012. "Localization of Wireless Sensor Networks in the Wild: Pursuit of Ranging Quality." *IEEE/ACM Transactions on Networking* 21 (1): 311–323.

Zhao, Yue, Zan Li, Benjian Hao, Pengwu Wan, and Linlin Wang. 2019. "How to Select the Best Sensors for TDOA and TDOA/AOA Localization?" *China Communications* 16 (2): 134–145.

Zhong, Ziguo, and Tian He. 2009. "Achieving Range-Free Localization beyond Connectivity." In *Proceedings of the 7th ACM Conference on Embedded Networked Sensor Systems*, Berkeley, CA, 281–294.

Zhou, Junyi, and Jing Shi. 2011. "A Comprehensive Multi-Factor Analysis on RFID Localization Capability." *Advanced Engineering Informatics* 25 (1): 32–40.

# 8 Distributed Intelligent Networks
## *Convergence of 5G, AI, and IoT*

M. Z. Shamim, M. Parayangat,
V. P. Thafasal Ijyas, and S. J. Ali
King Khalid University

## CONTENTS

- 8.1 Introduction .................................................................................................. 137
- 8.2 Global Impact of 5G and AIoT ..................................................................... 139
- 8.3 Requirements for Next Generation Distributed Intelligence Wireless Networks ....................................................................................................... 140
- 8.4 Enabling Technological Uses Cases for 5G and AIoT Systems ................... 141
  - 8.4.1 Industry 4.0 ...................................................................................... 141
    - Use Case 1: Automated Factories and Remote Inspection ............... 142
  - 8.4.2 Transportation and Logistics ............................................................ 142
    - Use Case 1: Connected Intelligent Vehicles ..................................... 142
  - 8.4.3 HealthCare 5.0 ................................................................................. 142
    - Use Case 1: Precision Medicine ........................................................ 143
    - Use Case 2: Remote Diagnosis and Surgeries .................................. 143
  - 8.4.4 Security and Safety .......................................................................... 143
    - Use Case 1: Intelligent Surveillance .................................................. 143
    - Use Case 2: Border/Immigration Control and Emergency Services ............................................................................................. 144
  - 8.4.5 Entertainment and Retail ................................................................. 144
    - Use Case 1: 5G Cloud Gaming .......................................................... 144
    - Use Case 2: Personalized Shopping Experience ............................... 144
  - 8.4.6 Smart Cities ...................................................................................... 144
- 8.5 Conclusion .................................................................................................... 145
- References ............................................................................................................. 146

## 8.1 INTRODUCTION

The advent of 5G, Artificial Intelligence (AI), and Internet of Things (IoT) being the nascent technologies of this decade are leading toward what is being coined as the era of Distributed Intelligence (DI) (Fu et al., 2018; Wang et al., 2019; Ioannou et al., 2020; Song et al., 2020). AI and Big Data have already enabled a host of new applications across different verticals and is now at a crossroads at migrating its powerful cloud-centric processing capabilities closer to edge IoT devices, paving the way for Artificial Intelligence of Things (AIoT) edge computing applications. AIoT in simple terms will allow IoT devices to perform intelligent decision-making tasks with the aid of integrated AI algorithms independently and no human intervention. Traditionally, an array of different sensors, cameras, and IoT devices collect data and transmit them to centralized data centers in the cloud for contextualization using AI. By 2020, more than 50 billion IoT edge devices will be connected to the World Wide Web (Davis, 2018). By 2025, these devices will generate nearly 150 zettabytes (ZB) of data annually (Reinsel, Gantz, and Rydning, 2018). Transmitting and processing this huge amount of data to the cloud poses several challenges. First, it is not cost effective to transmit this huge amount of data from edge IoT devices to data centers in terms of energy consumed, bandwidth capacity, and required computational power. Second, estimates show that only 12% of this transmitted data is analyzed by organizations. Third, power utilization for data transmission and processing is enormous and would require finding new ways to minimize this energy wastage and related transmission cost. Edge computing (EC) has the potential to address these challenges by utilizing AI techniques nearer to the edge (Santoyo Gonzalez and Cervello Pastor, 2019). This has the added benefit of improving end-to-end service latency, reducing energy usage by large data centers, minimizing cybersecurity risks with regard to data privacy, and also increasing network reliability, as applications would now be able to function even during network outages. Hence most local server installations are now utilizing low power systems on chips (SoCs) with integrated GPU-assisted hardware acceleration for AI contextualization.

In DI vision, digital data acquired by sensors, devices, and machines making up the IoT would be analyzed and contextualized using AI technologies at the edge, thanks to exponential growth in computational resources. Deployment of 5G is the key missing ingredient linking AI with IoT for real-time use cases, making the DI vision ubiquitous. Ultra-fast, ultra-low latency and high bandwidth connectivity provided by 5G networks, combined with a large amount of data acquired and contextualized using AIoT devices, will enable transformation in virtually every sector and will require a high level of flexibility and adaptability that cannot be guaranteed by existing network design paradigms. The fusion of 5G and EC has the potential to transform the way we live and work by accelerating technological development and thereby enabling new disruptive digital applications and services. It would also establish new markets for telecom providers and semiconductor companies and develop new software ecosystems. To achieve this vision, network transformation will be vital, as service providers would need to adapt their existing wireless networks to accommodate the exponential increase in new use cases once they become mainstream. This would require considerable flexibility in terms of enhanced mobile broadband (eMBB), ultra-reliable low-latency communication (URLLC), and massive machine type communication (mMTC) enabling new cloud native platforms based on network orchestration (Fu et al., 2018). These transformations are essential to take advantage of new business opportunities created by the convergence of 5G and AIoT technologies as illustrated in Figure 8.1 (Lowman, 2020).

# Distributed Intelligent Networks

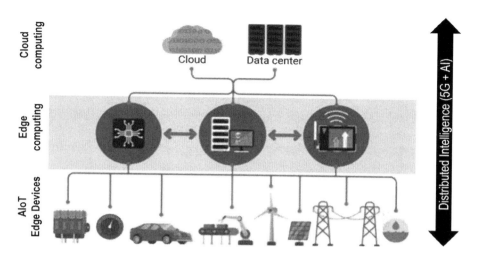

**FIGURE 8.1** Schematic illustration of future distributed intelligent networks utilizing 5G and AIoT.

## 8.2 GLOBAL IMPACT OF 5G AND AIoT

The convergence of 5G and AIoT will enable proliferation of fully automated processes across many verticals. It will serve as a catalyst to several technology sectors (e.g., industrial manufacturing, healthcare, automotive) by creating innovative business models capable of generating economic value globally. Market research indicates that this convergence will generate nearly US$ 4 trillion worth of revenue by 2025 (Figure 8.2) (Saadi and Mavrakis, 2020). This is forecasted to grow to nearly US$ 20 trillion by 2036, or 9.7% of global gross domestic product (GDP) when both the technologies reach maturity (Saadi and Mavrakis, 2020).

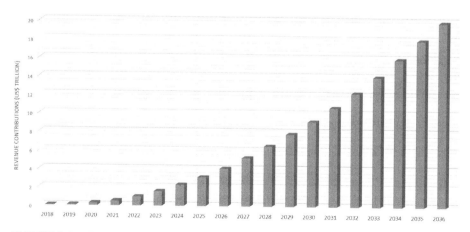

**FIGURE 8.2** Contribution of 5G and AI to Global GDP.

The current COVID-19 crisis is also bound to accelerate the adoption of 5G and AIoT technologies. It has changed our perspective of the value of technology in healthcare and remote working (Shen, 2020). DI can address the pressing challenge of tracking the pandemic and mitigating its spread globally and also play a vital role in various aspects of detection. For example, AIoT cameras with integrated thermal sensors can detect infected individuals and communicate automatically with the concerned health authorities in real time over 5G networks. Post-COVID will see a significant shift in public sentiment toward adoption of AIoT solutions for ensuring a healthy and secure lifestyle (Shen, 2020). Dissenting opinions about surrendering our privacy rights for security and well-being will become the new norm. For example, consider the use of implants or facial recognition technologies for contactless ID and security systems. Wearable AIoT solutions will become the entry point for consumers to gain insight into their personal health. It is believed that 5G will provide the much-needed infrastructure to support this growing number of intelligent connected devices and will finally find its feet.

## 8.3 REQUIREMENTS FOR NEXT GENERATION DISTRIBUTED INTELLIGENCE WIRELESS NETWORKS

The move toward distributed intelligent networks using 5G and AIoT systems would force the entire telecommunications industry to evolve to cater to the growing demand across all verticals. The convergence of 5G and AI will allow the development of distributed intelligent infrastructure that permeates autonomous decision-making process across various industries and applications in real time. Such flexible and adaptable intelligent networks would enable:

1. Scalability: Using network slicing and orchestration techniques, 5G intelligent networks would be able to cater to different service types and applications with different quality of service requirements. For example, network slice dedicated to healthcare services can be given utmost priority among all the use cases.
2. Decentralized Intelligence: By shifting AI compute capabilities away from central nodes toward the edge, intelligent macro and micro cells will bring about a paradigm shift of cloud computing closer to the end user. Intelligent 5G networks would be expected to operate in such highly dense deployments, thereby improving performance by enabling intelligence to be distributed across the entire network from the cloud to the IoT edge. This would have the added benefit of improving service experiences and would enhance the overall infrastructure efficiency.
3. Enhanced Operational Efficiency: Utilizing AI-assisted intelligent 5G networks would allow for several infrastructure processes to be automated, thus, enabling industries to minimize waste, make smart decisions and thereby increase production and operational efficiency by reducing human intervention and simplifying tech complexities.
4. Improved Network Security: Utilizing distributed intelligence in 5G networks will allow analysis of a large quantity of data, thereby enabling much effective detection and defense against malicious network attacks. For example, AI

# Distributed Intelligent Networks

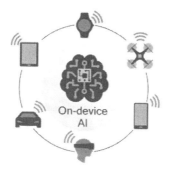

**FIGURE 8.3** On-device contextual and environmental sensing to reduce network access overhead and latency.

could detect anomalies in network traffic (e.g., flooding, jamming) by analyzing unusual spectrum usage patterns and automatically take corrective steps without human intervention.

5. Radio Awareness: On-device AI inference would allow AIoT devices contextual and environmental sensing in complex RF environments that would improve 5G end-to-end systems by reducing network data traffic for efficient wireless mobility, spectrum utilization and improved radio security, shown in Figure 8.3 (Smee and Hou, 2020). Such devices would be capable of intelligent beamforming and power management, thereby improving data throughput and increase battery life and robustness.

## 8.4 ENABLING TECHNOLOGICAL USES CASES FOR 5G AND AIoT SYSTEMS

Distributed intelligent wireless networks have the potential to become the technological platform for a new wave of innovations and use cases. Utilizing ultra-low-latency and high-bandwidth 5G spectrum and on-device AI inference, wireless AIoT edge devices will enable flexible solutions for different verticals and use cases. Such an architecture would be able to adapt to different network environments and allow for appropriate economic and performance trade-offs to achieve the optimum requirements for different applications. The convergence of 5G and AIoT is expected to play a major role in five key domains, namely industrial manufacturing, transportation and logistics, healthcare, security and the entertainment industry. Below we discuss a few use cases in each domain based on how the convergence of 5G and AIoT can enhance productivity.

### 8.4.1 INDUSTRY 4.0

In the industrial manufacturing segment, distributed intelligent 5G networks will be instrumental for implementing the next industrial revolution coined as Industry 4.0 or even as Industrial Internet of Things (IIoT) (Wollschlaeger, Sauter, and Jasperneite, 2017; Aijaz and Sooriyabandara 2018). Enabling optimized logistics, automated factories, and remotely operated machinery would significantly trim production downtime, allow for automated real-time preventive maintenance, and enhance quality assurance, thereby increasing productivity.

**Use Case 1: Automated Factories and Remote Inspection**

Amalgamation of ultra-fast and ultra-low-latency 5G connectivity with intelligent AIoT devices will enable enhanced automation of several industrial processes and machinery (Wollschlaeger et al., 2017). For example, connected sensors and cameras with onboard AI could automatically predict manufacturing defects in a production line and take corrective decisions in real time. 5G connectivity would allow seamless remote supervision and control of industrial robots with real-time visual and haptic feedback using interactive tools such as smart gloves and smart headsets with virtual/augmented reality (Aijaz and Sooriyabandara, 2018). This would result in lowering costs and minimizing operational risks in hazardous areas such as nuclear power plants, off-shore oil rigs, etc.

### 8.4.2 Transportation and Logistics

In the automotive industry, AI is already transforming vehicular automation and transportation and logistics management. Using onboard sensors and cameras in conjunction with AI-assisted machine vision models, autonomous vehicles are now capable of self-driving, navigation, and even traffic monitoring in complex and dynamically altering environments (Zhao et al., 2018; Tanwar et al. 2019). Using 5G in conjunction with AIoT would allow these applications to reach their full potential.

**Use Case 1: Connected Intelligent Vehicles**

Connected vehicles with onboard AIoT systems will be capable of self-navigation by contextually communicating wirelessly with each other and roadside systems using vehicle-to-everything (V2X) communication protocol over 5G networks in real time (Zhao et al., 2018). This would also enable automated traffic management and eventually lead to the development of driverless public transport models that will improve public safety by minimizing traffic accidents. Fleets of connected and Unmanned Aerial Vehicles (UAVs) can also be used for logistics operations (Tanwar et al., 2019; Ullah et al., 2019). UAVs are currently being adopted for delivering goods in challenging terrains and even urban congestions. Using 5G networks and onboard AIoT systems would allow swarms of UAVs to self-coordinate, avoid collisions with each other, and avoid other obstacles along their path. Such systems would provide significant cost savings to government authorities and the end user as they would be much cheaper to maintain than manned delivery systems.

### 8.4.3 Healthcare 5.0

Presently, AI in healthcare is deployed in diagnosis, drug synthesis, and patient screening. The next medical revolution, coined as Healthcare 5.0, will utilize superlow latency and high bandwidths and data rates of 5G networks in conjunction with

AIoT medical devices, for innovative applications such as real-time personalized patient monitoring and care (also known as precision medicine) and even aid medical practitioners to perform remote AI-assisted surgeries to name a few (Soldani et al., 2017; Mohanta, Das, and Patnaik, 2019; Ullah et al., 2019).

### Use Case 1: Precision Medicine

The current generation of medical-grade edge devices such as electrocardiograms (ECGs) and glucose monitoring devices utilize trained AI algorithms to detect anomalies in patients' vital signs for early diagnosis of any potential medical condition (Ullah et al., 2019). Utilizing 5G connectivity, it is possible for these AI-enabled devices to provide more effective care by facilitating remote monitoring and diagnosis. This technology has the potential to revolutionize medical care, which in some instances is limited to geographical location of medical practitioners. Adoption of wearable semi- or noninvasive biometric devices such as fitness trackers and on-skin sensors have gained popularity (Mohanta, Das, and Patnaik, 2019). The current generation of these medical devices can only perform data measurement and transmission to the user's smartphone for data processing and notification. Future versions with integrated 5G radio and trained AI models would capable of continuously monitoring patients' vital signs for detecting any medical anomalies and alert the concerned authorities for immediate medical care in real time.

### Use Case 2: Remote Diagnosis and Surgeries

There are instances when patients are either too far away from medical institutions or are in too critical a condition to be moved. The tangible nature of distributed intelligent networks enabled using ultra-reliable, ultra-high-speed and ultra-low-latency 5G networks would enable doctors to remotely diagnose such cases using audio video connectivity with real-time haptic feedback. Using 5G and AIoT-based robotic systems, it would also be possible for medical specialists to perform remote surgeries in the near future (Soldani et al., 2017).

### 8.4.4 Security and Safety

Distributed intelligent 5G networks have the potential to make our localities and cities much safer by allowing government institutions to fight crime. This can be achieved by improving security surveillance systems and emergency services while minimizing incurred cost (Horn and Schneider, 2015; Ahmad et al., 2018).

### Use Case 1: Intelligent Surveillance

Distributed intelligent 5G networks will facilitate deployment of AI-integrated connected surveillance systems such as security alarms, cameras, and sensors. This would allow for real-time surveillance and automated assessment (Horn and Schneider, 2015). For example, AIoT-enabled security cameras will be capable of automatically analyzing human behavior (e.g., body language, facial expressions) to detect suspects in real time and alert the relevant authorities. In the future such systems would be able to predict offences well in advance and thereby aid in optimizing the use of crime-prevention systems.

## Use Case 2: Border/Immigration Control and Emergency Services

Swarms of autonomous or remotely operated drones with onboard AI inference cameras connected over 5G networks can aid deployment of controlled emergency services and disaster management and can also be used for border control (Ahmad et al., 2018). Using ultra-fast 5G networks would aid emergency service personnel in operations in hazardous environments such as sites of toxic contamination and forest fires. They can also be used at border control sites to detect unauthorized trespassers and prevent crime.

### 8.4.5 Entertainment and Retail

The amalgamation of 5G networks and AIoT devices introduces several advantages for the media and entertainment industry. Innovative applications such as immersive augmented and/or virtual reality (Schmoll et al., 2018) and 8K resolution content (Inoue et al., 2017) have very high computational requirements. This inherently points toward the need for high-bandwidth 5G networks combined with AIoT edge devices to provide consumers with personalized entertainment experiences.

## Use Case 1: 5G Cloud Gaming

Current online gaming experiences have been poor due to limited bandwidth and delays associated with our current 4G/LTE networks. Online gaming is an area where even a millisecond delay affects playability. It is predicted that intelligent 5G networks will play a revolutionary role in meeting these requirements (Braun et al., 2017). Companies like Google (with Stadia) and Microsoft (with X-cloud) are rolling out their online gaming platforms based on the backbone of intelligent 5G networks. Utilizing ultra-low latency data transmission technologies, 5G networks will be able to reduce transmission delays down to less than 1 millisecond. Service providers could dedicate network slices dedicated for online gaming, thereby guaranteeing reliable service to gamers as per their contractual agreement with service providers. Distributed intelligence in 5G networks would then automatically allocate network resources to guarantee reliable performance to consumers in different peak times and days.

## Use Case 2: Personalized Shopping Experience

Targeted advertising can be a highly effective form of advertising by providing every consumer with a tailored shopping experience, thereby increasing effectiveness of the promotions (Park and Farr, 2007; Kshetri, 2018; Meani and Paglierani, 2018). 5G and AIoT penetration can engage customers effectively, hence increasing revenue. For example, facial recognition technology utilizing AI augmented machine vision for payment of goods is slowly gaining traction. This would help decrease checkout times, decrease lost sales in physical stores and improve customer experiences.

### 8.4.6 Smart Cities

All of the use cases discussed above and many more pave the foundation toward forming a digital society, thereby establishing smart cities. Current infrastructure in our established cities are prone to several problems such as traffic congestion, pollution, limited public resources, etc., to mention a few. These are challenges that low-latency 5G networks with distributed intelligence can address reliably. To address this challenge, AIoT-enabled devices should be able to analyze data at a much faster rate than what is possible with current network and hardware technologies. Advancement in sensor technology, AI inference at the network edge, and machine learning are adding value toward the dream of establishing intelligent societies. Real-time data processing at the network edge using AI would allow AIoT edge devices to make decisions automatically without human intervention. For example, self-driving vehicles would need to make instant decisions in milliseconds when dealing with potential road hazards, which our current network infrastructure is unable to sustain if the data is to be transmitted to the cloud for processing and returned back to the vehicle. Such mission-critical applications are only possible by bringing the power of cloud computing near to the network's edge. 5G distributed intelligent networks' gigabit speed, ultra-low latency, and ultra-high reliability are essential for this effort. They would enable seamless machine-to-machine communication between AIoT sensors and devices. The potent amalgamation of edge computing and 5G would enable cities to optimize every aspect of its operations, from waste disposal management to traffic management, environmental monitoring, etc., thus enabling innovative services.

However, future systems would need to guarantee data privacy, security, and integrity and not be prone to hacking before they can be deployed in critical city infrastructures such as power plants. The end-to-end security integrated into 5G networks would allow diverse AIoT devices with customized security parameters and tolerances to be safely plugged into the network. 5G intelligent networks should allow for mMTC for transmission and reception of small data blocks over low-bandwidth pipelines for general applications such as environmental sensing, logistics, etc. At the same time, these intelligence networks should also cater to critical machine type communications for delay-intolerant, secure, and reliable transmission applications such as autonomous vehicles, healthcare, traffic control, power plants, etc. Automated intelligent management of connected resources and operations citywide can provide an efficient and cost-effective solution.

## 8.5 CONCLUSION

It is understood that 5G in conjunction with AIoT systems will disrupt our present technological infrastructure and create innovative business and operational models. It has the potential to serve as an enabling catalyst for several other technologies capable of establishing and sustaining new use cases. 5G speeds will allow intelligence to be distributed across the entire communication infrastructure, from the central cloud to edge computing servers to the AIoT-enabled edge devices and sensors. This radical transformation would allow service providers and new businesses to adopt AIoT-enabled devices without compromising data security and privacy. It is predicted that the gradual convergence of 5G technology and AIoT edge systems will enable many industries to improve productivity and enhance the value for goods and services offered to consumers.

## REFERENCES

Ahmad, Ijaz, Tanesh Kumar, Madhusanka Liyanage, Jude Okwuibe, Mika Ylianttila, and Andrei Gurtov. 2018. "Overview of 5G Security Challenges and Solutions." *IEEE Communications Standards Magazine* 2 (1): 36–43. doi:10.1109/MCOMSTD.2018.1700063.

Aijaz, Adnan, and Mahesh Sooriyabandara. 2018. "The Tactile Internet for Industries: A Review." *Proceedings of the IEEE* 107 (2): 414–435. doi:10.1109/JPROC.2018.2878265.

Braun, Patrik J., Sreekrishna Pandi, Robert-Steve Schmoll, and Frank H. P. Fitzek. 2017. "On the Study and Deployment of Mobile Edge Cloud for Tactile Internet Using a 5G Gaming Application." In *2017 14th IEEE Annual Consumer Communications and Networking Conference, CCNC 2017*, Las Vegas, NV, 154–159. IEEE. doi:10.1109/CCNC.2017.7983098.

Davis, Gary. 2018. "2020: Life with 50 Billion Connected Devices." In *2018 IEEE International Conference on Consumer Electronics (ICCE)*, Las Vegas, NV, 1-1). doi:10.1109/icce.2018.8326056.

Fu, Yu, Sen Wang, Cheng Xiang Wang, Xuemin Hong, and Stephen McLaughlin. 2018. "Artificial Intelligence to Manage Network Traffic of 5G Wireless Networks." *IEEE Network* 32 (6): 58–64. doi:10.1109/MNET.2018.1800115.

Horn, Günther, and Peter Schneider. 2015. "Towards 5G Security." *Proceedings – 14th IEEE International Conference on Trust, Security and Privacy in Computing and Communications, TrustCom 2015*, Helsinki, Finland, 1: 1165–1170. doi:10.1109/Trustcom.2015.499.

Inoue, Yuki, Shohei Yoshioka, Yoshihisa Kishiyama, Satoshi Suyama, Yukihiko Okumura, James Kepler, and Mark Cudak. 2017. "Field Experimental Trials for 5G Mobile Communication System Using 70 GHz-Band." *2017 IEEE Wireless Communications and Networking Conference Workshops, WCNCW 2017*, San Francisco, CA. doi:10.1109/WCNCW.2017.7919092.

Ioannou, Iacovos, Vasos Vassiliou, Christophoros Christophorou, and Andreas Pitsillides. 2020. "Distributed Artificial Intelligence Solution for D2D Communication in 5G Networks." *IEEE Systems Journal*, 1–10. doi:10.1109/jsyst.2020.2979044.

Kshetri, Nir. 2018. "5G in E-Commerce Activities." *IEEE IT Professional* 20 (4): 73–77. doi:10.1109/MITP.2018.043141672.

Lowman, R. 2020. "How AI in Edge Computing Drives 5G and the IoT." *Synopsys Technical Bulletin*. https://www.synopsys.com/designware-ip/technical-bulletin/ai-edge-computing-5g-iot.html.

Meani, Claudio, and Pietro Paglierani. 2018. "Enabling Smart Retail through 5G Services and Technologies." In *European Conference on Networks and Communications (EuCNC)*, Lubljana.

Mohanta, Bhagyashree, Priti Das, and Srikanta Patnaik. 2019. "Healthcare 5.0: A Paradigm Shift in Digital Healthcare System Using Artificial Intelligence, IOT and 5G Communication." *Proceedings – 2019 International Conference on Applied Machine Learning, ICAML 2019*, Bhubaneswar, India, 191–196. doi:10.1109/ICAML48257.2019.00044.

Park, Nam Kyu, and Cheryl A. Farr. 2007. "Retail Store Lighting for Elderly Consumers: An Experimental Approach." *Family and Consumer Sciences Research Journal* 35 (4): 316–337. doi:10.1177/1077727X07300096.

Reinsel, David, John Gantz, and John Rydning. 2018. "The Digitization of the World – From Edge to Core." IDC White Paper, no. November: US44413318. https://www.seagate.com/files/www-content/our-story/trends/files/idc-seagate-dataage-whitepaper.pdf.

Saadi, Malik, and Dimitris Mavrakis. 2020. *5G AND AI the Foundations for the Next Societal and Business Leap*. New York: ABIresearch. www.abiresearch.com.

Santoyo Gonzalez, Alejandro and Cristina Cervello Pastor. 2019. "Edge Computing Node Placement in 5G Networks: A Latency and Reliability Constrained Framework." *Proceedings – 6th IEEE International Conference on Cyber Security and Cloud Computing, CSCloud 2019 and 5th IEEE International Conference on Edge Computing and Scalable Cloud, EdgeCom 2019*, Paris, France, 183–189. doi:10.1109/CSCloud/EdgeCom.2019.00024.

Schmoll, Robert Steve, Sreekrishna Pandi, Patrik J. Braun, and Frank H. P. Fitzek. 2018. "Demonstration of VR/AR Offloading to Mobile Edge Cloud for Low Latency 5G Gaming Application." *CCNC 2018 – 2018 15th IEEE Annual Consumer Communications and Networking Conference*, Las Vegas, NV, January 1–3, 2018. doi:10.1109/CCNC.2018.8319323.

Shen, Jijay. 2020. "The Importance of 5G, AI and Embracing New Technologies in a Post-Covid World." Silicon Republic, June 24, 2020. https://www.siliconrepublic.com/machines/5g-ai-huawei-ireland.

Smee, J. E., and J. Hou. 2020. "5G+AI: The Ingredients Fueling Tomorrow's Tech Innovations." Qualcomm Webinar, 2020. https://www.qualcomm.com/news/onq/2020/02/04/5gai-ingredients-fueling-tomorrows-tech-innovations.

Soldani, David, Fabio Fadini, Heikki Rasanen, Jose Duran, Tuomas Niemela, Devaki Chandramouli, Tom Hoglund, et al. 2017. "5G Mobile Systems for Healthcare." *IEEE Vehicular Technology Conference*, Sydney, NSW, Australia, June 2017. doi:10.1109/VTCSpring.2017.8108602.

Song, Hao, Jianan Bai, Yang Yi, Jinsong Wu, and Lingjia Liu. 2020. "Artificial Intelligence Enabled Internet of Things: Network Architecture and Spectrum Access." *IEEE Computational Intelligence Magazine* 15 (1): 44–51. doi:10.1109/MCI.2019.2954643.

Tanwar, Sudeep, Sudhanshu Tyagi, Ishan Budhiraja, and Neeraj Kumar. 2019. "Tactile Internet for Autonomous Vehicles: Latency and Reliability Analysis." *IEEE Wireless Communications* 26 (4): 66–72. doi:10.1109/MWC.2019.1800553.

Ullah, Hanif, Nithya Gopalakrishnan Nair, Adrian Moore, Chris Nugent, Paul Muschamp, and Maria Cuevas. 2019. "5G Communication: An Overview of Vehicle-to-Everything, Drones, and Healthcare Use-Cases." *IEEE Access* 7: 37251–37268. doi:10.1109/ACCESS.2019.2905347.

Wang, Dan, Bin Song, Dong Chen, and Xiaojiang Du. 2019. "Intelligent Cognitive Radio in 5G: AI-Based Hierarchical Cognitive Cellular Networks." *IEEE Wireless Communications* 26 (3): 54–61. doi:10.1109/MWC.2019.1800353.

Wollschlaeger, Martin., Thilo Sauter, and Juergen Jasperneite. 2017. "The Future of Industrial Communication: Automation Networks in the Era of the Internet of Things and Industry 4.0." *IEEE Industrial Electronics Magazine*, March 2017. doi:10.1109/MIE.2017.2649104.

Zhao, Liang, Xianwei Li, Bo Gu, Zhenyu Zhou, Shahid Mumtaz, Valerio Frascolla, Haris Gacanin, et al. 2018. "Vehicular Communications: Standardization and Open Issues." *IEEE Communications Standards Magazine* 2 (4): 74–80. doi:10.1109/MCOMSTD.2018.1800027.

# 9 Antenna Design Challenges for 5G
## *Assessing Future Direction*

*S. Arif Ali and M. Wajid*
Aligarh Muslim University

*M. Shah Alam*
College of Engineering

## CONTENTS

9.1 Introduction .................................................................................. 150
9.2 Antenna Design Flow ................................................................. 151
9.3 Antenna Integration with Radio Transceiver ICs ................... 151
    9.3.1 Antenna-in-Package (AiP) ............................................. 152
    9.3.2 Antenna-on-Chip (AoC) ................................................. 152
9.4 Multiple Beam Antenna Patterns and Their Characterization ....... 154
    9.4.1 Antenna Pattern Characterization ................................. 154
        9.4.1.1 Gain ($G$) ............................................................ 155
        9.4.1.2 Directivity ($D$) ................................................. 155
        9.4.1.3 Reflection Coefficient ...................................... 156
        9.4.1.4 Radiation Pattern .............................................. 156
        9.4.1.5 Effective Isotropic Radiated Power (EIRP) ...... 156
        9.4.1.6 Antenna Coverage Efficiency ......................... 158
        9.4.1.7 Effective Beam-Scanning Efficiency ............. 159
9.5 The Antenna Design Challenges for 5G Communication ............ 160
    9.5.1 High Gain Arrays ............................................................ 160
    9.5.2 Beamforming and Beamsteering ................................. 162
    9.5.3 Massive MIMO ................................................................. 163
    9.5.4 Multiband with Backward Compatibility ................... 163
    9.5.5 Compactness ................................................................... 164
    9.5.6 Diversity Performance – MIMO ................................... 165
        9.5.6.1 Envelope Correlation Coefficient (ECC) ........ 166
        9.5.6.2 Mean Effective Gain (MEG) .......................... 166
        9.5.6.3 Channel Capacity ........................................... 167
        9.5.6.4 Total Active Reflection Coefficient (TARC) .... 167
    9.5.7 Antenna Placement ........................................................ 168
    9.5.8 Antenna Environment ................................................... 169

9.5.8.1 Surroundings within Mobile ............................................... 169
9.5.8.2 Influence of User's Hand .................................................. 169
9.5.8.3 Propagation Channels ....................................................... 169
9.6 Radiation Exposure in 5G Communication ............................................. 170
  9.6.1 Specific Absorption Rate (SAR) .............................................. 170
  9.6.2 Power Density (PD) ........................................................... 171
  9.6.3 Antenna Measurement ......................................................... 171
9.7 Conclusion ........................................................................ 172
References ............................................................................ 173

## 9.1 INTRODUCTION

The world has witnessed four generations of wireless mobile communication in the last four decades. Initially, the first generation (1G), started with only analog transmission, has now entered into the fifth-generation (5G). The new 5G technology is expected to offer innovative services, broadly in three domains: enhanced mobile broadband, ultra-reliable and low-latency communication, and massive machine-type communication (Shafi et al. 2017); see Figure 9.1 (Santo 2017).

The details of various services offered and their characteristics are provided below (Shafi et al. 2017):

a. High data rates and high traffic volumes, possible through enhanced mobile broadband (eMBB) services.
b. Very low latency, very high reliability, and availability, promising through ultra-reliable low-latency communications (URLLC) services.
c. Massive number of device connectivity, low cost and low energy consumption, achievable through massive machine-type communications (mMTC) services.

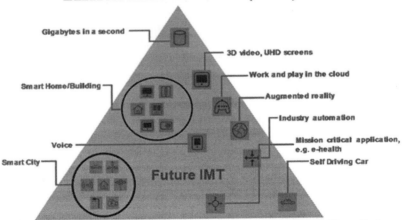

**FIGURE 9.1** Various 5G Use Cases.

… Antenna Design Challenges for 5G

There are substantial challenges involved, i.e., keeping high performance and security at a low cost while implementing 5G technology. However, in the future, health hazards (primarily due to radiation exposure) may become a severe issue due to the high frequency of operation, e.g., mmWave, terahertz (THz) used for the antenna development for 5G technology. Both academia and industry are actively involved in addressing the design challenges involved for this critical front-end component (e.g., antenna) at a low cost (Chin, Fan, and Haines 2014; Shafi et al. 2017). The challenges to the antenna designer are to cater to the 5G requirement, e.g., several Gb/s, beamforming, and beamsteering capability for interference-free and reliable communication. Moreover, due to the increased number of smaller cells in 5G networks, the number of transmitters will increase by manifold, which may cause human health and fauna disappearance (Ali, Wajid, and Alam 2020).

In addition to the issues discussed above, there is a strong need to address the thermal effects, which require a multiphysics environment to develop the thermally robust design. So there is a need to develop novel antennas, which are hazard free, high efficient, and low cost (Hong, Baek, and Ko 2017; Alibakhshikenari et al. 2019; Parchin et al. 2019). The benefit of innovative antenna design must reach to 5G industries involved by reducing the cost of fabrication.

The antenna provides the backbone of wireless connectivity for creating a global society. Furthermore, society has become more dependent on wireless devices due to COVID-19. This pandemic posed a greater use of wireless connectivity, whether it is working from home, online teaching, and learning, operating big/small business houses, and functioning of government organizations. Therefore, wireless connectivity is going to transform our society and remove the barriers caused due to an unanticipated pandemic. So, there will be a compulsion on 5G developers to accelerate the present pace of antenna design and delivering its solution at a low cost.

Given the 5G technology details provided above, this chapter begins with how the antenna is designed (see Figure 9.2). The two approaches for antenna integration, e.g., antenna-in-package (AiP) and antenna-on-chip (AoC), are discussed. The performance metrics and the design challenges involved for 5G antenna design are dealt with, followed by how the radiation exposure is quantified. Finally, antenna measurement challenges and concluding remarks are given.

## 9.2 ANTENNA DESIGN FLOW

The antenna is the last link to overcome in the design of the radio transceiver of mobile phones and follows the procedure as elaborated in Figure 9.3.

## 9.3 ANTENNA INTEGRATION WITH RADIO TRANSCEIVER ICs

The earlier radio systems utilize discrete antennas. However, due to high insertion losses, their use in the newer systems was abandoned and replaced by integrated antennas (Song 1986). However, as CMOS technology matures, an effort to integrate antenna with radio transceiver into a single chip had started (Song 1986). There are two approaches to incorporate antenna with the radio transceiver, named as

**FIGURE 9.2** Chapter Organization.

antenna-in-package (AiP), and antenna-on-chip (AoC) (Zhang 2019). The next section provides their details.

### 9.3.1 Antenna-in-Package (AiP)

The AiP is a cost-effective technology to integrate an antenna with a radio transceiver IC (e.g., front-end and base-band circuits) using the standard surface-mount package on a PCB (Zhang 2019) (see Figure 9.4). This approach requires only the addition of antenna with a radio transceiver chip at the IC packaging stage. While packaging the antenna, its impedance $Z_{ANT}$ must be optimized to optimum noise impedance $Z^{OPT}_{LNA}$ of front-end block of a radio receiver to minimize the noise performance, reduce insertion losses, and minimize the PCB area. This facilitates the codesign of antenna and front-end without using traditional 50 Ω matching network (MN), which helps to improve the integration level and reduce the cost (see Figure 9.4).

AiPhas been the most widely accepted approach for radio development for mobile phones at RF, microwave (μwave), and mmWave) bands (Zhang 2019). However, at the terahertz (THz) (0.1–10 THz) band (Cherry 2004), the AiP antenna performance degrades, and thus AoC is used instead.

### 9.3.2 Antenna-on-Chip (AoC)

In AoC technology, the antenna and radio transceiver circuits, designed using the back end of the line (BEOL) technology (Thayyil et al. 2018), are integrated on a

# Antenna Design Challenges for 5G

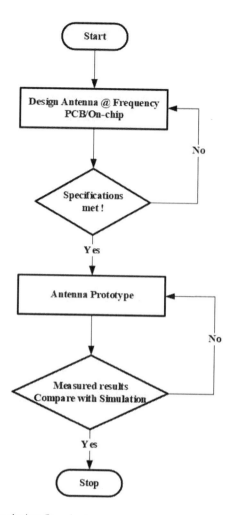

**FIGURE 9.3** Antenna design flowchart.

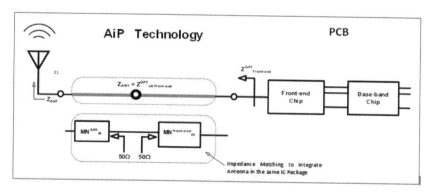

**FIGURE 9.4** AiP Technology: Radio transceiver, antenna, and matching network utilizing a single PCB.

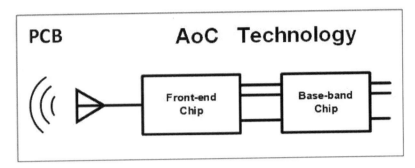

**FIGURE 9.5** AoC Technology: Radio transceiver integrated with the antenna on a single chip.

single-chip (see Figure 9.5). The AoC is the preferred design approach due to lower cost and improved performances in the THz band (Cherry 2004).

Thus, AiP technology offers a ubiquitous system-on-chip (SoC) solution (Zhang 2019). Most radio transceiver developers preferred AiP as it critically balances cost, size, and performances. However, with the launch of 6G technology in the year 2030 (Yang et al. 2019), when THz bandwidth will become a norm, then AoC will become a mainstream technology to develop radio transceiver on a single chip.

## 9.4 MULTIPLE BEAM ANTENNA PATTERNS AND THEIR CHARACTERIZATION

With the launch of fifth-generation (5G) technology to cater to the demand for high data rate requirements, advanced antennas are required to deliver eMBB, URLLC, and mMTC services (Saunders 2018; Americas 2019; Sayidmarie et al. 2019). Therefore, upgraded antenna technology of massive MIMO (in contrast to the passive and limited MIMO of an earlier generation) with beamforming is an effective way to construct antenna patterns to achieve the desired goals of 5G communication. This technology requires driving $M$ densely active antenna elements, where $M = 64$ or higher to build antenna beam patterns in different directions at the base station to communicate with the various mobile user equipment (UE) (see Figure 9.6a [Marzetta 2010]). Similar antenna arrays arrangements are used in the mobile UE to communicate with the base station and Wi-Fi user (see Figure 9.6b [Ojaroudiparchin, Shen, and Pedersen 2016])

### 9.4.1 Antenna Pattern Characterization

5G communications utilizes massive antenna arrays to generate several antenna beam patterns and steering ability. However, for characterization, each radiation pattern is classified in terms of gain, directivity, radiation patterns, return loss, and bandwidth. In contrast, a group of beam patterns is classified by effective isotropic radiated power, antenna coverage efficiency, and effective beam-scanning efficiency.

# Antenna Design Challenges for 5G

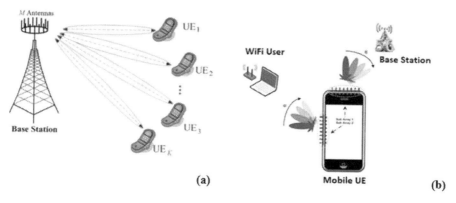

**FIGURE 9.6** (a) M×M antenna arrays in the base station to construct radiation patterns in different directions to communicate with various user equipment (UE1, UE2, UE3…UEK) (b) Similar, antenna arrays arrangement are used in mobile UE.

### 9.4.1.1 Gain (G)

The antenna gain $G$ is defined as the ratio of radiation intensity $U$ of the antenna to the radiation intensity of the isotropic antenna at a given distance. In contrast, the radiation intensity of the isotropic antenna is equal to the power transmitted by the antenna divided by solid angle $4\pi$ (Balanis 1997). Therefore, the gain $G$ expressed as:

$$G = 4\pi \frac{\text{radiation intensity of the antenna}}{\text{total received or transmitted power}} \qquad (9.1)$$

The antenna gain is the product of efficiency and directivity (Balanis 1997). The efficiency accounts for various losses, especially at the input and within the antenna structure. To the extent that directivity is concerned, it relates to directional properties and the antenna radiation pattern (Garg et al. 2001).

### 9.4.1.2 Directivity (D)

The directivity signifies the antenna's electromagnetic energy flow in a specific direction. It is defined as a ratio of radiation intensity of the antenna in a given direction to the average radiation density of the isotropic antenna. The average radiation intensity of the isotropic antenna is equal to the total radiated power divided by solid angle $4\pi$ (Fang 2010). Therefore, directivity $D$ is defined as:

$$D = 4\pi \left( \frac{U}{P_{rad}} \right), \qquad (9.2)$$

where $U$ represents the radiation intensity of the antenna (watt/solid angle), and $P_{rad}$ represents the total radiated power of an isotropic antenna.

### 9.4.1.3 Reflection Coefficient

The reflection coefficient $S_{11}$ (Alam 2015) indicates the ratio of reflected power $P_{ref}$ to the incident power $P_{inc}$. It merely exemplifies how well the impedance at the antenna input is matched to the feed-line (feeding power to the antenna). A good impedance match signifies a low reflection coefficient. A low value of this is always desirable as it results in a low power loss. Generally, in an antenna, the bandwidth (BW) is determined as the frequency range for which $S_{11} = -10$ dB. At $S_{11} = -10$ dB, the antenna accepts 90% of the power and reflects back the remaining 10% (Balanis 1997; Garg et al. 2001).

### 9.4.1.4 Radiation Pattern

It is a graphical representation of radiation properties in the far-field region of the antenna as a function of space coordinates (θ,Φ) (Balanis 1997). It includes radiation intensity and field strength in a given direction. As shown in Figure 9.7, its various constituent elements are the main lobe, side lobes, and back lobe. Depending on the applications, the radiation patterns could be omnidirectional and bidirectional (see Figure 9.8).

### 9.4.1.5 Effective Isotropic Radiated Power (EIRP)

EIRP denotes the gain of a transmitting antenna in a particular direction multiplied by the power delivered to the antenna (Zhao et al. 2018; Remcom 2020). It is essentially the power given to an isotropic antenna, which produces the same signal power in the given direction. For example, if an antenna takes 3.0 dBm of power from the transmitter and the antenna gain is 5.0 dB in a given direction, then the EIRP in that direction is equal to 8.0 dBm. Therefore, the signal in that direction is equal to that of an isotropic antenna supplied with a power input of 8.0 dBm. Because the EIRP ($E$) accounts for the direction of radiation, $E(\theta,\phi)$, therefore, $\theta$ and $\phi$ are taken in the range of $0 \leq \theta \leq \pi$, and $0 \leq \phi < 2\pi$, respectively, to cover the entire spherical area (see Figure 9.9 [Zhao et al. 2018]).

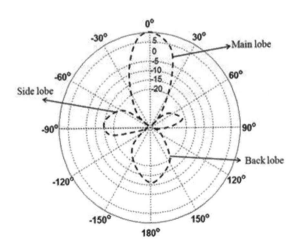

**FIGURE 9.7** Antenna radiation pattern.

# Antenna Design Challenges for 5G 157

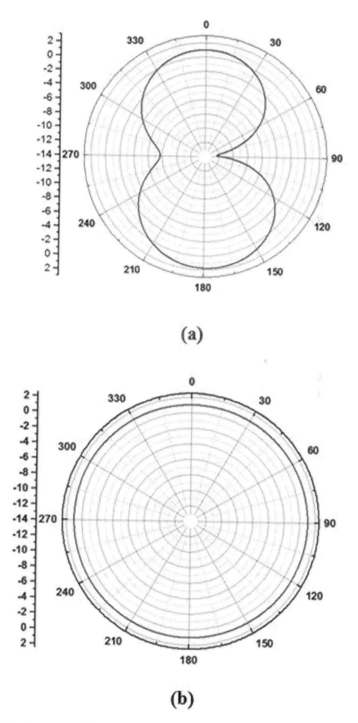

**FIGURE 9.8** Antenna radiation pattern: (a) bidirectional and (b) omnidirectional.

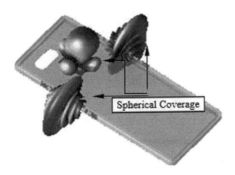

**FIGURE 9.9** Spherical coverage of a mobile UE using multiple beams.

The probability density function of EIRP $f(E(\theta, \phi))$ is defined (Remcom 2020):

$$\int_{-\infty}^{\infty} f(E)dE = \int_{E_{min}}^{E_{max}} f(E)dE = 1 \qquad (9.3)$$

The CDF $F_E(x)$ is the probability that EIRP is $\leq x$. When $F_E(x < E_{min})$, the CDF is 0, and when $F_E(x \geq E_{max})$, the CDF is 1. For $E_{min} \leq x \leq E_{max}$, $F_E(x)$ gives the fraction of all possible directions, or fraction of $4\pi$ steradians, for which $E \leq x$. When EIRP over the entire spherical coverage is sampled at the number of finite directions, then CDF is approximated as:

$$F_E(x) = \text{CDF} \cong \frac{\text{number of directions such that } E \leq x}{\text{total number of direction}}. \qquad (9.4)$$

The max hold $M(\theta, \phi)$ is the envelope of a set of several radiation patterns

$$M(\theta,\phi) = \max(E1(\theta,\phi), E2(\theta,\phi),...,En(\theta,\phi)) \qquad (9.5)$$

The CDF of $M$ denoted as $F_M(x)$ is taken when there are several radiation patterns, which is similar to $F_E(x)$ used for a single radiation pattern.

### 9.4.1.6 Antenna Coverage Efficiency

The coverage efficiency, $\eta_C$, denotes the beam-scanning capability of array antennas, and defined as (Xu et al. 2018):

$$\eta_C = \text{coverage solid angle total solid angle} = \frac{\Omega_C}{\Omega_0}, \qquad (9.6)$$

where

$$\Omega_C = \int_{\Omega_0} h(G_{TS}(\Omega))d\Omega \qquad (9.7)$$

# Antenna Design Challenges for 5G

$$h(G_{TS}) = \begin{cases} 1 & G_{TS} \geq G_{min} \\ 0 & G_{TS} < G_{min,} \end{cases} \quad (9.8)$$

and $h$ denotes a step function, $G_{TS}$, is higher than the minimum required gain, $G_{min}$, satisfying the system requirements, and $\Omega_o$ is equal to $4\pi$ steradian for full-spherical coverage. In order to achieve full-spherical coverage, use of multiple subarray arrangement have been reported in the literature for 5G mobile UE (Xu et al. 2018).

### 9.4.1.7 Effective Beam-Scanning Efficiency

The effective beam-scanning efficiency $\eta_{EBS}$ for an ensemble of $N$ antenna beams formed using a dense array of the antenna is defined as (Xu et al. 2018):

$$\eta_{EBS}(G_{min}, \Omega_d) = \frac{\int_{\Omega_d} |F_{TS}(\Omega)|^2 h(G_{TS}(\Omega)) d\Omega}{\int_{\Omega_d} |F_{TS}(\Omega)|^2 d\Omega}, \quad (9.9)$$

where

$|F_{TS}(\Omega)|$ = farfield strength of total scan pattern = $\max |F_i(\Omega)|, i = 1, 2, \ldots, N$

$|F_i| = \sqrt{|F_{i,\theta}|^2 + |F_{i,\Phi}|^2}$ is the farfield strength of each scan of beam pattern.

If a beam-scanning has less distorted coverage will lead to high $\eta_{EBS}$ (see Figure 9.10a). In contrast, the severely distorted coverage gives a low $\eta_{EBS}$, shown in Figure 9.10b. If minimum required gain $G_{min}$ is obtained within the specified $\Omega_d$, $\eta_{EBS}$ becomes 1. However, practically no array can converge all energy within specified $\Omega_d$, therefore $\eta_{EBS}$ is always less than 1.

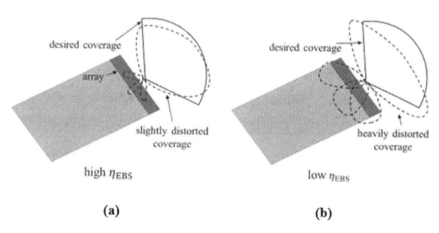

**FIGURE 9.10** Effective beam-scanning efficiency $\eta_{EBS}$: (a) less distorted and (b) more distorted.

## 9.5 THE ANTENNA DESIGN CHALLENGES FOR 5G COMMUNICATION

The antenna is an integral part of any wireless device. There are several design challenges for antenna design to satisfy the requirements of 5G new radio (NR). The 3GPP documents (5G Americas 2019) provided two parallel paths to establish wireless connectivity between the radios as described below:

- **Path I:** Non-stand-alone mode, where the UE will have dual connectivity for LTE and 5G NR.
- **Path II:** Stand-alone mode, where both data and control will use the 5G new radio link. In this mode, mmWave and sub-6 GHz bands will be combined with carrier aggregation (CA) (5G Americas 2019).

The frequency ranges (FR1, FR2) used for 5G NR as per the latest release (Radio, 2020b) is provided in Table 9.1.

Further, details of FR1 and FR2 for 5G NR systems are given in Table 9.2 (Radio, 2020b) and Table 9.3 (Radio, 2020c). Several bands are used, which are designated by letter $n$ as given in Table 9.2. For example, n65–n256 is reserved for NR FR1, whereas n257–n512 is for NR FR2. The antenna designers utilize these bands, which are popularly known as sub-6 GHz and mmWave bands, respectively. In the sections from 9.5.1 to 9.5.8, various challenges involved in the 5G antenna development are discussed, and their possible, solutions are provided.

### 9.5.1 HIGH GAIN ARRAYS

In consumer electronics, omnidirectional antennas always remained a preferable choice because they radiate in all horizontal directions equally and receive signals from any direction. These antennas are preferred below 3 GHz spectrum. However, in the case of antennas operating in the mmWave spectrum, atmospheric signal attenuation becomes more severe as compared to the sub-6 GHz band (see Figure 9.11 (FCC Report 1997)). Molecules of water vapors and gases like oxygen may create resonances and absorb energy in some portions of the mmWave spectrum.

While low-frequency (sub-6 GHz band) signals can travel long distances and are not absorbed by building walls, mmWave signals can only propagate for few kilometers and are prone to absorption by walls. However, small and dense networks (Pico cells) based on mmWave bands can provide the advantage of spectrum reuse and

**TABLE 9.1**
**Designation of Frequency Ranges**

| Designation | Frequency Range |
|---|---|
| FR1 | 410–7125 MHz |
| FR2 | 24.25–52.6 GHz |

### TABLE 9.2
### Glimpse of New Radio FR1 Bands

| NR Band | Uplink Band (MHz) | Downlink Band (MHz) | Mode |
|---|---|---|---|
| n1 | 1920–1980 | 2110–2170 | FDD |
| n2 | 1850–1910 | 1930–1990 | FDD |
| n3 | 1710–1785 | 1805–1880 | FDD |
| n5 | 824–849 | 869–894 | FDD |
| n7 | 2500–2570 | 2620–2690 | FDD |
| ⋮ | ⋮ | ⋮ | ⋮ |
| n29 | N/A | 717–728 | SDL |
| n30 | 2305–2315 | 2350–2360 | FDD |
| n34 | 2010–2025 | 2010–2025 | TDD |
| n38 | 2570–2620 | 2570–2620 | TDD |
| n39 | 1880–1920 | 1880–1920 | TDD |
| ⋮ | ⋮ | ⋮ | ⋮ |
| n77 | 3300–4200 | 3300–4200 | TDD |
| n78 | 3300–3800 | 3300–3800 | TDD |
| n79 | 4400–5000 | 4400–5000 | TDD |
| ⋮ | ⋮ | ⋮ | ⋮ |
| n93 | 880–915 | 1427–1432 | FDD |
| n94 | 880–915 | 1432–1517 | FDD |
| n95 | 2010–2025 | N/A | SUL |

### TABLE 9.3
### New Radio FR2 Bands

| NR Band | Uplink/downlink Band (GHz) | | | Mode |
|---|---|---|---|---|
| n257 | 26.5 | – | 29.5 | TDD |
| n258 | 24.25 | – | 27.5 | TDD |
| n260 | 37 | – | 40 | TDD |
| n261 | 27.5 | – | 28.35 | TDD |

increased security (FCC Report 1997). So now, mmWave systems require antennas with much higher gain to compensate for the atmospheric losses. With limited energy available, to get higher gain at the mmWave band, the concept of antenna arrays is extended for the user terminals as well (Hong et al. 2014).

Within the NR sub-6 GHz bands (FR1), because atmospheric losses are minimum, there is no need to use array topology. However, at mmWave bands (FR2 bands, especially n257, n258, n260, and n261), antenna designers have to utilize

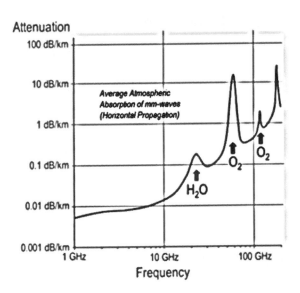

**FIGURE 9.11** Average atmospheric absorption of mm Wave signals.

array topology to maximize the gain (Hussain et al. 2017; Kurvinen et al. 2018; Yu et al. 2018).

### 9.5.2 BEAMFORMING AND BEAMSTEERING

The radio communication is much more challenging at mmWave than at FR1 bands (sub-6 GHz), and antenna designers have to deal with it. There is a finite amount of energy available, which cannot be utilized for spherical coverage using omnidirectional radiation patterns at mmWave bands due to high atmospheric losses. Therefore, with the help of massive antenna arrays, the beam is appropriately reshaped, which is called beamforming, to concentrate finite energy in a particular direction. Furthermore, the beam is provided with a full sweeping capability, which is called beamsteering, for fast scanning in 3D space for communicating the signals (Hong et al. 2014). Consequently, the beamforming gain will compensate for the substantial atmospheric losses that occurred when FR2 is used for signal transmission. However, for FR1, the capacity enhancement is essential due to the high interference-limited environment, but for the higher spectrum, FR2, coverage enhancement is needed. Hence, beamforming and beamsteering technologies provide enhanced coverage. Figure 9.12 (Forouzmand and Mosallaei 2016) depicts how the electronic beamsteering process is achieved.

In this arrangement, with digital signal processing, a pencil-type radiation pattern is generated (beamforming), and they are steered both in elevation and azimuthal planes (beamstreering). The steering is performed either by switching antenna elements or by creating phase difference between antenna elements (Forouzmand and Mosallaei 2016).

# Antenna Design Challenges for 5G 163

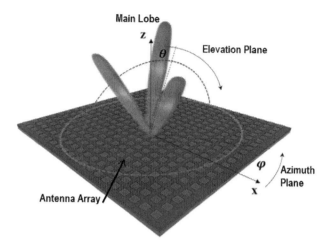

**FIGURE 9.12** Mechanism of beamsteering.

## 9.5.3 Massive MIMO

In the 5G technology, massive-MIMO technology is utilized at the base station, where a large number of phased arrays are used to generate multiple beam patterns simultaneously, to serve the number of user equipment (UE) or given geographical area (see Figure 9.13). The dynamic positioning of these beams is required to follow each user as they move in the given geographical area. Designing phased array antennas to meet these requirements is a challenging task (White Paper, Ansys 2020).

As an example, the maximum number of allowed antennas for 5G wireless indoor communication at 70 GHz are 1024 and 64 for the base station and user equipment, respectively (Busari et al. 2017). Massive-MIMO technology deals with heterogeneous networks, consisting of small and macro cell base stations, communicating at sub-6 GHz and mmWave bands (see Figure 9.13) (Busari et al. 2017).

## 9.5.4 Multiband with Backward Compatibility

5G NR technology is not an evolved version of LTE; instead it is defined by the used cases to be supported. However, NR reuses many of the features of LTE as well. It means the evolution of LTE and the development of NR are going on simultaneously. From the antenna designer perspective, the transceiver must be able to communicate both at NR and LTE bands in the present scenario (5G NR Release 15). So the NR devices preferably are backward compatible with LTE. Therefore, to deal with the current requirement, the antenna must operate in multiband. A single antenna may possibly operate either in multiband mode or multi antennas with different operating frequencies. The multiband may be from LTE and NR bands or only from NR's sub-6 GHz and mmWave bands (stand-alone mode with CA) supporting different frequencies. The main challenge in such a design scenario is to preserve the radiation performance of the antennas from deterioration due to coupling effects, and each operating band must also be independently tuned.

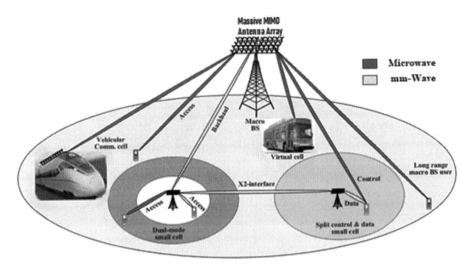

**FIGURE 9.13** 5G heterogeneous networks.

A backward LTE-compatible NR mobile user-integrated antenna (Hussain et al. 2017) was developed in 2017 when the first version of NR specifications was released. In this work planar, MIMO antenna for LTE bands (1870–2530 MHz) and array for the mmWave band (28 GHz) were integrated, simultaneously, in the smartphone. In a more practical approach (Kurvinen et al. 2018), a metal-rimmed mobile handset is used. This antenna, operating in 25–30 GHz mmWave band, supports 4G LTE lower band (700–960 MHz) and upper band (1710–2690 MHz), without compromising their performance as well as keeping sufficient isolation. Also, meta-material antennas, which exhibits negative permeability and permittivity and generates extra resonances, are explored for multiband operation. The meta-material concept is used to develop a 4G backward-compatible antenna for 5G operating in the sub-6 GHz band (Sarkar and Srivastava 2017).

### 9.5.5 Compactness

The overall size of the wireless systems is reducing due to advancements in IC fabrication technologies. Therefore, the antennas connected with RF modules also need to be compact. As the antennas in 5G new radio operated in mmWave band, their sizes are reduced significantly. However, integrating them with lower-band antennas within the limited space available in UEs is a challenging task for the antenna designers.

As discussed in Section 9.3, AiP and AoC are the two techniques to integrate antenna with RF circuitry (see Figure 9.14 [Cheema and Shamim 2013]). AoCs is a future trend for next generation communication technology. The AoCs are expected to be mainstream technology for antenna development in the future due to their low cost and high efficiency (Cheema and Shamim 2013; 5G Americas, 2019). A monolithic antenna (i.e., AoC) for 5G applications was recently reported (Hedayati,

# Antenna Design Challenges for 5G

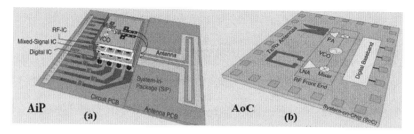

**FIGURE 9.14** (*a*) Antenna-in-package and (*b*) antenna-on-chip.

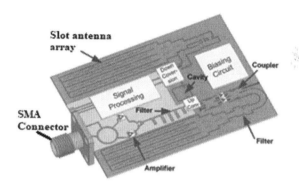

**FIGURE 9.15** SIW-based platform integrating antenna with all other RF circuitry.

Cetintepe, and Staszewski 2019). In this work, an active integrated antenna achieved a very high rate of 14 dBi, which is part of a nano-scale CMOS receiver. Substrate integrated waveguide (SIW) planar fabrication technology also offers antenna designers an innovative way to develop a complete 5G RF module along with antenna module on a single substrate (see Figure 9.15 [Djerafi, Doghri, and Wu 2015]). An effective way of miniaturizing antenna size is by using meta-material structures. While designing antennas for AiP, meta-material structures are utilized to generate negative or zeroth-order resonances at lower bands while designing mmWave antennas to make the overall size more compact (Dong and Itoh 2010).

## 9.5.6 Diversity Performance – MIMO

In order to further enhance high data rates capacity of 5G NR radio, massive MIMO antenna configuration is used. In this technique, multiple data streams are sent from multiple antennas at the transmitter, and multiple antennas receive the same at the receiver end. This arrangement provides a linear increase in the data rates as a function of the number of antennas utilized (Sharawi 2013). Thus, designing a compact MIMO antenna with reduced mutual coupling having less polarization mismatch, and immune to noisy and fading channels, is a big challenge. Therefore, the diversity performance of MIMO antennas is necessary to figure out the isolated behavior of each of the individual antenna elements present in the MIMO system (Sharawi 2013).

There are various diversity performance parameters of a MIMO system, as discussed below.

### 9.5.6.1 Envelope Correlation Coefficient (ECC)

The correlation coefficient ($\rho$) reflects how much the communication channels are isolated from each other. ECC indicates the coupling of radiation characteristics of the antenna system, which is the square of the correlation coefficient (Sharawi 2013). The ECC requirements for the 5G MIMO is less than 0.5 (Li et al. 2018). The ECC ($\rho_e$) is determined using the following relation:

$$\rho_e = \frac{\left|\iint_{4\pi}\left[\vec{F_1}(\theta,\phi) * \vec{F_2}(\theta,\phi)\right]d\Omega\right|^2}{\iint_{4\pi}\left|\vec{F_1}(\theta,\phi)\right|^2 d\Omega \iint_{4\pi}\left|\vec{F_2}(\theta,\phi)\right|^2 d\Omega}, \quad (9.10)$$

In a particular case for the lossless and single-mode antenna, ECC is calculated as:

$$\rho_{eij} = \left|\frac{\left|S_{ii}^* S_{ij} + S_{ji}^* S_{jj}\right|}{\sqrt{\left(1-\left|S_{ii}\right|^2-\left|S_{ji}\right|^2\right)\left(1-\left|S_{jj}\right|^2-\left|S_{ij}\right|^2\right)}\eta_{radi}\eta_{radj}}\right|^2 \quad (9.11)$$

### 9.5.6.2 Mean Effective Gain (MEG)

In practical applications, the antenna is not used in an anechoic chamber; instead, it works in a certain environment for a given application. Therefore, taking into account the effect of environment on the radiation characteristics is important while measuring gain of an antenna. Then the only true picture of antenna performance will come out (Sharawi 2013).

A time-consuming and costly procedure for this is to test the designed antenna in the actual environmental conditions with a standard antenna. Alternatively, one can combine radiation patterns measured in an ideal environment (i.e., anechoic chamber) with the proposed statistical model of specific environment, to get mean effective gain (MEG) numerically using Equations (9.12) and (9.13a–9.13b) (Sharawi 2013).

$$MEG = \int_0^{2\pi}\int_0^{2\pi}\left[\frac{XPD}{1+XPD}G_\theta(\theta,\varphi)P_\theta(\theta,\varphi) + \frac{1}{1+XPD}G_\varphi(\theta,\varphi)P_\varphi(\theta,\varphi)\right]\sin\theta d\theta d\varphi \quad (9.12)$$

$$\int_0^{2\pi}\int_0^{2\pi}\left[G_\theta(\theta,\varphi) + G_\varphi(\theta,\varphi)\right]\sin\theta d\theta d\varphi = 4\pi \quad (9.13a)$$

$$\int_0^{2\pi}\int_0^{2\pi}\left[P_\theta(\theta,\varphi)\right]\sin\theta d\theta d\varphi = \int_0^{2\pi}\int_0^{2\pi}\left[P_\varphi(\theta,\varphi)\right]\sin\theta d\theta d\varphi = 1 \quad (9.13b)$$

$$XPD = \frac{P_V}{P_H}, s \quad (9.14)$$

# Antenna Design Challenges for 5G

where

$XPD$ = cross-polarization of the power ratio, which is the ratio of vertical and horizontal mean incident powers indicating the distribution of incoming power.
$G_\theta(\theta,\varphi)$ and $G_\varphi(\theta,\varphi)$ = antenna gain components
$P_\theta(\theta,\varphi)$ and $P_\varphi(\theta,\varphi)$ = statistical distribution of the incoming waves in the environment, assuming that they are not correlated.

Equations (9.13a–9.13b) represent the conditions required for evaluating Equation (9.12).

### 9.5.6.3 Channel Capacity

In a multipath environment the MIMO antenna system is advantageous as it improves channel capacity. The maximum channel capacity is also considered one of the performance metrics for a MIMO system. The channel capacity Equation (9.15) reduces to Equation (9.16) in case of uncorrelated transmitting waves with antenna elements having zero correlation coefficients at both transmitter and receiver. Hence, Equation (9.15) sets the ideal limit for a MIMO system. However, practically this limit cannot be achieved due to the presence of coupling between antenna elements and channel correlation (Sharawi 2013).

The channel capacity in bits/sec/Hz is given by

$$C = \log_2\left[\det\left(I_N + \frac{\rho}{N} HH^T\right)\right], \quad (9.15)$$

where $\rho$ denotes the average SNR, $H$ is the channel matrix, $I_N$ is an $N \times N$ identity matrix, and $N$ is the number of antenna elements at both the receiver and transmitter.

For ideal channel capacity, Equation (9.14) reduces to the following equation

$$C = N \times \log_2\left[\left(1 + \frac{\rho}{N}\right)\right] \quad (9.16)$$

The goal of the MIMO antenna designer will be to bring the channel capacity as near to the ideal limit as possible given by Equation (9.15) for good performance.

### 9.5.6.4 Total Active Reflection Coefficient (TARC)

TARC is used to characterize a MIMO antenna system other than scattering parameters, which are not sufficient for a multiport antenna system. It indicates the ratio of the square root of the total reflected power divided by the square root of the total incident power. For an $N$-element antenna, the TARC is given by

$$\Gamma_a^t = \frac{\sqrt{\sum_{i=1}^{N}|b_i|^2}}{\sqrt{\sum_{i=1}^{N}|a_i|^2}} \quad (9.17)$$

where $a_i$ and $b_i$ are the incident signals and reflected signals, respectively. These can be computed from the measured $S$ parameters. Their vector relationship is given by

$$b = Sa \qquad (9.18)$$

After solving Equations (9.17) and (9.18), one can obtain TARC, whose value lies between zero and one. Zero corresponds to all power radiated, while one corresponds to all incident power reflected back, nothing radiated. The TARC is commonly presented in decibels (dB).

### 9.5.7 ANTENNA PLACEMENT

The current deployment of mmWave systems brought to light many issues related to form factor for UE design. One such issue is to achieve spherical coverage by proper placement of antennas. Therefore, one solution is to use the number of antenna array modules at multiple locations on a UE. However, an engineer has to compromise between increased device cost power usage and beam management overhead. The position of antenna array modules on UE design is a crucial parameter to be considered during the antenna design optimization phase. A recent study (Raghavan et al. 2019) revealed the impact of the positioning of antenna modules to obtain good spherical coverage and eventually minimum specific array gains. Additionally, the study reported to achieve a good trade-off in beam management overhead, complexity, and cost.

In this study, the authors considered two cases, namely face and edge models (see Figure 9.16) and compared their performances. The CDF analysis reveals that the edge design is more attractive as compared to the face design approach (Raghavan et

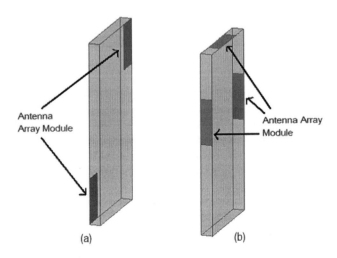

**FIGURE 9.16** Antenna array placement on (*a*) face-model and (*b*) edge-model.

# Antenna Design Challenges for 5G

al. 2019). An example of face antenna and edge antenna designs for the 5G FR1 band have been recently reported (Li et al. 2019).

### 9.5.8 Antenna Environment

Antenna performance is susceptible to the surroundings and environment at mmWave FR2 bands as the wavelength becomes comparable to the size of various surroundings. Sensitivity analysis is a crucial aspect that emerges with FR2 applications.

#### 9.5.8.1 Surroundings within Mobile

The antenna is surrounded by various electrical components like LCD, battery, frame, and circuitry. These surroundings affect the final performance. It is, therefore, essential to study the effects of the surrounding environment, as well as effect of temperature.

Shorting pins with a battery store the antenna performance (Li et al. 2019). LCD detunes the antenna, but it is manageable. Plastic frames in UEs offer dielectric loading resulting in shifting of resonant modes (Li et al. 2019). The thermal stability of miniature wireless assemblies in a 5G-capable smartphone or other antenna-enabled 5G user equipment is essential to ensure that the entire system conforms to its expected behavior. Power-hungry applications in varying environmental conditions can cause swings in device temperatures, leading to thermal cycling effects, which detune the antenna. Overheating is risky, as it can adversely impact various components of the mobile phone and degrades their performance. A detailed multiphysics simulation study is required to identify the potential issues and their possible solutions to ensure a reliable high performance (White Paper, Ansys 2019).

#### 9.5.8.2 Influence of User's Hand

When a smartphone is held by a user, the position of the fingers and palm can affect the antenna's performance. Therefore, by taking human body models, the impact of this blockade on antenna performance is thoroughly examined. These simulation studies help to obtain a better understanding of how different positions of the fingers influence the antenna performance. Similar simulations predict the antenna performance for various phone orientations and hand positions (White Paper, Ansys 2019). The study in a single-hand mode and dual-hand mode, depending on how the user holds the equipment and their impact on radiation performances, was carried out recently (Li et al. 2019).

#### 9.5.8.3 Propagation Channels

The 5G massive MIMO arrays have to be tested for both indoor and outdoor propagation mode scenarios. Propagation scenarios of mmWave massive MIMO networks are different in many aspects with sub-6 GHz cellular systems. The reason is that at mmWaves wavelength becomes of the order of the obstacles, i.e., rain, foliage, therefore, more affected by shadowing, diffraction, and blockage. Additionally, atmospheric absorption is significant in mmWave range (Busari et al. 2017). Therefore, channel modeling in mmWave scenarios and their simulations with massive-MIMO systems is a challenging task before the installation of the network.

## 9.6 RADIATION EXPOSURE IN 5G COMMUNICATION

Any wireless consumer electronics device must comply with various criteria and meet FCC or European Union (EU) regulatory standards and rules to ensure user safety. By the year 2022, mobile networks will reach >95% population across the globe. Out of the total mobile user, the percentage of LTE users will be >55% and 5G $\cong$ 15% (Ericsson Mobility Report, 2020). With the use of the Internet of Things and artificial intelligence, mobile, and portable devices, the radio network coverage will reach every metropolitan city and urban area. The use of dense pico and atto-cells in the 5G networks, there will several sources radiating RF energy simultaneously (Shikhantsov et al. 2019). Therefore, mobile users will be under continuous radiofrequency (RF) exposure, which is usually nonionizing radiation. Various internationally recognized bodies, namely the Federal Communications Commission (FCC), the International Commission on Non-Ionizing Radiation Protection (ICNIRP), and the IEEE International Committee on Electromagnetic Safety (ICES), have laid down guidelines for human RF exposure in terms of specific absorption rate (SAR) and power density (PD). As an example, the FCC recommends the use of SAR for wireless devices operating in the sub-6 GHz band. However, in the mmWave band, where the wave energy is mainly confined at the surface (due to skin effect), power density (PD) is a more suitable performance metric, keeping in view the FCC guidelines.

### 9.6.1 SPECIFIC ABSORPTION RATE (SAR)

Specific absorption rate (SAR) is a measure of the volume of RF energy absorption by the humanoid from the radiating source. It offers a straightforward way to quantify the RF contact due to base stations and/or UEs to safeguard that they are operating within the stated limits. Furthermore, its analysis involves standardized models of the humanoid packed with specialized fluids that simulate RF properties of different human tissues (FCC, 2019). Likewise, it is generally averaged either over the whole body or over a small sample volume (usually volume occupied by 1 g or 10 g of tissue). The value mentioned is then the extreme level observed in the body. From the electric field penetrating the body, *SAR* (for sub-6 GHz band) can be calculated using the following expression:

$$SAR = \frac{1}{V} \int_{\text{sample}} \frac{\sigma(r)|E(r)|^2}{\rho(r)} dr, \qquad (9.19)$$

where $\sigma$ and $\rho$ are the electrical conductivity and density of the sample, respectively, and $E$ is the RMS electric field inside $V$, which is the volume of the sample. Finally, the FCC requires that for general public usage, UEs sold should not cross the stated

# Antenna Design Challenges for 5G

limit of *SAR*, i.e., 1.6 watts per kilogram (W/kg). However, in the EU, the limit of *SAR* is more relaxed to 2 W/kg.

## 9.6.2 Power Density (PD)

PD is preferred over SAR in mmWave range. Therefore, keeping in view the massive MIMO used in mmWave frequency range, the power density (PD) corresponding to each beam for a selected surface on the human body is computed as follows:

$$PD = \frac{\iint_A E \times H * ds}{A}, \qquad (9.20)$$

where, $E$ and $H$ denote the vector electric and magnetic fields, respectively, and $A$ is the surface area. As per the ICNIRP guidelines, the PD limit for the general public mobile user is set at 20 watts per square meter (Wm²) (Physics 2020).

## 9.6.3 Antenna Measurement

Following the 3GPP standards documents (Radio, 2020a), the RF tests for the user equipment operating in at mmWave bands must take into account the following facts:

- The measurements should be carried out over-the-air (OTA).
- Allowed test methods are:
    a. Indirect far-field (IFF)
    b. Direct far-field (DFF)
    c. Near-field to far-field transform (NFFFT)

OTA testing is one of the most challenging aspects of 5G device development. In the OTA test environment, it is necessary to visualize, characterize, and validate 5G device beam patterns and performance in a variety of real-world scenarios. The three main challenges that came up with antenna measurements at mmWave (White Paper: First Steps in 5G 2019) are

a. Excessive path loss at mmWave frequencies
b. mmWave OTA test methods not well calibrated
c. Measuring device performance in real-world channel conditions

The far-field distance and associated path loss grow as the frequency of operation increases. This increase in distance results in the use of large-sized far-field test chambers and the high path loss and to make accurate measurements at mmWave frequencies very challenging. Therefore, the 3GPP approves the use of an indirect far-field (IFF) test method within the compact antenna test range (CATR). The other two methods for antenna measurement performances are DFF and NFFFT (Radio, 2020a).

The IFF test method creates the far-field environment using a transformation with a parabolic reflector (see Figure 9.17 [Hurtarte and Wen 2019]). Inside the

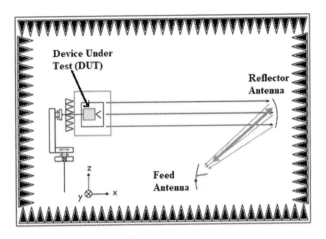

**FIGURE 9.17** CATR OTA chamber.

CATR chamber, the device under test (DUT) radiates a wavefront to the parabolic reflector that then collimates the radiated wavefront into a receiver-feed antenna. Enough distance is maintained between the DUT and the receiver so that the originating spherical wave converts to a plane wave. However, the CATR method is 10 times costlier than the DFF method. For small antenna apertures (< 3 cm), DFF offers approximately the same path loss as the CATR method (Hurtarte and Wen 2019). Therefore, DFF provides a cost-effective solution for small aperture antennas.

## 9.7 CONCLUSION

This chapter illustrated an overview of 5G technology antennas. A general antenna design flow along with the antenna integration with radiofrequency transceiver ICs is covered. Because the 5G technology will cater to the demand for high data rate requirements, advanced antenna technology of massive multiple input multiple output (MIMO) is needed. It improves not only the data rate capacity but also the data delivery by employing antenna pattern steering at low energy requirements. Several specified terms that are used to characterize the antenna patterns for 5G New Radio are covered.

Because the antenna is an integral part of wireless communication both at the user and the base station, several design challenges that are to be overcome for the antenna design to satisfy the requirements of the 5G new radio are covered. The various 5G deployment scenarios, along with the performance metrics and service requirements for the 5G radio, are described in this chapter. Toward the end, the chapter provides a detailed picture of how radiation exposures are quantified and finally concludes with a discussion of the challenges involved in the antenna performance measurement.

# REFERENCES

Alam, M. S. 2015. "Analytical Modelling and Design of CMOS Low-Noise Amplifier (LNA) with Electro-Static Discharge Protection." *IETE Technical Review* 32 (3): 227–235.

Ali, Sayyed Arif, Mohd Wajid, and M. S. Alam. 2020. "Mobile Communication and Threat to Human Health." In *Proceedings of ACAPE 2020, AMU,* Aligarh, India.

Alibakhshikenari, Mohammad, Mohsen Khalily, Bal Singh Virdee, Chan Hwang See, Raed A. Abd-Alhameed, and Ernesto Limiti. 2019. "Mutual Coupling Suppression between Two Closely Placed Microstrip Patches Using EM-Bandgap Metamaterial Fractal Loading." *IEEE Access* 7: 23606–23614.

Americas, 5G. 2019. White Paper: Advanced Antenna Systems for 5G. 5G Americas. https://www.5gamericas.org/advanced-antenna-systems-for-5g.

Balanis, C. A. 1997. *Antenna Theory, Analysis and Design.* New York: Wiley.

Busari, Sherif Adeshina, Kazi Mohammed Saidul Huq, Shahid Mumtaz, Linglong Dai, and Jonathan Rodriguez. 2017. "Millimeter-Wave Massive MIMO Communication for Future Wireless Systems: A Survey." *IEEE Communications Surveys & Tutorials* 20 (2): 836–869.

Cheema, Hammad M., and Atif Shamim. 2013. "The Last Barrier: On-Chip Antennas." *IEEE Microwave Magazine* 14 (1): 79–91.

Cherry, Steven. 2004. "Edholm's Law of Bandwidth." *IEEE Spectrum* 41 (7): 58–60.

Chin, Woon Hau, Zhong Fan, and Russell Haines. 2014. "Emerging Technologies and Research Challenges for 5G Wireless Networks." *IEEE Wireless Communications* 21 (2): 106–112.

Djerafi, Tarek, Ali Doghri, and Ke Wu. 2015. *Handbook of Antenna Technologies.* Singapore: Springer.

Dong, Yuandan, and Tatsuo Itoh. 2010. "Miniaturized Substrate Integrated Waveguide Slot Antennas Based on Negative Order Resonance." *IEEE Transactions on Antennas and Propagation* 58 (12): 3856–3864.

Ericsson. 2020. *Ericsson Mobility Report.* Stockholm, Sweden: Author.

Fang, D. G. 2010. *Antenna Theory and MIcrostrip Antenna.* Boca Raton, FL: CRC Press.

FCC Report. 1997. "Millimeter Wave Propagation: Spectrum Management Implications." Federal Communications Commission, Office of Engineering and Technology.

FCC. 2019. "Specific Absorption Rate (SAR) for Cell Phones: What It Means for You." https://www.fcc.gov/consumers/guides/specific-absorption-rate-sar-cell-phones-what-it-means-you (accessed July 3, 2020).

Forouzmand, A., and H. Mosallaei. 2016. "Tunable Two Dimensional Optical Beam Steering with Reconfigurable Indium Tin Oxide Plasmonic Reflectarray Metasurface." *Journal of Optics* 18 (12): 125003.

Garg, R., P. Bhartia, I. Bahal, and A. Ittipiboon. 2001. *Microstrip Antenna Design Hand Book.* Norwood, MA: Artech House Antenna and Propagation Library.

Hedayati, Mahsa Keshavarz, Cagri Cetintepe, and Robert Bogdan Staszewski. 2019. "Challenges in On-Chip Antenna Design and Integration with RF Receiver Front-End Circuitry in Nanoscale CMOS for 5G Communication Systems." *IEEE Access* 7: 43190–43204. doi:10.1109/ACCESS.2019.2905861.

Hong, Wonbin, Kwang-Hyun Baek, Youngju Lee, Yoongeon Kim, and Seung-Tae Ko. 2014. "Study and Prototyping of Practically Large-Scale MmWave Antenna Systems for 5G Cellular Devices." *IEEE Communications Magazine* 52 (9): 63–69.

Hong, Wonbin, Kwang-Hyun Baek, and Seung-Tae Ko. 2017. "Millimeter-Wave 5G Antennas for Smartphones: Overview and Experimental Demonstration." *IEEE Transactions on Antennas and Propagation* 65 (12): 6250–6261.

Hurtarte, Jeorge S., and Middle Wen. 2019. *Over-the-Air Testing for 5G MmWave Devices: DFF or CATR?* Microwaves & RF. https://cdn.baseplatform.io/files/base/ebm/mwrf/document/2019/03/mwrf_10332_0319_30c_pdflayout.pdf.

Hussain, Rifaqat, Ali T. Alreshaid, Symon K. Podilchak, and Mohammad S. Sharawi. 2017. "Compact 4G MIMO Antenna Integrated with a 5G Array for Current and Future Mobile Handsets." *IET Microwaves, Antennas & Propagation* 11 (2): 271–279.

Kurvinen, Joni, Henri Kähkönen, Anu Lehtovuori, Juha Ala-Laurinaho, and Ville Viikari. 2018. "Co-Designed Mm-Wave and LTE Handset Antennas." *IEEE Transactions on Antennas and Propagation* 67 (3): 1545–1553.

Li, Yixin, Chow Yen Desmond Sim, Yong Luo, and Guangli Yang. 2018. "Multiband 10-Antenna Array for Sub-6 GHz MIMO Applications in 5-G Smartphones." *IEEE Access* 6: 28041–28053. doi:10.1109/ACCESS.2018.2838337.

Li, Yixin, Chow Yen Desmond Sim, Yong Luo, and Guangli Yang. 2019. "High-Isolation 3.5 GHz Eight-Antenna MIMO Array Using Balanced Open-Slot Antenna Element for 5G Smartphones." *IEEE Transactions on Antennas and Propagation* 67 (6): 3820–3830. doi:10.1109/TAP.2019.2902751.

Marzetta, Thomas L. 2010. "Noncooperative Cellular Wireless with Unlimited Numbers of Base Station Antennas." *IEEE Transactions on Wireless Communications* 9 (11): 3590–3600.

Ojaroudiparchin, Naser, Ming Shen, and Gert Frolund Pedersen. 2016. "Multi-Layer 5G Mobile Phone Antenna for Multi-User MIMO Communications." In *23rd Telecommunications Forum, TELFOR 2015*, Belgrade, Serbia, 559–562. doi:10.1109/TELFOR.2015.7377529.

Parchin, Naser Ojaroudi, Yasir Ismael Abdulraheem Al-Yasir, Ammar H. Ali, Issa Elfergani, James M. Noras, Jonathan Rodriguez, and Raed A. Abd-Alhameed. 2019. "Eight-Element Dual-Polarized MIMO Slot Antenna System for 5G Smartphone Applications." *IEEE Access* 7: 15612–15622.

Physics, Health. 2020. "Guidelines for Limiting Exposure to Electromagnetic Fields (100 KHz to 300 GHz)." *Health Physics*. 118. doi:10.1097/HP.0000000000001210.

Radio, New. 2020a. "Study on Test Methods (Release 16)," 3GPP TS 38810-G50." Technical Specification.

Radio, New. 2020b. "User Equipment (UE) Radio Transmission and Reception Part 1: Range 1 Standalone (Release 16)," 3GPP TS 38101-1-G30." Technical Specification.

Radio, New. 2020c. "User Equipment (UE) Radio Transmission and Reception Part 2: Range 2 Standalone (Release 16)," 3GPP TS 38101-2-G31." Technical Specification.

Raghavan, Vasanthan, Mei-li Clara Chi, M. Ali Tassoudji, Ozge H. Koymen, and Junyi Li. 2019. "Antenna Placement and Performance Tradeoffs with Hand Blockage in Millimeter Wave Systems." *IEEE Transactions on Communications PP (c)*, 1. doi:10.1109/TCOMM.2019.2891669

Remcom. 2020. "CDF of EIRP & Max Hold |REMCOM." https://support.remcom.com/xfdtd/knowledge-articles/post-processing/cdf-of-eirp-max-hold.html (accessed July 3, 2020).

Santo, Brian. 2017. "The 5 Best 5G Use Cases | EDN." https://www.edn.com/electronics-blogs/5g-waves/4458756/The-5-best-5G-use-cases (accessed July 3, 2020).

Sarkar, D., and K. V. Srivastava. 2017. "Compact Four-Element SRR-Loaded Dual-Band MIMO Antenna for WLAN/WiMAX/WiFi/4G-LTE and 5G Applications." *Electronics Letters* 53 (25): 1623–1624.

Saunders, Jake. 2018. *White Paper: The Rise & Outlook of Antennas in 5G*. New York: ABI Research. Shanghai, China. https://go.abiresearch.com/lp-rise-and-outlook-of-antennas-in-5g.

Sayidmarie, Khalil H., Neil J. McEwan, Peter S. Excell, Raed A. Abd-Alhameed, and Chan H. See. 2019. *"Antennas for Emerging 5G Systems." International Journal of Antennas and Propagation* vol. 2019, Article ID 9290210, 3 pages. doi:10.1155/2019/9290210

Shafi, Mansoor, Andreas F. Molisch, Peter J. Smith, Thomas Haustein, Peiying Zhu, Prasan De Silva, Fredrik Tufvesson, Anass Benjebbour, and Gerhard Wunder. 2017. "5G: A Tutorial Overview of Standards, Trials, Challenges, Deployment, and Practice." *IEEE Journal on Selected Areas in Communications* 35 (6): 1201–1221.

Sharawi, Mohammad S. 2013. "Printed Multi-Band MIMO Antenna Systems and Their Performance Metrics [Wireless Corner]." *IEEE Antennas and Propagation Magazine* 55 (5): 218–232.

Shikhantsov, Sergei, Arno Thielens, Günter Vermeeren, Piet Demeester, Luc Martens, Guy Torfs, and Wout Joseph. 2019. "Statistical Approach for Human Electromagnetic Exposure Assessment in Future Wireless Atto-Cell Networks." *Radiation Protection Dosimetry* 183 (3): 326–331.

Song, B. 1986. "CMOS RF Circuits for Data Communications Applications." *IEEE Journal of Solid-State Circuits* 21 (2): 310–317.

Thayyil, Manu Viswambharan, Paolo Valerio Testa, Corrado Carta, and Frank Ellinger. 2018. "A 190 GHz Inset-Fed Patch Antenna in SiGe BEOL for On-Chip Integration." In *2018 IEEE Radio and Antenna Days of the Indian Ocean (RADIO)*, Grand Port, Mauritius, 1–2.

White Paper: ANSYS 5G Mobile / UE Solutions. 2019. ANSYS Inc. https://www.ansys.com/resource-library/white-paper/ansys-5g-mobile-ue-solutions.

White Paper: First Steps in 5G. 2019. Keysight. https://www.keysight.com/in/en/assets/7018-06219/white-papers/5992-3082.pdf.

White Paper: ANSYS 5G Antenna Solutions. 2020. ANSYS Inc. https://www.ansys.com/resource-library/white-paper/5g-antenna-solutions.

Xu, Bo, Zhinong Ying, Lucia Scialacqua, Alessandro Scannavini, Lars Jacob Foged, Thomas Bolin, Kun Zhao, Sailing He, and Mats Gustafsson. 2018. "Radiation Performance Analysis of 28 GHz Antennas Integrated in 5G Mobile Terminal Housing." *IEEE Access* 6: 48088–48101.

Yang, Ping, Yue Xiao, Ming Xiao, and Shaoqian Li. 2019. "6G Wireless Communications: Vision and Potential Techniques." *IEEE Network* 33 (4): 70–75.

Yu, Bin, Kang Yang, Chow-yen-desmond Sim, and Guangli Yang. 2018. "A Novel 28 GHz Beam Steering Array for 5G Mobile Device with Metallic Casing Application." *IEEE Transactions on Antennas and Propagation* 66 (1): 462–466. doi:10.1109/TAP.2017.2772084

Zhang, Yueping. 2019. "Antenna-in-Package Technology: Its Early Development [Historical Corner]." *IEEE Antennas and Propagation Magazine* 61 (3): 111–118.

Zhao, Kun, Shuai Zhang, Zuleita Ho, Olof Zander, Thomas Bolin, Zhinong Ying, and Gert Frølund Pedersen. 2018. "Spherical Coverage Characterization of 5G Millimeter Wave User Equipment with 3GPP Specifications." *IEEE Access* 7: 4442–4452.

# 10 Design and Simulation of New Beamforming *Based Cognitive Radio for 5G Networks*

Tanzeela Ashraf, Javaid A. Sheikh,
Sadaf Ajaz Khan, and Mehboob-ul-Amin
University of Kashmir

## CONTENTS

10.1 Introduction .................................................................................................. 177
10.2 Evading the Eergy Crunch: Cognitive Radio ............................................. 179
    10.2.1 Cognitive Radio Architecture ......................................................... 180
10.3 Introduction to Beamforming ..................................................................... 181
    10.3.1 Beamforming System Approaches ................................................. 181
    10.3.2 Beamforming Signals ..................................................................... 181
    10.3.3 Types of Beamforming ................................................................... 182
    10.3.4 IEEE 802.11 ad Beamforming Protocol ......................................... 183
10.4 System Model .............................................................................................. 184
10.5 Optimization Problem ................................................................................. 185
10.6 Framework for Optimal Solution ................................................................ 187
10.7 Results ......................................................................................................... 187
10.8 Conclusion ................................................................................................... 193
References ............................................................................................................ 193

## 10.1 INTRODUCTION

The origin of wireless technology occurred in the 1970s. The number of smart devices connected to networks is ever increasing. According to the survey conducted by *Forbes*, roughly 24 billion devices would be connected to the network in 2020. This may outmode the present 4G technology. Therefore, a next generation technology is required for faster and efficient data transfer. The probable solution to meet the escalating data rates is 5G technology (De and Singh 2016). 5G is a revolutionary technology that will be successor of 4G. The 5G rationale is poles apart from its forerunner technologies, which involve setting up the mobile broadband. The existing infrastructure and architecture cannot help to achieve the objective of 5G.

Therefore, everything has to be redefined (Tudzarov and Janevski 2011; Modi, Patel, and Gohil, 2013; Andrews et al. 2014; Chen and Zhao 2014; Mitra, and Agrawal 2015; Ji et al. 2018.). A major concern in network load balancing is the offloading of data on unlicensed bands with an increase in the number of devices connected to the Internet. Hence, to satisfy the high demand of real-time traffic and smooth connectivity, 5G is expected to work seamlessly with dense heterogeneous networks (Sheikh et al. 2019). Internet of Everything (IoT) being a significant part in 5G devices can sustain enormous objects that are used to access the network. The fundamental trait carried by IoT is the limited power of the sensors, as they are battery powered. The escalation in IoT has raised the consumption of energy drastically (Tutuncuoglu, Yener, and Ulukus 2015). Thus, the dearth of resources poses a decisive challenge to the existing spectrum, which is scarce, and scores of solutions are considered to deal with it (White and Reil 2016). Energy efficiency concept has evolved as a key feature for future 5G networks, where focus has shifted from capacity enhanced to such a communication that is efficient in utilization of energy. The resources that a communication system exhibit shouldn't be exclusively amended for transmitted information, but rather the amount of transmitted information per Joule of energy consumed (Rappaport et al. 2015). The resource allocation works on the concept of maximization of energy efficiency and requires the use of novel mathematical tools. The traditional frequencies used for mobile communications are deficient. Therefore, an alternative is to use mmWave bands. At these frequencies, the large bandwidths proffer with such a transfer speed with the aim to quench 5G requirements. The mobile communication setup is complicated at mmWave bands as compared to the current frequencies in use (Sayeed and Raghavan 2007). With these conditions, mmWave technology has come into sight as a new leading edge for high-speed data links. There has been a successful establishment of mmWave for indoor communication. As a result, various standards have materialized such as the IEEE 802.15.3c wireless personal area network (WPAN), wireless high definition (WiHD), European computer manufacturers association (ECMA), and IEEE 802.11ad wireless local area network (WLAN). To make use of the mmWave band economically, the key enabling technology is the usage of multiple antennas. By exploiting communication that is directional in nature, capacity of the link can be improved. Array architectures with compact form factors have been entrenched into portable devices due to smaller wavelengths making beamforming a potent solution (Sayeed and Raghavan 2007).

In MIMO systems, large bandwidths and high dimensionalities of antennas result in large multipath effects. Thus mmWave systems have meager MIMO channels comparatively (Sayeed and Raghavan 2007; Hariharan et al. 2008; Kutty and Sen 2015). Beamsteering at either end in a communication link ensures high directivity. In 5G implementations, beamforming antenna arrays will play an imperative role as mobile handsets can house numerous antennas at mmWave frequencies. These antennas can present multifarious beamforming capabilities besides higher directive gain. Aiming directly to the user groups results in boosting the signal to interference ratio (SIR), which can improve the mobile radio network. The interference faced by the wireless environment is reduced by the fine transmit beam, and ample signal power is sustained at the receiver. Appropriate antenna configurations are given to multiantenna beamformers to diminish RMS delay spread and elevate SNR and Rician gain factor owing to multipath effects (Raghavan et al. 2016). In this study, the authors have studied codebook-based MIMO beamforming, which aims to improve the spectral efficiency of the system with restricted feedback. Owing to the heavy feedback load, a new technique has been proposed that uses beamforming-based vectors that are pseudo-random in nature and threshold at different angles in the process of feedback. In this idea, there is no need to store a codebook by base station or the user. For each access route, a pseudo-random beamforming vector is generated by the base station. The user calculates the angle between the beamforming vector and channel state information (CSI) vector. The indicators of channel quality are sent back by the user if angle is less than a referenced angle-threshold; otherwise it doesn't send anything. The projected idea can condense the rate of feedback to a great extent with insignificant loss in the throughput. However, a proper threshold is required to achieve steadiness in the system throughput and feedback rate. In this chapter, an mmWave-based homogeneous network is considered that is multi-user and multicellular. Four hybrid beamforming techniques are compared with each other. This comparison is based on the assumption that complete CSI is known by the transmission points and can exchange it within each other. This gives the transmission points an account of intracell as well as intercell interference for devising the precoding matrices. In addition, interference and SNR level can influence the performance of these techniques, which themselves are affected by the number of users per cell, cell radius, and the number of streams per user. Distinctively, high spectral efficiency per user can be achieved by only some users per cell and petite cell radius. In Sun, Rappaport, and Shafi (2018), the authors have established the concepts related to beamforming to realize linear array architectures along with their radiation patterns. In Sun et al. (2018), the authors have elucidated mmWave beamforming and its various trends. The various trends include resolving the intricate issues of beam search process and its optimization, polarization diversity, hybrid beamforming, synchronized protocols for beamforming, vigorous adaptive beamforming, and beamforming in 3D. The above specified solutions should consider the requirements of less computational expenditure, adequate quality of service during the process of beamforming with delay, and power consumption constraints. In this study, the authors typify a low-complexity beamforming solution to detect the loss in SNR in received signal. It requires phase control in each antenna in the array to steer the beam in the direction of the dominant path at both the ends of the link. The loss in the SNR of received signal is negligible for large realizations of channel emphasizing directional beamforming as a way out for mm-MIMO-based systems.

## 10.2   EVADING THE EERGY CRUNCH: COGNITIVE RADIO

In 1999, cognitive radio (CR) was proposed by Joseph Mitola. Cognitive radio is an intelligent radio in which transceivers adjust themselves to new network condition. There are two important areas that are covered by CR: fully cognitive radio and spectrum sensing. Fully cognitive radio has the information of network parameters, and those parameters are modified where change is required for optimization of network. In spectrum sensing, researches notice location and time are imperative for proficient utilization of spectrum that is available. Cognitive radio is the best choice for detecting and allocation of vacant spectrum. This is termed as dynamic spectrum access. When the detection of vacant band radio occurs, spectrum pooling policy will be adopted where OFDMA sub bands will dwell in the actual bands. The various expressions those are involved in CR concept are

Spectrum sensing: To avail best frequency band cognitive terminal departs from existing to best presented frequency.

Spectrum sharing: Licensed user (PU) occupies sharing of resource, neighbor licensed users are not affected.

Sensing-based spectrum: Methods are used to see if the certified user is legitimate or not. Then CR user transmits the data.

### 10.2.1 Cognitive Radio Architecture

Researchers faced two problems before suggesting the design of CR. First, ambiguity in CR algorithms may be caused due to CR devices changing their transmission mode anytime. Second, change in bandwidth occurs as the users are mobile and are capable of moving about their location corresponding to different ranges in bandwidth. So it required an adjustment. These problems led to architecture of cognitive radio which is shown in Figure 10.1.

The architecture comprises of three basic, which are

- Decision making, which is done by cognitive unit on the basis of input provided.
- SDR (software-defined radio) unit, which has operating software to provide the operating environment.
- To find signal and user characteristics another component is required.

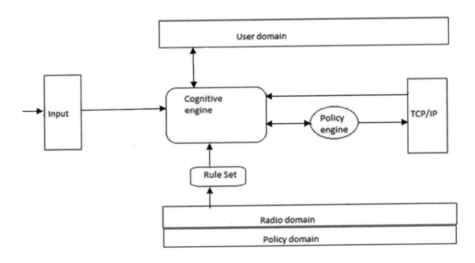

**FIGURE 10.1** CR architecture.

… Design and Simulation of New Beamforming

## 10.3 INTRODUCTION TO BEAMFORMING

The data storm triggered by the social media and mobile-phone uprising has made the communication field pervasive, indicating that the data rates will grow at an exponential rate. Using channels with wider bandwidths, greater data rates can be accomplished using mmWave cellular systems in fifth-generation (5G) (Xu, Li, and Shengqun 2012). The foremost challenge in dense networks, which needs attention, is intercell interference. Two rudimentary methods for regulating multi-user interference are power control and antenna array beamforming. In power control, the focus lies on enhancing the standard of fragile links by balancing the signal-to-interference-plus-noise ratio (SINR) evenly in a cell. Yet, signal quality can be improved by modifying the beam patterns while mitigating interference using array beamforming. Antenna array beamforming is more convincing for mmWave systems, as arrays of antennas are used at both transmitting and receiving ends to deliver array gain to recompense the loss in the first meter of transmission.

### 10.3.1 Beamforming System Approaches

Fundamentally, beamforming is a spatial filtering technique that uses a group of radiators to emit energy within a particular direction along the aperture. There is a huge margin of improvement in the transmit/receive gain as compared to that of unidirectional transmission/reception. Diversity gain combined with array gain plus interference extenuation is considered nowadays by deploying smart antennas, which in addition can escalate the capacity of a link. Electronic beamsteering is used to accomplish this by using a phased array, which has a specific geometric configuration and is a device with multiple radiative elements. The resultant of the fields given out by individual elements, gives spatial distribution of power that is labeled as array radiation pattern.

### 10.3.2 Beamforming Signals

In a broad sense, beamforming is compatible with simple candidate wave (CW) signals along with complex waveforms. Currently, CW is of growing interest for 5G as some disadvantages are experienced at mmWave bands for many of today's implementations. A vital precondition for beamforming analysis is the phase coherent signal. This indicates that all RF carriers are defined with a distinct and stable phase relationship. Therefore, to guide the main lobe to a desired direction, a preset delta phase is defined among the carriers, which can be seen in Figure 10.2.

To achieve phase coherence, varied generators are coupled by means of an ordinary orientation (i.e., 20 MHz). However, instability in immediate differential phase ("delta phase") in the RF signals is in the view of:

- Two synthesizers accompanied by phase noise
- A long synthesis chain for the RF output and "weak" coupling at 10 MHz
- Change in effective parameters of synthesizers due to temperature differences

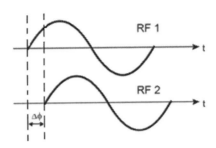

**FIGURE 10.2** Phase coherent signals with phase offset.

### 10.3.3 Types of Beamforming

1. Fixed-weight beamforming: To steer the main beam, antenna weights that are unvarying (amplitude/phase) are applied in digital or analog field to the antenna array elements.
2. Adaptive beamforming: Adaptive beamforming is a more authoritative technique than the fixed weight beamforming as it is able to acclimatize the pattern of RF radiation. This technique is predominantly suitable for mobile communications, which requires a weight vector to be updated recursively due to the unpredictable direction-of-arrival (DoA). Signal-processing algorithms that are well planned and structured are employed with adaptive arrays to constantly deduce the desired, multipath, and interfering signals. The main beam is vigorously directed in a direction that is looked-for by continuously updating the significant vectors. Implementation of beamforming for an application essentially relies on the proficiencies of the transmitter part and the receiver part, characteristics of the channel that is wireless, and bandwidth of the signal that is being transmitted.
3. Analog baseband beamforming: In the temporal domain, by using the elements that introduce time delay or shifting the phase of signal unvaryingly prior to radio frequency up conversion stage, antenna weights can be obtained.
4. Analog RF beamforming: Here the antenna weights are applied after the radio frequency up conversion stage.
5. Digital beamforming: Using a digital signal processor, the processing of beamforming is done, which gives greater flexibility in terms of degrees of freedoms to implement efficient beamforming algorithm. Here digital baseband acts as start off for the control signals.
6. Frequency domain beamforming: With the help of transform domain tools followed by inverse transforming, frequency domain is used to operate the signal.
7. Transmit beamforming: By the side of the transmitter, beamformers are used between the source and the components that radiate to govern the emitted electromagnetic field in 3D spaces formed on localization principle of receiver,

# Design and Simulation of New Beamforming 183

and this whole practice is named as transmit beamforming. The beamforming procedure usually depends on the potential of the transmitting end and the receiving end, attributes of radio channel, and the bandwidth of the transmitted signal.

8. Receive beamforming: A type of beamforming that is achieved between the range of antenna arrays and the receiver elements in the receivers to manage the relative spatial sensitivity of antenna to the signals is termed as receive beamforming. It usually entails the estimation of DoA, which is done by ascertaining the localized maxima of the spectrum.
9. Implicit (open loop) beamforming: It is often referred to as open loop beamforming. Depending on the channel reciprocity hypothesis, transmitter implements a channel sounding procedure. Due to multipath effects, phase-shift differences are regulated. Training sequences are directed to compute the modules of the steering matrix. However, due to the underlying reciprocity assumption, the problem is to precisely accomplish channel calibration.
10. Explicit (closed loop) beamforming: It is often referred as closed loop beamforming. In this technique, the channel is being assessed by the receiver, and the information is given to the transmitter. The three modes of explicit beamforming that are used universally are
    a. uncompressed mode where a steering matrix is used as a feedback, which is premeditated by the receiving component,
    b. compressed mode, where a compressed steering matrix is sent by the receiver to the transmitting component, and
    c. The channel state information (CSI), which runs in a feedback mode and raw channel estimates are sent by the receiver to the transmitter in order to calculate the steering matrix.

## 10.3.4 IEEE 802.11 AD Beamforming Protocol

The three phases of beamforming protocol framed by IEEE 802.11ad are given as:

1. Sector level sweep (SLS) phase: Here, quasi-omni patterns are used by transmit and receive antennas to choose the paramount transmitting and receiving sector.
2. Beam refinement phase (BRP): BRP comes into action once best sector pair is acknowledged. The transmitting and receiving arrays of antenna are guided to sort the beam patterns that have narrow beam widths.
3. Beam tracking (BT) phase: This phase is discretionary and can be used to fiddle with channel changes during data transmission. Training fields are appended by data packets to perform beam tracking. Automatic gain control (AGC) field is used within which AGC gain is calculated by changing the direction of set of beams sequentially for a receiver. For channel estimation (CE), each beam is allocated with training field associated with CE chain that improves the precision of tap delay approximation.

## 10.4 SYSTEM MODEL

A 5G cellular system is taken into consideration with PUs and SUs. A macro base station (MBS) serves PU at each $K^{th}$ position. The total number of users distributed in the cell is given by m ∈ M, with SUs following Gaussian distribution for ample variables. $h_{PU}$ signifies channel between MBS and the $K^{th}$ position PU and $h_{SU}$ is the channel between the $K^{th}$ PU and the $m^{th}$ SU with the assumption that MBS and PU have information about CSI. The link formed amid MBS and PU is backhaul link, and joining PU with channel is the access link. The signal that is received at the $K^{th}$ PU in the initial time notch is conferred by

$$Y_{mk} = \sum_{m=1}^{M}\sum_{K=1}^{N} h_{PU}^{m} X + n_m, \qquad (10.1)$$

where $n_m$ is AGWN noise, and $K = 1, 2, 3, 4, 5 \ldots N$ entails $N$ positions for PU. The PU at the $K^{th}$ position undertakes the acquired symbol linearly by exploiting weighting matrix "$w_{mk}$" and sends out: $T_{mk} = Y_{mk} v_{mk}$, in the succeeding time notch to the access link. $v_{mk}$ Is the beamforming vector with the aim of disseminating the signal in the necessary direction and restrains any interference amid PUs and SUs. Therefore, the signal output obtained in the succeedingtime slot is as:

$$Z_{mk} = \sum_{m=1}^{M}\sum_{k=1}^{N} h_{SU}^{m} T_{mk} + n_m. \qquad (10.2)$$

The total power restraint at $K^{th}$ PU, $m^{th}$ SU is specified by:

$$\sum_{k=1}^{Kmax} E\{E_r(Y_{mk}Y_{mk}^*)\} \le P_{PU}^k \qquad (10.3)$$

$$\sum_{m=1}^{m} E\{E_r(Z_{mk}Z_{mk}^*)\} \le P_{SU}^m. \qquad (10.4)$$

To define the optimization problem, contour is set, and overall channel capacities across it are defined, i.e., PU and SU. In totality, throughput of the system is

$$C_{Tout} = \sum_{m=1}^{M}\sum_{k=1}^{N} R_{m,k}(V_m^-), \qquad (10.5)$$

where $R_{m,k}(V_m^-) = \sum_{m=1}^{M}\sum_{n=1}^{k}(C_{PU}, C_{PU})$

Here $C_{PU}$ and $C_{SU}$ represent the throughput of PU and SUs, respectively, and are given as:

$$C_{PU} = \log_2\left(1 + \frac{h_{PU}^m (h_{PU}^m)^H Y_{mk}(Y_{mk})^H V_{PU}^k (V_{PU}^k)^H}{n_m}\right) \qquad (10.6)$$

# Design and Simulation of New Beamforming

$$C_{SU} = \log_2\left(1 + \frac{h_{SU}^m \left(h_{SU}^m\right)^H Z_{mk} \left(Z_{mk}\right)^H V_{SU}^k \left(V_{SU}^k\right)^H}{n_m}\right) \quad (10.7)$$

Here beamforming vector $V_k$ is based on numerous antenna receivers, which is defined for both PU and SUs as

$$V_{PU}^k = \frac{\left(I_{Nj} + \sum_{m=1}^{M}\sum_{k=1}^{6N} \left(Y_{mk}\right)^H h_{PU}^m \left(h_{PU}^m\right)^H \left(Y_{mk}\right)^H\right)^{-1}}{\left(I_{Nj} + \sum_{m=1}^{M}\sum_{k=1}^{N} \left(Y_{mk}\right)^H h_{PU}^m \left(h_{PU}^m\right)^H \left(Y_{mk}\right)^H\right)^{-1}} \quad (10.8)$$

$$V_{SU}^k = \frac{\left(I_{Nr} + \sum_{m=1}^{M}\sum_{k=1}^{N} \left(Z_{mk}\right)^H h_{SU}^m \left(h_{SU}^m\right)^H \left(Z_{mk}\right)^H\right)^{-1}}{\left(I_{Nr} + \sum_{m=1}^{M}\sum_{k=1}^{N} \left(Z_{mk}\right)^H h_{SU}^m \left(h_{SU}^m\right)^H \left(Z_{mk}\right)^H\right)^{-1}}. \quad (10.9)$$

$N_j, N_r$ are antennas at the receiving side at PU and SU. The overall power consumed for PU and SUs is as under:

$$P_{PU} = \sum_{m=1}^{M}\sum_{k=1}^{N} p_{PU} \quad (10.10)$$

$$P_{SU} = \sum_{m=1}^{M}\sum_{k=1}^{N} p_{SU}. \quad (10.11)$$

$p_{PU}$ and $p_{SU}$ are subject to $\sum p_{PU} \leq 1$ and $\sum p_{CU} \leq 1$ with $p_{PU}$ and $p_{SU}$ being the power allocation coefficient intended for PU and SU.

As a result, power consumed entirety in the given outline is

$$P_{Total} = p_{PU} + p_{SU}. \quad (10.12)$$

The energy efficiency is given as:

$$EE = \frac{C_{Tout}}{P_{Total}}. \quad (10.13)$$

The system model is shown in Figure 10.3.

## 10.5 OPTIMIZATION PROBLEM

The augmentation of the problem is devised as under:

$$\max p_{PU}, p_{SU}, C_{Tout} \quad (10.14)$$

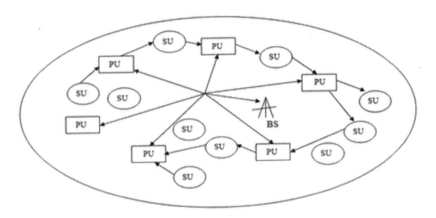

**FIGURE 10.3** System model.

$$\max EE, p_{PU}, p_{SU}, \tag{10.15}$$

subjected to the following constraints:
F1:

$$\sum_{m=1}^{M}\sum_{k=1}^{N} p_{PU} \leq P_{PU} \tag{10.16}$$

$$\sum_{m=1}^{M}\sum_{k=1}^{N} p_{SU} \leq P_{SU} \tag{10.17}$$

$$p_{PU} \leq 1 \tag{10.18}$$

$$. p_{SU} \leq 1 \tag{10.19}$$

$$\sum_{m=1}^{M}\sum_{k=1}^{N} p_{PU} h_{PU} \leq J_{th}^{PU} \tag{10.20}$$

$$\sum_{m=1}^{M}\sum_{k=1}^{N} p_{SU} h_{SU} \leq J_{th}^{SU}, \tag{10.21}$$

where, $C_{Tout} = \sum_{m=1}^{M}\sum_{k=1}^{N} R_{m,k}\left(V_m^-\right)$ is the aggregate throughput of the system, and $J_{th}^{PU}$ and $J_{th}^{SU}$ are interference threshold for $K^{th}$PU and $m^{th}$ SU.

## 10.6 FRAMEWORK FOR OPTIMAL SOLUTION

F1 being a problem of Fractional Programming (FP) has a fractional nonconcave function and constraints that are nonlinear, making it tricky to uncover the most favorable solution. Therefore, F1 is transformed in a diminution problem, which is defined as:

F2:

$$\max\left(C_{Tout} - \lambda\left(p_{PU} + p_{SU} + \mu_{mk}C_{Tout}\right)\right) \quad (10.22)$$

S.T.

$$\left(C_{PU}, C_{SU}\right) \quad (10.23)$$

Where, $\lambda = (\lambda_{1k}, \lambda_{2k}, \lambda_{3k}, \ldots \ldots \lambda_{mk})$ is a vector known as Lagrangian multiplier vector, and $\mu_{mk}$ is the Lagrangian multiplier related to $m^{th}$ constant on the two links. The Lagrangian function is articulated as:

$$L = \sum_{m=1}^{M}\sum_{k=1}^{6}\left[\log_2\left(1\frac{h_{PU}^m\left(h_{PU}^m\right)^H Y_{mk}\left(Y_{mk}\right)^H V_{PU}^k\left(V_{PU}^k\right)^H}{n_m}\right)\right.$$
$$+\log_2\left(1+\frac{h_{SU}^m\left(h_{SU}^m\right)^H Z_{mk}\left(Z_{mk}\right)^H V_{SU}^k\left(V_{SU}^k\right)^H}{n_m}\right) \quad (10.24)$$
$$\left.+\mu_{mk}P_{Total}\sum\left(p_{PU}-P_{PU}\right)\sum\left(p_{SU}-P_{SU}\right)\right]$$

Employing KKT, the problem's concavity is resolved to arrive at the finest solution.

Solving it after: $\dfrac{\partial L}{\partial p_{PU}} = 0$ and $\dfrac{\partial L}{\partial p_{SU}} = 0$ by including (Karush Kuhn Tucker) KKT constraint an optimal solution is derived as:

$$p_{PU} = \mu_{mk}\left[\sum_{m=1}^{M}\sum_{k=1}^{N}\frac{\left(P_{Total}\mu_{mk} + V_{PU}^k - n_m\right)}{\sum P_{PU}}\right] \quad (10.25)$$

$$p_{SU} = \mu_{mk}\left[\sum_{m=1}^{M}\sum_{k=1}^{N}\frac{\left(P_{Total}\mu_{mk} + V_{SU}^k - n_m\right)}{\sum P_{SU}}\right] \quad (10.26)$$

## 10.7 RESULTS

Beamforming will play a vital part for the enactment of networks used in next generation. Countless 5G techniques are in the zone of enduring exploration. This work objects at reviewing the future generation networks over future wireless networks. The prime emphasis is to enhance the spectrum efficiency, increase network coverage, and diminish end-to-end delay. The implementation deals with divesting the traffic at base station (BS) so that the users can communicate without any deviation. Multiple antennas at transmitter and receiver are used with directing the beam in a particular direction. The implementation has been conceded in Matlab software, and simulation results are deliberated.

### CASE 1  WHEN ONLY ONE PRIMARY USER IS PRESENT AND THE REST OF THE SLOTS ARE ABSENT (P AAA A).

Figure 10.4 shows the plot of cdf versus spectral efficiency where a lone PU is existing and all others are missing.

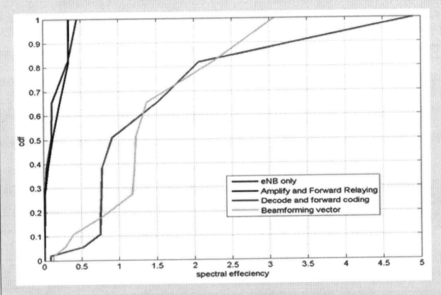

**FIGURE 10.4**   Plot of cdf vs. spectral efficiency for case 1.

From the figure, it can be seen that eNB (enhanced node B) has minimum spectral efficiency. The spectral efficiency increases when we apply amplify and forward (AF) relaying. Decode and forward (DF) coding further enhances the spectral efficiency. The beamforming vector (BFV) employed provides the highest spectral efficiency at various percentiles of cdf. At the 5th percentile of cdf, DF coding has a maximum spectral efficiency of 0.5 bits/sec/Hz, and that provided by BFV is 0.3 bits/sec/Hz. At the 50th percentile of cdf, maximum spectral efficiency provided by DF coding is 0.9 bits/sec/Hz, and that provided by BFV is 1.2 bits/sec/Hz. The percentage increase in spectral efficiency at the 50th percentile is 50%.

# Design and Simulation of New Beamforming

## CASE 2   WHEN ONE PRIMARY USER AND ONE SECONDARY USER ARE PRESENT AND THREE SLOTS ARE LEFT BEHIND VACANT (P PAA A).

Figure 10.5 shows the plot of cdf versus spectral efficiency where one PU and one SU are present.

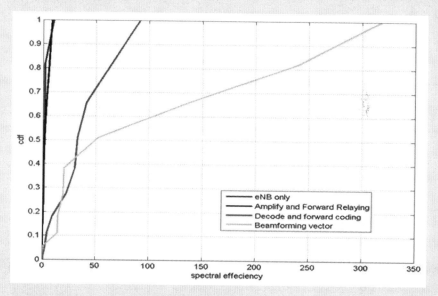

**FIGURE 10.5**   Plot of cdf vs. spectral efficiency for case 2.

Spectrum is being sensed by the SU and occupies the available empty slot leaving behind three vacant slots. From the figure, it can be seen that eNB (enhanced node B) has minimum spectral efficiency. The spectral efficiency increases when we apply amplify and forward (AF) relaying. Decode and forward (DF) coding further enhances the spectral efficiency. The beamforming vector (BFV) employed provides the highest spectral efficiency at various percentiles of cdf. At the 5th percentile of cdf, DF coding has a maximum spectral efficiency of 10 bits/sec/Hz, and that provided by BFV is 15 bits/sec/Hz. At the 50th percentile of cdf, maximum spectral efficiency provided by DF coding is 35 bits/sec/Hz, and that provided by BFV is 50 bits/sec/Hz. The percentage increase in spectral efficiency at the 50th percentile is 42.85%.

## CASE 3 : WHEN ONE PRIMARY USER AND TWO SECONDARY USERS ARE PRESENT AND TWO SLOTS ARE LEFT BEHIND VACANT (P PP A A).

Figure 10.6 shows the plot of cdf versus spectral efficiency where one PU and two SUs are present.

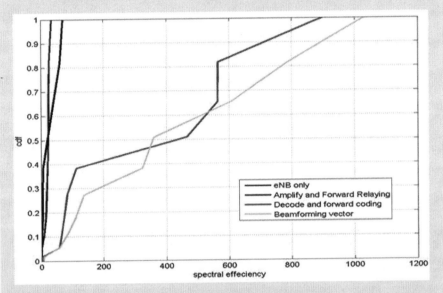

**FIGURE 10.6** Plot of cdf vs. spectral efficiency for case 3.

Spectrum is being sensed by the SUs, and they occupy the available empty slot leaving behind two vacant slots. From the figure, it can be seen that eNB (enhanced node B) has minimum spectral efficiency. The spectral efficiency increases when we apply amplify and forward (AF) relaying. Decode and forward (DF) coding further enhances the spectral efficiency. The beamforming vector (BFV) employed provides the highest spectral efficiency at various percentiles of cdf. At the 5th percentile of cdf, DF coding has a maximum spectral efficiency of 70 bits/sec/Hz, and that provided by BFV is 90 bits/sec/Hz. At the 50th percentile of cdf, maximum spectral efficiency provided by DF coding is 450 bits/sec/Hz, and that provided by BFV is 370 bits/sec/Hz. The percentage increase in spectral efficiency at the 50th percentile is 21.62%.

# Design and Simulation of New Beamforming 191

## CASE 4 : WHEN ONE PRIMARY USER AND THREE SECONDARY USERS ARE PRESENT AND ONE SLOT IS LEFT VACANT (P PP P A).

Figure 10.7 shows the plot of cdf versus spectral efficiency where one PU and three SU's are present.

Spectrum is being sensed by the SUs and they occupy the available empty

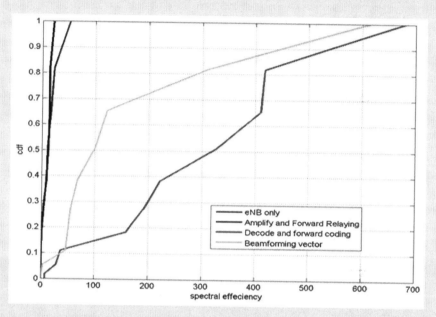

**FIGURE 10.7** Plot of cdf vs. spectral efficiency for case 4.

slot, leaving behind a vacant slot. From the figure, it can be seen that eNB (enhanced node B) has minimum spectral efficiency. The spectral efficiency increases when we apply amplify and forward (AF) relaying. Decode and forward (DF) coding further enhances the spectral efficiency. The beamforming vector (BFV) employed provides the highest spectral efficiency at various percentiles of cdf. At the 5th percentile of cdf, DF coding has a maximum spectral efficiency of 35 bits/sec/Hz, and that provided by BFV is 25 bits/sec/Hz. At the 50th percentile of cdf, maximum spectral efficiency provided by DF coding is 220 bits/sec/Hz, and that provided by BFV is 100 bits/sec/Hz. The percentage increase in spectral efficiency at the 50th percentile is 120%.

## CASE 5  WHEN NO SLOT IS LEFT VACANT (P PP P P).

Figure 10.8 shows the plot of cdf versus spectral efficiency where one PU and three SUs are present.

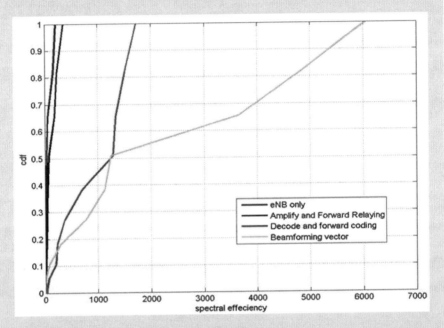

**FIGURE 10.8**  Plot of cdf vs. spectral efficiency for case 5.

Spectrum is being sensed by the SUs, and they occupy all the available slots. From the figure, it can be seen that eNB (enhanced node B) has minimum spectral efficiency. The spectral efficiency increases when we apply amplify and forward (AF) relaying. Decode and forward (DF) coding further enhances the spectral efficiency. The beamforming vector (BFV) employed provides the highest spectral efficiency at various percentiles of cdf. At the 5th percentile of cdf, DF coding has a maximum spectral efficiency of 200 bits/sec/Hz, and that provided by BFV is 300 bits/sec/Hz. At the 50th percentile of cdf, maximum spectral efficiency provided by DF coding is 700 bits/sec/Hz, and that provided by BFV is 1500 bits/sec/Hz. The percentage increase in spectral efficiency at the 50th percentile is 114.28%.

The results that have been achieved are summarized in Table 10.1.

### TABLE 10.1
### Summary of the Derived Results for Spectral Efficiency

| Case | 5th Percentile | 50th Percentile |
|---|---|---|
| PAAAA | 66.6% | 50% |
| PPAAA | 50% | 42.85% |
| PPPAA | 28.57% | 21.62% |
| PPPPA | 40% | 120% |
| PPPPP | 50% | 114.28% |

## 10.8 CONCLUSION

The work presented in this chapter has aimed to improve the spectral efficiency of the MIMO system by using beamforming and CR algorithm. Capacity enhancement is the key focus here besides other parameters. Mathematical model has been provided to support the results acquired using the MATLAB software. It is seen that the spectral efficiency obtained using beamforming is better as compared to the enB, AF relaying and DF coding. Also when only one PU is present, the efficiency at the node is 50%. If we increase the number of users along with beamforming and CR algorithm, the efficiency ascends to 114.28%. The required results that are close to Shannon limit are attained when CR algorithm is combined with beamforming.

The future scope of this chapter is that the massive MIMO (m-MIMO) along with NOMA can be used to apportion the resources and to solve the optimization problem and see its effect on the spectral efficiency of various users.

## REFERENCES

Andrews, Jeffrey G, Stefano Buzzi, Wan Choi, Stephen Hanly, Angel Lozano, Anthony CK Soong, and Jianzhong Charlie Zhang. 2014. "What Will 5G Be?" *IEEE JSAC Special Issue on 5G Wireless Communication Systems* 32 (6): 1065–1082. Ieeexplore.Ieee.Org.

Chen, S, and J Zhao. 2014. "The Requirements, Challenges, and Technologies for 5G of Terrestrial Mobile Telecommunication." *IEEE Communications Magazine* 52 (5): 36–43. Ieeexplore.Ieee.Org.

De, P, and S Singh, 2016. "Journey of Mobile Generation and Cognitive Radio Technology in 5G." *Journal of Mobile Network Communications* 52 (5): 36–43. Papers.Ssrn.Com.

Hariharan, Gautham, Vasanthan Raghavan, and Akbar M Sayeed. 2008. "Capacity of Sparse Wideband Channels with Partial Channel Feedback." *European Transactions on Telecommunications* 19 (4): 475–493.

Ji, B, K Song, C Li, W Zhu, and L Yang. 2018. "Energy Harvest and Information Transmission Design in Internet-of-Things Wireless Communication Systems." *AEU–International Journal of Electronics, and Communications* 87 (April): 124–127.

Kumar, Railmajra Puneet, J K Sharma, S B S Nagar, and M Singh. 2012. "5G Technology of Mobile Communication." *International Journal of Electronics and Computer Science Engineering* 2 (4):1265–1275.

Kutty, S, and D Sen. 2015. "Beamforming for Millimeter Wave Communications: An Inclusive Survey." *IEEE Communications Surveys & Tutorials* 18 (2): 949–973, Ieeexplore.Ieee.Org.

Mitra, RN, DP Agrawal, 2015. "5G Mobile Technology: A Survey." *ICT Express* 1 (3): 132–137, Amsterdam: Elsevier.

Modi, Hardik, Shobhit K Patel, and Asvin Gohil. 2013. "5G Technology of Mobile Communication: A Survey Liquid Metamaterial Antenna View Project Numerical Investigation of Liquid Metamaterial-Based Superstrate Microstrip Radiating Structure View Project 5G Technology of Mobile Communication: A Survey." *Ieeexplore.Ieee.Org*. doi:10.1109/ISSP.2013.6526920.

Raghavan, Vasanthan, and Akbar M. Sayeed. 2011. "Sublinear Capacity Scaling Laws for Sparse MIMO Channels Cognitive MIMO View Project Distributed Sensing and Communication in Wireless Sensor Networks View Project Vasanthan Raghavan Sub-Linear Capacity Scaling Laws for Sparse MIMO Channels." *Ieeexplore.Ieee.Org*. doi:10.1109/TIT.2010.2090255.

Raghavan, Vasanthan, Juergen Cezanne, Sundar Subramanian, Ashwin Sampath, and Ozge Koymen. 2016. "Beamforming Tradeoffs for Initial UE Discovery in Millimeter-Wave MIMO Systems." *IEEE Journal of Selected Topics in Signal Processing* 10 (3): 543–549. *Ieeexplore.Ieee.Org*.

Rappaport, T S, R W Heath Jr, R C Daniels, and J N Murdock. 2015. *Millimeter Wave Wireless Communications*. New York: Pearson.

Sayeed, Akbar M, and Vasanthan Raghavan. 2007. "Maximizing MIMO Capacity in Sparse Multipath with Reconfigurable Antenna Arrays." *IEEE Journal of Selected Topics in Signal Processing* 1 (1): 156–166. doi:10.1109/JSTSP.2007.897057.

Sheikh, Javaid A., Mehboob-ul Amin, Shabir A. Parah, and G. Mohiuddin Bhat. 2019. "Resource Allocation in Co-Operative Relay Networks for IOT Driven Broadband Multimedia Services." In *Handbook of Multimedia Information Security: Techniques and Applications*, 703–721. Springer International Publishing, USA. doi:10.1007/978-3-030-15887-3_34

Sun, Shu, Theodore S Rappaport, and Mansoor Shafi. 2018. "Hybrid Beamforming for 5G Millimeter-Wave Multi-Cell Networks." *IEEE INFOCOM 2018 – IEEE Conference on Computer Communications Workshops (INFOCOM WKSHPS)*, Honolulu, HI, pp. 589–596. *Ieeexplore.Ieee.Org*.

Tudzarov, A, and T Janevski. 2011. "Functional Architecture for 5G Mobile Networks." *International Journal of Advanced Science* 32 (July): 65–78. *Academia.Edu*.

Tutuncuoglu, Kaya, Aylin Yener, and Sennur Ulukus. 2015. "Optimum Policies for an Energy Harvesting Transmitter under Energy Storage Losses." *IEEE Journal on Selected Areas in Communications* 33 (3): 467–481.

White, P, and GL Reil. 2016. "Millimeter-Wave Beamforming: Antenna Array Design Choices & Characterization White Paper." *Rohde-Schwarz-Ad.Com*. https://doi.org/10.2016.

Xu, Dong, Ying Li, and Wei Shengqun. 2012. " Basic Theories in Cognitive Wireless Networks." *China Science Bulletin* 57 (28–29): 3698–3704. doi:10.1007/s11434-012-5102-6.

Yarrabothu, R S, J Mohan, G Vadlamudi, and G Vadlamudi. 2015. "A Survey Paper on 5G Cellular Technologies – Technical & Social Challenges." *International Journal of Emerging Trands in Electrical and Electronics* 11 (2): 98–100. *Researchgate.Net*.

# 11 Image Transmission Analysis Using MIMO-OFDM Systems

*Akanksha Sharma, Lavish Kansal, Gurjot Singh Gaba*
Lovely Professional University, India
*Mohamed Mounir*
El Gazeera High Institute for Engineering and Technology, Egypt

## CONTENTS

11.1 Introduction .................................................................................................. 195
11.2 MIMO Systems ............................................................................................. 197
    11.2.1 Spatial Multiplexing (SM) ................................................................ 197
    11.2.2 Beamforming ..................................................................................... 197
    11.2.3 Spatial Diversity (SD) ....................................................................... 198
        11.2.3.1 Diversity Combining Techniques ..................................... 198
        11.2.3.2 Selection Combining (SC) ................................................ 199
        11.2.3.3 Maximal Ratio Combining (MRC) ................................... 199
        11.2.3.4 Equal Gain Combining (EGC) ......................................... 199
11.3 Simulation Results ........................................................................................ 199
    11.3.1 Beamforming ..................................................................................... 200
    11.3.2 Maximal Ratio Combining ............................................................... 201
    11.3.3 Selection Combining ......................................................................... 203
11.4 Conclusion .................................................................................................... 208
References ............................................................................................................. 208

## 11.1 INTRODUCTION

Image transmission in the arena of wireless communication is considered a big challenge. Today, a large amount of focus is on the quality transmission of visual images rather than the increased data rate and reliability through wireless means. Image transmission often gets depraved due to difficulties like cochannel interference, distortions in signals, limitations in bandwidth, and channels that vary according to time causes much more decrement in image transmission performance. So, various improvement factors are added, and perfect visual images can be seen through the M-PSK scheme proposed below. MIMO-OFDM overcomes the challenges that are being faced by all future wireless communications by providing full data and quality of service. It protects the users from interference by other users and having high spectral efficiency and reliability through various multiplexing and antenna diversity gains (Banelli and Cacopardi 2000). A coherent combination is used to realize diversity gain and array gain, and interference cancellation is performed by use of multiple antennas, i.e., MIMO. The use of a transmit diversity scheme, i.e., beamforming, along with various other diversity-combining techniques added in MIMO, are used in this chapter are maximal ratio combining (MRC) and selection combining (SC).

Study of transmission of the image was done over many years and showed a great impact on the current wireless scenario. Embedded image transmission is done over MIMO channels, which guarantees the quality of service (QoS) and optimizing transmit power (Chen and Wang 2002). Transmit antenna selection based on MRC is used in MIMO systems with enhancement of orthogonal space–time block code (OSTBC). The analysis is being done under Rician fading channel for the combination of different transmit and receive channels in bit error rate (BER) (Bighieri et al. 2007). Use of MIMO for providing large information data and high rates along with high communication energy and bandwidth is provided for high-quality image transmission and communication. Various MIMO communication techniques like ODQ, BST, OBST, RO, and CO are used to improve image quality for transmission over 4G wireless networks (Li et al. 2007). The study of large MIMO-CDMA systems is done using Rayleigh fading channels. Evaluation of the system has been done by MSSIM index, which is used to compare the quality of images that are received by keeping the original image as a reference (Maré and Maharaj 2008). Transmission of images progressively in wireless communication systems is done by combining three things, i.e., joint source-channel coding, space-time coding (STC), and orthogonal frequency division multiplexing (OFDM). The adaptive modulation scheme is proposed to pick constellation size that represents the best quality of image along with average SNR (Alex et al. 2008). WiMAX, when compared with Wi-Fi technology by adding the issues like security, seamless handover, and QoS, has more security and reliability compared to Wi-Fi (Sezginer et al. 2009). Alamouti STC and MRC are combined at the receiver side. Both MIMO choices with adaptive modulation and coding by deciding the SNR edges of working areas is done in a legitimate way (Kobeissi et al. 2009; Mehlführer et al. 2009). Results showed Matrix A performs superior to Matrix B, and throughput is augmented by consolidating MIMO with adaptive modulation and coding (Tan et al. 2010; Murty et al. 2012). Performance analysis using noncoherent modulations is done in SC in double rice fading channels, which result in better BER performance and high usage of such type of techniques with the proposed method (Tilwari and Kushwah 2013). In MRC the weights must be picked as about the individual signals level for maximizing the combined carrier-to-noise ratio (CNR). The connected weighting to the diversity branches must be balanced bestowing to the SNR (Hemalatha et al. 2013).

The main goal of this chapter is to investigate the performance of high-quality image transmission through different M-PSK (QPSK, 16-PSK, 64-PSK) schemes used with various SNR values in dB. MIMO wireless and its areas are discussed in Section 11.2. Section 11.3 includes a brief introduction to diversity combining techniques. Finally, conclusions are incorporated in Section 11.4.

# Image Transmission Analysis Using MIMO-OFDM Systems

## 11.2 MIMO SYSTEMS

MIMO makes use of space dimension and multipath to improve wireless system capacity range, reliability, and efficiency. There can be much of MIMO configurations like 2*2 MIMO where two antennas are used to transmit the signal, and two antennas are used to receive. Here multiple antennas are used to send multiple parallel signals as shown in Figure 11.1. First, beamforming is performed to maintain the beam on the side so that check on the particular signal can be retained excluding other signals. Same data to transmitter and receiver is being sent so that they can experience different fading, which is popularly known as spatial diversity (Wang and Chu 2016; Yang et al. 2016).

### 11.2.1 Spatial Multiplexing (SM)

In spatial multiplexing, the same bandwidth is provided along with power, where several independent subchannels are drawn having better scattering in the environment. Improvement in bit rate is done by transmitting data at a higher rate, and no expansion in bandwidth is done here in the same frequency channel. It increases its capacity by introducing layered architectures in it, i.e., V-BLAST. The technique named as the V-BLAST technology is mainly used to enhance the system's spectral efficiency and was developed by Bell Laboratories (Divya et al. 2015). It organizes channel limit at high SNR and involves complexity and costs at both transmitter and receiver side (Zerrouki and Feham 2014).

### 11.2.2 Beamforming

Beamforming provides a better and destined way to the channels rather than observing them and guessing what is being received at the receiver. It chooses a particular beam and uses its much of the energy in maintaining it, which is to be reflected to the receiver for confirmation of received signal as shown in Figure 11.2. Better

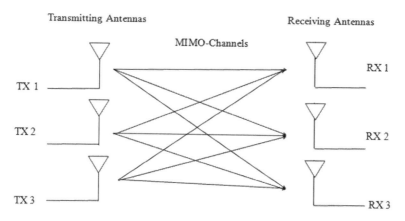

**FIGURE 11.1** Block diagram of MIMO system.

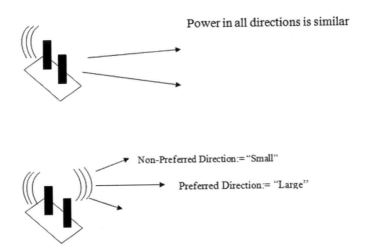

**FIGURE 11.2** Beamforming (with and without).

connections are to be maintained from both transmitter and receiver side. Usually, a closely spaced antenna is used to transmit the phased signal so that power radiated toward a particular beam can be maximized and excluding the interference from other directions. Beamforming permits better robustness and less sensitivity to the other beams by increasing the transmission speed. Improvement in gain and range is done and provides a maximum advantage in nonsymmetric systems.

### 11.2.3 Spatial Diversity (SD)

SD is an improved topic, where independent or different channels are being sent to avoid fading and interference in communication systems. If we send different copies of data to the receiver, then the total addition of fade suffered by the individual copy of data will be different and results in providing us with the better copy of data that is less faded among all. It increases the opportunity of getting the proper data that is being sent from transmitter side due to changes in quality of the channel, the stream of data can be lost or corrupted that cannot be recovered by the receiver. Here links do not depend on each other for better reception quality because if one is less faded, it goes to receiver fast compared to the one being more faded. For Alamouti code in a symbol period, only one independent symbol is being sent as can be seen from Figure 11.3.

#### 11.2.3.1 Diversity Combining Techniques

The diversity technique mentioned above tells that different signals or beams are received at receiver side having different and independent fading. This means that if one signal or beam undergoes more fade, then another beam will have less fade and stronger signal. So, to select the SNR by choosing more than one path is provided by appropriate diversity combining technique.

# Image Transmission Analysis Using MIMO-OFDM Systems 199

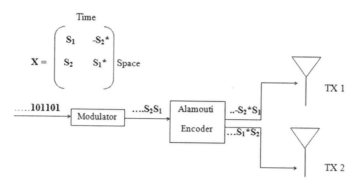

**FIGURE 11.3** Alamouti STBC.

### 11.2.3.2 Selection Combining (SC)

It follows the principle of selecting the best signal among other signals received from the receiver at the receiving end as shown in Figure 11.4. We can say in this combining, the receiver receives the information from many diversity branches, and the branch having a high signal to noise ratio gives proper information at the receiver side. It is tentatively demonstrated that the execution change accomplished by the selection combining is simply a little lower than execution enhanced accomplished by a perfect MRC.

### 11.2.3.3 Maximal Ratio Combining (MRC)

In maximal ratio combining, the diversity branches are weighted by their complex fading gains and combined at the receiver. Here, all values of signals are individually calculated or weighted according to their noise power ratios and then summed as shown in Figure 11.5. The most convenient thing is that this combining technique produces the adequate amount of SNR rather than other signals. Reduction in fading is done better in this case.

### 11.2.3.4 Equal Gain Combining (EGC)

MRC is the best-assorted qualities consolidating; the plan, however, requires an extremely costly outline at receiver circuit to modify the gain in each branch. It needs a proper pattern for following the mind for complex fading, which is extremely hard to accomplish for all intents and purposes. However, by utilizing a basic phase lock summing circuit, it is not much difficult to execute an equal gain combining. EGC is somewhat looking like MRC in which the only dissimilarity is that the diversity branches are not weighted in this case, as can be seen from Figure 11.6.

## 11.3 SIMULATION RESULTS

This chapter involves the performance of the image transmission in the best-possible way. The M-PSK (QPSK, 16-PSK, 64-PSK) modulation schemes are used, and accordingly BER is compared. This type of image transmission helps in sending a

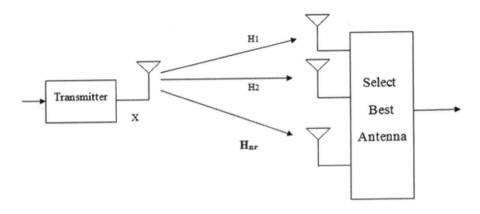

**FIGURE 11.4** Selection combining.

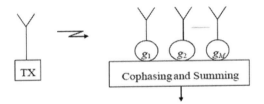

**FIGURE 11.5** Maximal ratio combining.

**FIGURE 11.6** Equal gain combining.

better-quality image along with better performance having various transmitters and receivers. Diversity combining techniques like MRC, SC, etc. is used along with beamforming of different values. Various values of SNR had been taken, i.e. (40, 50, 60) dB to get specified results. The original image, which has been taken as input, is shown in Figure 11.7.

### 11.3.1 BEAMFORMING

Analysis of beamforming has been done in this over various modulation rates, i.e., QPSK, 16-PSK, and 64-PSK, with different transmitter values, i.e., Ntx = 2, 3, 4, and

Image Transmission Analysis Using MIMO-OFDM Systems

**FIGURE 11.7** Original image.

various SNR values, i.e. (40, 50, 60), and results are plotted. Various images are received based on their performances and modulation rates provided to them as shown in Figure 11.8a–c. The plot in Figure 11.9 shows BER versus SNR performance of the image received by using various transmitters and with every modulation technique used. Overall results from the simulations shown tell that BER) is providing the best result in the case of QPSK and least is in the case of 64-PSK.

### 11.3.2 MAXIMAL RATIO COMBINING

Analysis of MRC has been done in this over various modulation rates, i.e., QPSK, 16-PSK, and 64-PSK with different transmitter values, i.e., Nrx = 2, 3, 4, and various

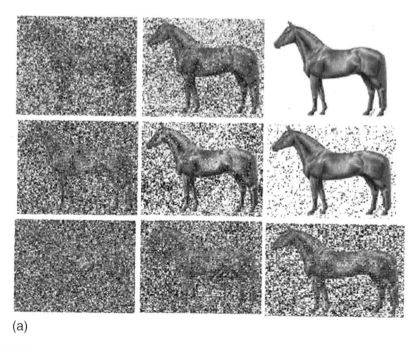

(a)

**FIGURE 11.8** Received images for MIMO-OFDM with beamforming for (a) Ntx = 2,

(*Continued*)

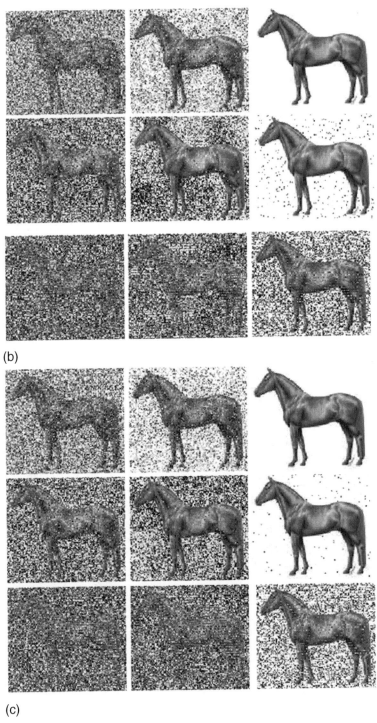

**FIGURE 11.8 (CONTINUED)**  (b) Ntx = 3, and (c) Ntx = 4.

# Image Transmission Analysis Using MIMO-OFDM Systems

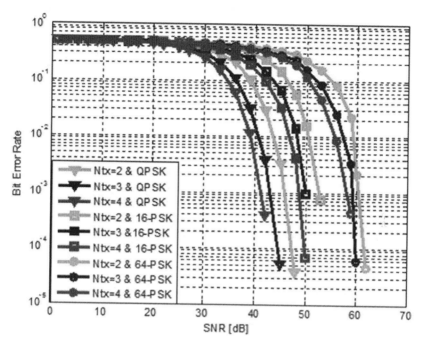

**FIGURE 11.9** BER vs. SNR for MIMO-OFDM with beamforming with modulation rates of 4-, 16-, 64-PSK.

SNR values, i.e. (40, 50, 60), and results are plotted. Various images are received based on their performances and modulation rates provided to them as shown in Figure 11.10a–c. The plot in Figure 11.11 shows BER versus SNR performance of the image received by using various receivers and with every modulation technique used. Overall results from the simulations shown tell that BER is providing the best result in the case of QPSK and least is in case of 64-PSK.

### 11.3.3 Selection Combining

Analysis of MRC has been done in this over various modulation rates, i.e., QPSK, 16-PSK, and 64-PSK with different transmitter values, i.e., Nrx = 2, 3, 4, and various SNR values, i.e. (40, 50, 60), and results are plotted. Various images are received based on their performances and modulation rates provided to them as shown in Figure 11.12. The plot in Figure 11.13 shows BER versus SNR performance of the image received by using various receivers and with every modulation technique used. Overall results from the simulations shown tell that BER is providing the best result in the case of QPSK and least is in case of 64-PSK.

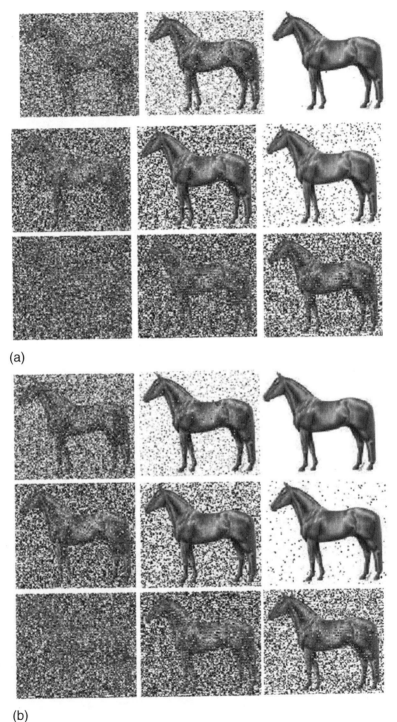

**FIGURE 11.10** Received images for MIMO-OFDM with MRC for (a) Nrx = 2, (b) Nrx = 3, and *(Continued)*

# Image Transmission Analysis Using MIMO-OFDM Systems

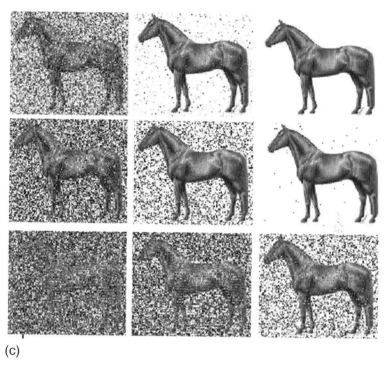

(c)

**FIGURE 11.10 (CONTINUED)**  (c) $N_{rx} = 4$.

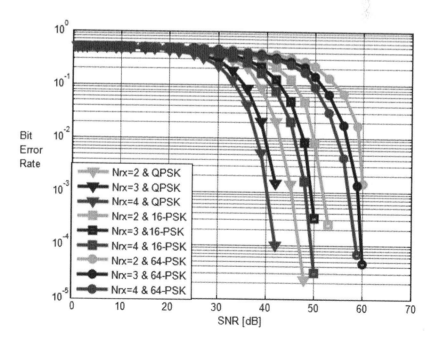

**FIGURE 11.11**  BER vs. SNR for MIMO-OFDM with MRC with modulation rates of 4-, 16-, 64-PSK.

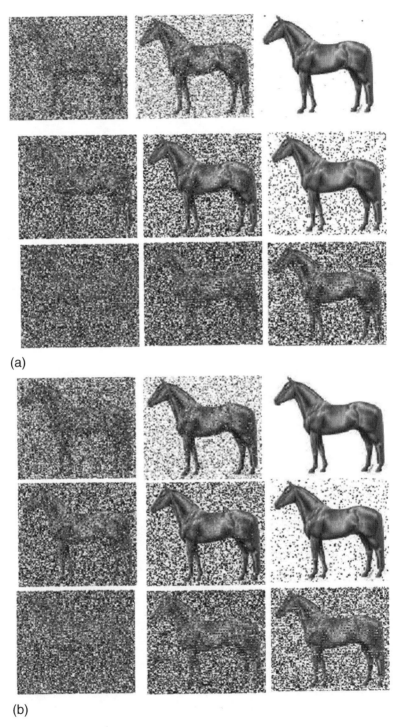

**FIGURE 11.12** Received images for MIMO-OFDM with SC for (a) Nrx = 2, (b) Nrx = 3, and *(Continued)*

# Image Transmission Analysis Using MIMO-OFDM Systems

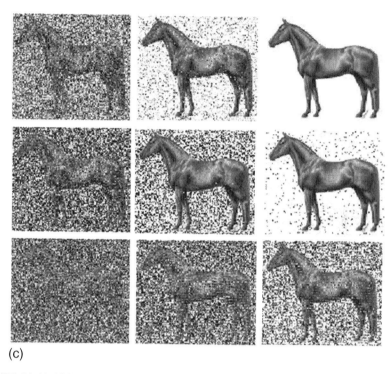

(c)

**FIGURE 11.12 (CONTINUED)**  (c) Nrx = 4.

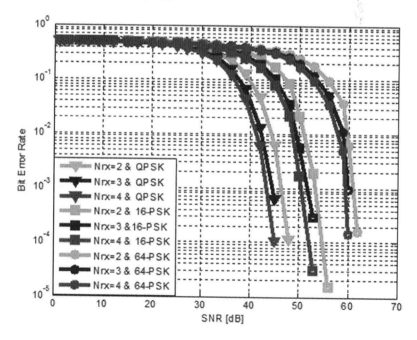

**FIGURE 11.13**  BER vs. SNR for MIMO-OFDM with SC with modulation rates of 4-, 16-, 64-PSK.

## 11.4 CONCLUSION

In this chapter, overall results have been demonstrated based on the transmission of images using M-PSK schemes with different transmitters and receivers having different SNR values. Different BER for different SNR values are performed, and their differences can be seen from the results shown above. Basically, QPSK > 16-PSK > 64 –PSK is the performance of the simulated results, which proves that the BER has the best configuration in case of QPSK and the least in the case of 64-PSK. Further, images received are of proper quality as they are transmitted well. Adding to this the results show that the proper quality images only contains less number of transmitters and receivers and fewer SNR values, which are the biggest advantage in this case. A less number of transmitters will cause less complexity on both sides, and the actual image that is to be received comes out in the proper way.

The noise effect can be measured from the images being received, and we can say that 64-PSK with SNR value of 60 has much noise in it. It is observed that the noise effect reduces, as we are increasing the SNR value, and clarity in the images can be seen. The above results have been checked and verified for the SNR values of (40, 50, 60). So, from the results it is clear that the above results have less BER in the case of QPSK and more in the case of 64-PSK, which also states that the requirement of signal power is minimum for QPSK and more for 64-PSK.

## REFERENCES

Alex, Sam P, and Louay M A Jalloul. 2008. "Performance Evaluation of MIMO in IEEE802. 16e/WiMAX." *IEEE Journal of Selected Topics in Signal Processing* 2 (2): 181–190.

Banelli, Paolo, and Saverio Cacopardi. 2000. "Theoretical Analysis and Performance of OFDM Signals in Nonlinear AWGN Channels." *IEEE Transactions on Communications* 48 (3): 430–441.

Bighieri, E, A Goldsmith, B Muquet, and Hikmet Sari. 2007. "Diversity, Interference Cancellation and Spatial Multiplexing in MIMO Mobile WiMAX Systems." In *2007 IEEE Mobile WiMAX Symposium*, Orlando, FL, 74–79.

Chen, Biao, and Hao Wang. 2002. "Maximum Likelihood Estimation of OFDM Carrier Frequency Offset." In *2002 IEEE International Conference on Communications. Conference Proceedings. ICC 2002*, New York, NY (Cat. No. 02CH37333), 1:49–53.

Divya, S, H Ananda Kumar, and A Vishalakshi. 2015. "An Improved Spectral Efficiency of WiMAX Using 802.16 G Based Technology." In *2015 International Conference on Advanced Computing and Communication Systems*, Cochin, India, 1–4.

Hemalatha, M, V Prithiviraj, S Jayalalitha, and Karuppusamy Thenmozhi. 2013. "Space Diversity Knotted with WiMAX – A Way for Undistorted and Anti-Corruptive Channel." *Wireless Personal Communications* 71 (4): 3023–3032.

Kobeissi, Rabi, Serdar Sezginer, and Fabien Buda. 2009. "Downlink Performance Analysis of Full-Rate STCs in 2x2 MIMO WiMAX Systems." In *VTC Spring 2009-IEEE 69th Vehicular Technology Conference*, Barcelona, Spain, 1–5.

Li, Qinghua, Xintian Eddie Lin, and Jianzhong Zhang. 2007. "MIMO Precoding in 802.16 e WiMAX." *Journal of Communications and Networks* 9 (2): 141–149.

Maré, Karel-Peet, and Bodhaswar T Maharaj. 2008. "Performance Analysis of Modern Space-Time Codes on a MIMO-WiMAX Platform." In *2008 IEEE International Conference on Wireless and Mobile Computing, Networking and Communications*, Avignon, France, 139–144.

Mehlführer, Christian, Sebastian Caban, José A Garcia-Naya, and Markus Rupp. 2009. "Throughput and Capacity of MIMO WiMAX." In *2009 Conference Record of the Forty-Third Asilomar Conference on Signals, Systems and Computers*, Pacific Grove, CA, 1426–1430.

Murty, M Sreerama, A Veeraiah, and Srinivas Rao. 2012. "Performance Evaluation of Wi-Fi Comparison with WiMax Networks." ArXiv Preprint ArXiv:1202.2634.

Sezginer, Serdar, Hikmet Sari, and Ezio Biglieri. 2009. "On High-Rate Full-Diversity $2\times 2$ Space-Time Codes with Low-Complexity Optimum Detection." *IEEE Transactions on Communications* 57 (5): 1532–1541.

Tan, Kefeng, Jean H Andrian, Hao Zhu, Frank M Candocia, and Chi Zhou. 2010. "A Novel Spectrum Encoding MIMO Communication System." *Wireless Personal Communications* 52 (1): 147.

Tilwari, Valmik, and Aparna Singh Kushwah. 2013. "Performance Analysis of Wi-Max 802.16 e Physical Layer Using Digital Modulation Techniques and Code Rates." *International Journal of Engineering Research and Applications (IJERA)* 3(4): 1449-1454.

Wang, Chin-Liang, and Kuan-Yu Chu. 2016. "An Improved Transceiver Design for Two-Relay SFBC-OFDM Cooperative Relay Systems." In *2016 IEEE 83rd Vehicular Technology Conference (VTC Spring)*, Nanjing, China 1–5.

Yang, Kai, Nan Yang, Chengwen Xing, Jinsong Wu, and Jianping An. 2016. "Space–Time Network Coding with Antenna Selection." *IEEE Transactions on Vehicular Technology* 65 (7): 5264–5274.

Zerrouki, Hadj, and Mohammed Feham. 2014. "A Physical Layer Simulation for WiMAX MIMO-OFDM System: Throughput Comparison between 2x2 STBC and 2x2 V-BLAST in Rayleigh Fading Channel." In *2014 International Conference on Multimedia Computing and Systems (ICMCS)*, Marrakesh, Morocco, 757–764.

# 12 Physical Layer Security in Two-Way Wireless Communication System

*Shashibhushan Sharma, Sanjay Dhar Roy, and Sumit Kundu*
National Institute of Technology, India

## CONTENTS

12.1 Introduction: Background and Motivations .................................................211
12.2 Secrecy at the Physical Layer in the TWC with a Number of Untrusted AF Relays That Harvest Energy from RF Sources and Operate in Half-Duplex Mode ....................................................214
    12.2.1 A System Model of TWC with a Number of Untrusted AF Relays ....................................................214
        12.2.1.1 Power Allocation to the Cognitive Nodes ...................216
        12.2.1.2 Secrecy Capacity and Relay Selection ........................217
    12.2.2 SOP Formulation with Untrusted AF Relays .............................219
    12.2.3 Numerical Results Based on the SOP in Two-Way Communication with AF Relays ..................................................222
12.3 Secrecy at the Physical Layer in TWC with Two Half-Duplex DF Relays in the Presence of an External Eavesdropper ......................225
    12.3.1 System Model of Two-Way Communication with DF Relays .....225
        12.3.1.1 Signal Strength ..............................................................226
        12.3.1.2 Global Secrecy Capacity ..............................................227
    12.3.2 SOP Formulation with DF Relays and Optimality of Source Power and Fraction of Relay Power ..............................................228
        12.3.2.1 SOP Formulation ..........................................................228
        12.3.2.2 Optimality of Source Power and Fraction of Relay Power ..............................................................229
    12.3.3 Numerical Results Based on the SOP and Observation of Optimal Value ..............................................................................231
12.4 Conclusion ....................................................................................................236
References ............................................................................................................237

## 12.1 INTRODUCTION: BACKGROUND AND MOTIVATIONS

Wireless communications have witnessed phenomenal growth in recent years due to an increase in the number of the communication users accessing different types of wireless applications. Due to the wireless broadcasting of information signals among a multiple number of communication users, the issues of secrecy of transmitted messages has become important. The physical layer security (PLS) (Wyner et al. 1975) has evolved as an interesting approach for maintaining the information security of the information signal in the wireless communication system of higher generation. Apart from the traditional approach to maintain security using cryptographic code, which involves complex key exchange, information security is also achieved at the physical layer as shown in Wyner et al. (1975). Mostly, the performance of security at the physical layer is measured in different metrics such as secrecy capacity (SC) and secrecy outage probability (SOP) (Pan et al. 2015). The SC has been analyzed in several sources (Sakran et al. 2012; Sun et al. 2012; Kalamkar and Banerjee 2017). In Sakran et al. (2012), SC has been evaluated under eavesdropper (EAV) attacking on the source and relay while in others (Sun et al. 2012; Kalamkar and Banerjee 2017), SC has been investigated with the use intermediate untrusted relay. The SOP metric has been analyzed (Pan et al. 2015; Son and Kong 2015) in energy-harvesting scenarios. In Son and Kong (2015), a relay network has been considered, and the EAV seeks the message from relay only, while in Pan et al. (2015), the EAV eavesdrops the message from the source in a single-input multiple-output system.

Further, due to a multiple number of users, the spectrum scarcity occurs, while the licensed spectrum remains underutilized in some cases. The CRN addresses the problem of spectrum scarcity and spectrum underutilization (Sakran et al. 2012). In Sakran et al. (2012), the PLS has been investigated in a CRN where the signal is eavesdropped by the EAV in both the hops of transmission. In the same, secrecy has been investigated under the interference limit of the primary receiver network. In Zou et al. (2014), the PLS of the cognitive network has been investigated where multiple cognitive users are attacked by multiple EAVs. Also, the power of the cognitive node is estimated under the interference level of the primary receiver. In Wang and Wang (2014), the security is investigated for a primary network that is attacked by the EAV where the secondary transmitter cooperates to the primary network by jamming the EAV. In Sharma et al. (2020), the security of information signal in the CRN has been investigated under the influence of untrusted amplify and forward relays.

Two-way communication (TWC) further improves bandwidth utilization (Sun et al. 2012). In some cases, the capacity sum of the TWC through the half-duplex relay network becomes equal to the capacity of the one-way full-duplex relay network (Sun et al. 2012). The PLS of a TWC network via the untrusted relay has been investigated with cloud jammers in Zhang et al. (2012). In Wang et al. (2013), the multiple amplify and forward (AF) relays provide TWC, which is secured by using jammer under the attack of the EAV. In Xu et al. (2015), a constellation-rotation aided approach has been proposed in a TWC to prevent the eavesdropping from the relaying. In Zhang et al. (2017), in a case of multiple EAV attacking, the PLS of TWC network has been investigated via the multiple AF relays. The relay selection scheme has been proposed based on the low-complexity method. This relay selection scheme is important for the 5G network. The secrecy performance has been measured in the average secrecy rate and the SOP metrics. In Khandaker et al. (2017), the secrecy rate has been evaluated in TWC when the EAV seeks the message from both the source and the relay nodes. In Jameel et al. (2018), the TWC has been studied through the half-duplex decode and forward (DF) relay, but communication is completed in three phases. In Sharma et al. (2020), TWC has been evaluated under the impact of multiple relays that are untrusted. In the same, secrecy performance has been measured in SOP metric. Thus TWC mostly takes the help of an intermediate relay.

The relay can be an energy-harvesting relay, as RF signal-based energy harvesting is an appealing approach to provide the energy to the communication nodes that are in remote areas (Sakran et al. 2012). There are basically two methods of energy harvesting; one is a time-splitting (TS) scheme (Salem et al. 2016; Kalamkar and Banerjee 2017), and another is a power-splitting (PS) scheme (Pan et al. 2015; Kalamkar and Banerjee 2017; Sharma et al. 2020). In Pan et al. (2015), the information receiver uses the PS scheme of energy harvesting. In Salem et al. (2016 and Kalamkar and Banerjee (2017), the authors use both schemes of harvesting energy and evaluate the secrecy performance in the ergodic secrecy capacity metric. In Sharma et al. (2020), the authors have used only the PS scheme of harvesting energy at the untrusted AF relay and investigate the secrecy performance in SOP metric.

The use of intermediate relay in communication networks has several benefits such as increasing the communication areas, the feasibility of two-way communication, and the utilization of green energy. Apart from these benefits, the intermediate relays have drawbacks also such as the relays may be untrusted (Sakran et al. 2012; Kalamkar and Banerjee 2017), and they may eavesdrop the message. In several sources (Sun et al. 2012; Zhang et al. 2012; Kalamkar and Banerjee 2017), the authors capture this issue and investigate the performance when an intermediate relay is untrusted. In Sun et al. (2012), directional antenna, on each of the selected untrusted relay, is used to prevent the eavesdropping by other untrusted nonselected relays.

The above work motivates us to present the secrecy performance of a relay network in TWC, where relay may be trusted and untrusted. The relay may harvest energy from RF sources. In the case of AF relay, two-way communication is easily possible with known channel state information (Sharma et al. 2019), but for the TWC utilizing DF relays, directional antennas are utilized in analyzing TWC via half-duplex DF relays (Sharma et al. 2019).

In this chapter, we will present a TWC in a CRN to improve the spectrum utilization with the help of intermediate AF or DF relay. Secrecy performance of such networks will be addressed when AF relays are untrusted and an external eavesdropper is present in the case of DF relay. Further, we will present some interesting results based on our research investigation on these topics.

The main contributions of presenting the secrecy performance of the TWC through the AF and DF relays are given below as:

- Describe a network architecture for efficient utilization of bandwidth using CRN with AF and DF relays and further, with two-way communication.
- Present secrecy performance of such a network when AF relays are untrusted and external eavesdropper is present in the case of DF relay.
- Depict the impact of energy harvesting on the secrecy performance at the untrusted AF relays.

- Detail mathematical model for evaluation of secrecy performance in SOP will be presented.
- Present some interesting results based on our original research to show the impact of several important physical parameters such as transmit power of signal and jamming, energy harvesting parameters, and relay power on the SOP.

We organize the chapter broadly in two sections. The outline of this chapter is indicated as:

- Secrecy at the physical layer in the TWC with a number of untrusted AF relays that harvest energy from RF sources and operate in half-duplex mode.
  o System model of TWC with a number of untrusted AF relays
  o SOP formulation with untrusted AF relays
  o Numerical results based on the SOP in TWC with AF relays
- Secrecy at the physical layer in TWC with half-duplex DF relays of number two in the attack of an external eavesdropper.
  o System model of TWC with DF relays
  o SOP formulation with DF relays and optimality of source power and fraction of relay power
  o Numerical results based on the SOP
- Chapter conclusions

## 12.2 SECRECY AT THE PHYSICAL LAYER IN THE TWC WITH A NUMBER OF UNTRUSTED AF RELAYS THAT HARVEST ENERGY FROM RF SOURCES AND OPERATE IN HALF-DUPLEX MODE

### 12.2.1 A System Model of TWC with a Number of Untrusted AF Relays

A cognitive TWC network through untrusted AF relays as in Sharma et al. (2020) is considered. All untrusted relays work as relays as well as eavesdroppers. However, at a time, a particular relay is in the active state, and therefore, only active untrusted relay eavesdrops the information signal. Only one information signal is eavesdropped at a time, while another information signal creates interference at the relay. In a CRN, the considered primary network consists of only primary transmitter (PT) and primary receiver (PR). It is also considered that the PT is at a far distance from both the sources and relays and therefore received interference signal from the PT is negligible at the receiving nodes of the cognitive network.

The complete system model is shown in Figure 12.1. In the network, the S1 represents the cognitive source one, S2 represents the cognitive source two, and $R_i$ represents the $i^{th}$ untrusted AF relay. Both the sources and all the relays transmit the signal satisfying the primary network outage constraint. The communication is completed in two-phases; the first phase is the broadcasting phase, and the second phase is the relaying phase as given in Figure 12.2. All nodes in the cognitive and primary

# Physical Layer Security in Two-Way Wireless Communication System

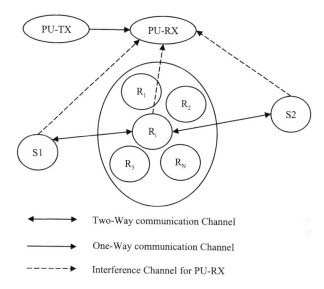

**FIGURE 12.1** System model of TWC of CRN with a number of untrusted AF relays that harvest energy and operate in half-duplex mode.

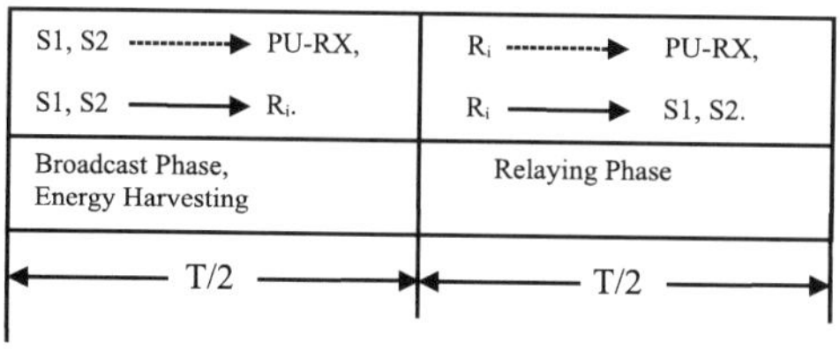

**FIGURE 12.2** Time frame of TWC with energy-harvesting relays.

network use an individual single omnidirectional antenna. In the broadcasting phase, both the sources send the information signal to a selected best relay. In the relaying phase, the relay follows the protocol of AF on received signal and broadcasts the signal toward both the sources using an omnidirectional antenna. As per knowledge of perfect channel state information (CSI) and prior transmitted message, both the sources first detect the self-interferer signal in the received signal and then delete that. After removal of the self-interference signal, both the sources easily decode the message without interference. Power corresponding to harvested energy by the selected relay from the RF signals in the broadcasting phase is used in the relaying phase.

All channels are faded with Rayleigh fading. All channel coefficients are random variables of independent nonidentically distributed type. We assume that the channel coefficient from source one (S1) to $i^{th}$ relay is $h_{S1Ri}$, and source two (S2) to $i^{th}$ relay is $h_{S2Ri}$, from $i^{th}$ the relay to source two is $h_{RiS2}$, and $i^{th}$ relay to S1 is $h_{RiS1}$, the channel from the S1 to the primary receiver is $h_{S1PR,}$ and source two to the primary receiver is

$h_{S2PR}$, and the channel from PU-TX to PU-RX is $h_{PP}$. The channel gains are indicated by $g_x$, where $x \in (S_1R_i, S_2R_i, R_iS_1, R_iS_2, S_1PR, S_2PR, R_iPR, PP)$. The mean channel gains of all channels are indicated by $\Omega_x$, where $x \in (S_1R_i, S_2R_i, R_iS_1, R_iS_2, S_1PR, S_2PR, R_iPR, PP)$. All channels have additive white Gaussian noise (AWGN) with mean zero and variance $N_0$. The probability distribution function (PDF) of channel gain is expressed as:

$$f_{g_x}(x) = \frac{1}{\Omega_x} \exp\left(-\frac{x}{\Omega_x}\right); \quad x \geq 0, \qquad (12.1)$$

and the cumulative distribution function (CDF) is expressed as:

$$F_{g_x}(x) = 1 - \exp\left(-\frac{x}{\Omega_x}\right); \quad x \geq 0. \qquad (12.2)$$

### 12.2.1.1 Power Allocation to the Cognitive Nodes

Assignment of power to the cognitive nodes is based on the outage probability constraint of the primary network (Tran et al. 2013; Sharma et al. 2020). We assume the sources' transmit power are equal. The primary network has the outage probability, which is expressed as:

$$P_{Out}^P = \left[ P\left\{ \left(1 + \frac{P_P g_{PP}}{P_{MS}(g_{S_1PR} + g_{S_2PR}) + N_0}\right) \leq R_P \right\} \right]$$

$$\leq \Delta = 1 - \frac{\exp\left(-\frac{\gamma_{TH}^P N_0}{P_P \Omega_{PP}}\right)}{\left(\frac{\gamma_{TH}^P P_{MS} \Omega_{S_1PR}}{P_P \Omega_{PP}} + 1\right)^2} \leq \Delta. \qquad (12.3)$$

where $\gamma_{TH}^P = 2^{2R_P}$ and $P_P$ is the transmit power of the PT, $R_P$ is the outage threshold rate of the primary network, $P_{MS}$ is the estimated power assigned to both sources under outage probability constraint of the primary network, AWGN has power $N_0$, and $\Delta$ is the primary network outage constraint. The assigned power to both cognitive sources ($P_{MS}$) is evaluated from Eq. (12.3) as:

$$P_{MS} = \frac{P_P \Omega_{PP}}{\gamma_{TH}^P \Omega_{SP}} \left[ \left\{ \frac{\exp\left(-\frac{\gamma_{TH}^P N_0}{P_P \Omega_{PP}}\right)}{1 - \Delta} \right\}^{0.5} - 1 \right], \qquad (12.4)$$

where $\Omega_{SP} = \Omega_{S_1PR} = \Omega_{S_2PR}$. In the relaying phase of communication, only the relay transmits the signals. Thus, the assignment of power to relay is only necessary.

Again, we evaluate the power for the relay in the same manner as in Eq. (12.4). The outage probability is expressed as:

$$P_{Out}^P = P\left\{\log_2\left(1 + \frac{P_P g_{PP}}{P_R g_{R_i PR} + N_O}\right) \le R_P\right\} = 1 - \frac{\exp\left(-\frac{\gamma_{TH}^P N_0}{P_P \Omega_{PP}}\right)}{\left(\frac{\gamma_{TH}^P P_R \Omega_{RP}}{P_P \Omega_{PP}} + 1\right)} \le \Delta. \quad (12.5)$$

The estimated power for the relay is (Tran et al. 2013; Sharma et al. 2020)

$$P_R = \frac{P_P \Omega_{PP}}{\gamma_{TH}^P \Omega_{RP}}\left[\left\{\frac{\exp\left(-\frac{\gamma_{TH}^P N_0}{P_P \Omega_{PP}}\right)}{1-\Delta}\right\} - 1\right]. \quad (12.6)$$

The cognitive nodes have limitation of transmit power due to constraint imposed by primary network. Both the cognitive sources have $P_{PK}$ as a maximum limit of transmit power, and the selected relay has the harvested power ($P_{Hi}$) as the maximum limit of transmit power. The final assigned power to both the cognitive sources is given as (Tran et al. 2013; Sharma et al. 2020)

$$P_S = \min(P_{MS}, P_{PK}) \quad (12.7)$$

The final assigned power to the selected relay is (Tran et al. 2013; Sharma et al. 2020)

$$P_{R_i} = \min(P_{H_i}, P_R) \quad (12.8)$$

### 12.2.1.2 Secrecy Capacity and Relay Selection

We will consider selection of relay based on maximizing the secrecy capacity, which will be shown by Eq. (12.22). First, we shall find the secrecy capacity and then consider the relay selection. The combined information signals received at the particular $i^{th}$ relay is

$$y_R' = \sqrt{P_s} h_{S_1 R_i} x_{S_1} + \sqrt{P_s} h_{S_2 R_i} x_{S_2} + n_R \quad (12.9)$$

where $n_0$ is the AWGN. The energy harvesting circuit, based on a power-splitting scheme, uses a "$\theta$" part of the combined received signals to harvest the energy, and another "$(1 - \theta)$" part of combined signal is used in signal processing (Pan et al. 2015; Kalamkar and Banerjee 2017; Sharma et al. 2020). The term "$\theta \in (0, 1)$" is the PS factor for harvesting energy. The harvested energy is expressed as

$$E_{H_i} = \eta \theta P_s \left(g_{S_1 R_i} + g_{S_2 R_i}\right)\frac{T}{2}, \quad (12.10)$$

where $\eta(0 < \eta < 1)$ is energy transfer efficiency, and $T$ is the complete communication time. The power, based on harvested energy, is

$$P_{H_i} = \frac{E_{H_i}}{\frac{T}{2}} = \eta\theta P_S\left(g_{S_1R_i} + g_{S_2R_i}\right). \tag{12.11}$$

After removal of harvesting part of the signal, the rest part of the signal is expressed as:

$$y_R = \sqrt{(1-\theta)P_S}\,h_{S_1R_i}x_{S_1} + \sqrt{(1-\theta)P_S}\,h_{S_2R_i}x_{S_2} + n_R \tag{12.12}$$

The signal strength is represented in signal-to-noise ratio (SNR). The SNR for each information signal at the relay is expressed as:

$$\gamma_{S_1R_i} = \frac{(1-\theta)P_S g_{S_1R_i}}{(1-\theta)P_S g_{S_2R_i} + N_0}; \gamma_{S_2R_i} = \frac{(1-\theta)P_S g_{S_2R_i}}{(1-\theta)P_S g_{S_1R_i} + N_0}, \tag{12.13}$$

where $\gamma_{S_1R_i}$ is the SNR for the first information signal one, and $\gamma_{S_2R_i}$ is the SNR for the second information signal. The selected $i^{th}$ relay amplifies the information signals with the amplification factor $\zeta_i$. The $\zeta_i$ is expressed as

$$\zeta_i = \sqrt{\frac{P_{R_i}}{(1-\theta)P_S\left(g_{S_1R_i} + g_{S_2R_i}\right) + N_0}} \tag{12.14}$$

The $i^{th}$ relay broadcasts the information signals. The source two receives the signal, which can be expressed as

$$y_{S_2} = \underbrace{\zeta_i\sqrt{(1-\theta)P_S}\,h_{S_1R_i}h_{R_iS_2}x_{S_1}}_{\text{Information Signal}} + \underbrace{\zeta_i\sqrt{(1-\theta)P_S}\,h_{S_2R_i}h_{R_iS_2}x_{S_2}}_{\text{Self interference Signal}} + \underbrace{\zeta_i n_0 h_{R_iS_2} + n_0}_{\text{Noise signal}}. \tag{12.15}$$

The source one receives the signal as

$$y_{S_1} = \underbrace{\zeta_i\sqrt{(1-\theta)P_S}\,h_{S_1R_i}h_{R_iS_1}x_{S_1}}_{\text{Self interference Signal}} + \underbrace{\zeta_i\sqrt{(1-\theta)P_S}\,h_{S_2R_i}h_{R_iS_1}x_{S_2}}_{\text{Information Signal}} + \underbrace{\zeta_i n_0 h_{R_iS_1} + n_0}_{\text{Noise signal}} \tag{12.16}$$

The self-interference is canceled at both the sources. The SNR at both the sources is represented as

$$\begin{aligned}\gamma_{R_iS_2} &= \frac{\zeta_i^2(1-\theta)P_S g_{S_1R_i}g_{R_iS_2}}{\zeta_i^2 N_0 g_{R_iS_2} + N_0} \approx \frac{\eta\theta(1-\theta)P_S g_{S_1R_i}g_{R_iS_2}}{\eta\theta g_{R_iS_2}N_0 + (1-\theta)N_0}.\\ \gamma_{R_iS_1} &= \frac{\zeta_i^2(1-\theta)P_S g_{S_2R_i}g_{R_iS_1}}{\zeta_i^2 N_0 g_{R_iS_1} + N_0} \approx \frac{\eta\theta(1-\theta)P_S g_{S_2R_i}g_{R_iS_1}}{\eta\theta g_{R_iS_1}N_0 + (1-\theta)N_0}.\end{aligned} \tag{12.17}$$

# Physical Layer Security in Two-Way Wireless Communication System

where $\gamma_{RiS2}$ is the SNR for the information signal one at the S2, and $\gamma_{RiS1}$ is the SNR for the information signal of S2 at the S1.

Using Shannon channel capacity formula, the channel capacities for both the information signal at both the sources are expressed as

$$C_{R_iS_2} = \left[\frac{1}{2}\log(1+\gamma_{R_iS_2})\right]; C_{R_iS_1} = \left[\frac{1}{2}\log(1+\gamma_{R_iS_1})\right], \quad (12.18)$$

The channel capacities of both the information signal at $i^{th}$ the untrusted relay are expressed as

$$C_{S_1R_i} = \left[\frac{1}{2}\log(1+\gamma_{S_1R_i})\right]; C_{S_2R_i} = \left[\frac{1}{2}\log(1+\gamma_{S_2R_i})\right]. \quad (12.19)$$

Now, the secrecy capacity of both the information signals is defined as (Pan et al. 2015; Kalamkar and Banerjee 2017)

$$C_{R_iS_2}^{SEC} = \left[\frac{1}{2}\log(1+\gamma_{R_iS_2}) - \frac{1}{2}\log(1+\gamma_{S_1R_i})\right]^+;$$

$$C_{R_iS_1}^{SEC} = \left[\frac{1}{2}\log(1+\gamma_{R_iS_1}) - \frac{1}{2}\log(1+\gamma_{S_2R_i})\right]^+ \quad (12.20)$$

where $[x]^+$ indicates maximum between 0 and $x$. The global secrecy capacity $\left(C_{G_i}^{SEC}\right)$ is defined as:

$$C_{G_i}^{SEC} = \min\left(C_{R_iS_2}^{SEC}, C_{R_iS_1}^{SEC}\right). \quad (12.21)$$

Based on maximum secrecy capacity, a particular relay is selected. The selected relay must follow the selection criteria

$$i^* = \arg\max_{1\leq i\leq N}\left(C_{G_i}^{SEC}\right). \quad (12.22)$$

A final secrecy capacity through a selected untrusted relay out of multiple untrusted relays is expressed as

$$C_{G_i}^{SEC*} = \max_{1\leq i\leq N}\left(C_{G_i}^{SEC}\right). \quad (12.23)$$

## 12.2.2 SOP Formulation with Untrusted AF Relays

The SOP is defined as the probability of an event in which secrecy capacity falls below a threshold secrecy rate $\left(R_{TH}^{SEC}\right)$. Numerically, it can be expressed as (Zhang et al. 2012; Zou et al. 2014)

$$P_{OUT}^{SEC} = P\left(C_{G_i}^{SEC*} < R_{TH}^{SEC}\right) = P\left(\max_{1\leq i\leq N} C_{G_i}^{SEC} < R_{TH}^{SEC}\right) \quad (12.24)$$

Equation (12.24) can be further simplified as

$$P_{OUT}^{SEC} = P\left[\max_{1\le i\le N}\left\{\min\left(C_{R_iS_2}^{SEC},C_{R_iS_1}^{SEC}\right)\right\} < R_{TH}^{SEC}\right] = \prod_N^{i=1}\left[P\left\{\min\left(C_{R_iS_2}^{SEC},C_{R_iS_1}^{SEC}\right) < R_{TH}^{SEC}\right\}\right]$$

$$= \left[1-\underbrace{\left\{1-P\left(C_{R_iS_2}^{SEC} < R_{TH}^{SEC}\right)\right\}}_{I_1}\times\underbrace{\left\{1-P\left(C_{R_iS_1}^{SEC} < R_{TH}^{SEC}\right)\right\}}_{I_2}\right]^N = \left[1-(1-I)^2\right]^N, \quad (12.25)$$

as per symmetry and independence about the relay, $I = I_1 = I_2$. The $I_1$ can be expressed as

$$I = P\left(C_{R_iSS_2}^{SEC} < R_{TH}^{SEC}\right) = \underbrace{P\left\{\left(C_{R_iS_2}^{SEC} < R_{TH}^{SEC}\right)|\left(P_{H_i} < P_R\right)\right\}P\left(P_{H_i} < P_R\right)}_{I_3}$$
$$+ \underbrace{P\left\{\left(C_{R_iS_2}^{SEC} < R_{TH}^{SEC}\right)|\left(P_{H_i} \ge P_R\right)\right\}P\left(P_{H_i} \ge P_R\right)}_{I_4} \quad (12.26)$$

Here, there are two cases, (i) $P_{Hi} < P_R$ and (ii) $P_{Hi} \ge P_R$. In the case of $P_{Hi} < P_R$, the $P(P_{Hi} < P_R)$ can be expressed in closed form as

$$P\left(P_{H_i} < P_R\right) = 1 - \left(1 + \frac{P_R}{\rho\theta P_S \Omega_{R_iS_2}}\right)\exp\left(-\frac{P_R}{\rho\theta P_S \Omega_{R_iS_2}}\right). \quad (12.27)$$

We consider $x = |h_{S2Ri}|^2$ and $y = |h_{RiS2}|^2$. The term $I_3$ can be simplified as

$$I_3 = P\left\{\left(C_{R_iSS_2}^{SEC} < R_{TH}^{SEC}\right)|\left(P_{H_i} < P_{MSR}\right)\right\}$$
$$= P\left\{\frac{1}{2}\log_2\left(1 + \frac{\eta\theta(1-\theta)P_S xy}{\eta\theta N_0 y + N_0(1-\theta)}\bigg/1+\frac{(1-\theta)P_S x}{(1-\theta)P_S y + N_0}\right) \le R_{TH}^S\right\} \quad (12.28)$$
$$= P\left\{v(y)x < (\delta-1)\right\} = 1 - \frac{1}{\Omega_{R_iS_2}}\int_\phi^\infty \exp\left(-\frac{(\delta-1)}{v(y)\Omega_{S_1R_i}} - \frac{y}{\Omega_{R_iS_2}}\right)dy,$$

where $\delta = 2^{2R_{TH}^{SEC}}$, $v(y) = (1-\theta)\left\{\frac{\eta\theta P_S y}{\eta\theta y N_0 + N_0(1-\theta)} - \frac{\delta P_S}{(1-\theta)P_S y + N_0}\right\}$,

$$\phi = \frac{\left(\frac{\delta-1}{1-\theta}\right) + \sqrt{\left(\frac{\delta-1}{1-\theta}\right)^2 + \frac{4\delta P_S}{N_0\eta\theta}}}{2\left(\frac{P_S}{N_0}\right)}.$$ In the case of $P_{Hi} \ge P_R$, $P_{Ri} = P_R$, which is constant.

The term $P(P_{Hi} \ge P_R)$ can be represented in closed form as

$$P\left(P_{H_i} \ge P_R\right) = 1 - P\left(P_{H_i} < P_R\right) = \left(1 + \frac{P_R}{\rho\beta P_S \Omega_{R_iS_2}}\right)\exp\left(-\frac{P_R}{\rho\beta P_S \Omega_{R_iS_2}}\right). \quad (12.29)$$

In the calculation of $I_4$, the SNR in Eq. (12.17) using Eq. (12.14), can be reexpressed as Sharma et al. 2020)

$$\gamma_{R_iS_2} = \frac{P_R}{(1-\beta)P_S + P_R}\left[\frac{\left\{\frac{(1-\theta)P_S}{N_0}x\right\}\left\{\frac{(1-\theta)P_S+P_R}{N_0}y\right\}}{1+\left\{\frac{(1-\theta)P_S}{N_0}x\right\}+\left\{\frac{(1-\theta)P_S+P_R}{N_0}y\right\}}\right] \quad (12.30)$$

$$\gamma_{R_iS_2} < A_1 \min(Q,R); \gamma_{R_iSS_2} \geq \frac{A_1}{2}\min(Q,R) \text{ where } Q = \frac{(1-\beta)P_S}{N_0}x;$$

$$R = \frac{(1-\beta)P_S + P_R}{N_0}y; A_1 = \frac{P_R}{(1-\beta)P_S + P_R}$$

From Eq. (12.30), the PDF of $\gamma_{RiS2}$ is expressed as

$$f_{\gamma_{R_iS_2}}(\tau) \approx \frac{1}{\Omega_D}\exp\left(-\frac{\tau}{\Omega_D}\right); \tau \geq 0, \quad (12.31)$$

where $\Omega_D = \frac{A_1}{2}T_{QR}, T_{QR} = \frac{T_Q T_R}{T_Q + T_R}, T_Q = \frac{(1-\theta)P_S\Omega_{S_1R_i}}{N_o}, T_R = \frac{\{(1-\theta)P_S+P_R\}\Omega_{R_iS_2}}{N_0}$,

The PDF of SNR at the selected relay is expressed as

$$f_{\gamma_{S_1R_i}}(\upsilon) = \frac{N_0}{(1-\theta)P_S\Omega_{R_iS_2}} \times \frac{\exp\left(-\frac{N_0\upsilon}{(1-\theta)P_S\Omega_{R_iS_2}}\right)}{\upsilon+1} + \frac{\exp\left(-\frac{N_0\upsilon}{(1-\theta)P_S\Omega_{R_iS_2}}\right)}{(\upsilon+1)^2}; \upsilon \geq 0. \quad (12.32)$$

The term $I_4$ in Eq. (12.26) is expressed in single integration form as

$$I_4 = P\left\{\left(C_{R_iS_2}^{SEC} < R_{TH}^{SEC}\right)|(P_{H_i} \geq P_R)\right\} = P\left\{\tau < \left(\delta(1+\upsilon)-1\right)\right\}$$

$$= \int_0^\infty \int_0^{(\delta(1+\upsilon)-1)} f_{\gamma_{R_iS_2}\gamma_{S_1R_i}}(\tau,\upsilon)d\tau d\upsilon = \int_0^\infty \left\{\int_0^{(\delta(1+\upsilon)-1)} f_{\gamma_{R_iS_2}}\left(\frac{\tau}{\upsilon}\right)d\tau\right\}f_{\gamma_{S_1R_i}}(\upsilon)d\upsilon$$
$$\quad (12.33)$$

$$= \int_0^\infty \upsilon\left[\left\{1-\exp\left(-\frac{(\delta(1+\upsilon)-1)}{\Omega_D\upsilon}\right)\right\}\left\{A_2\frac{\exp(-A_2\upsilon)}{\upsilon+1}+\frac{\exp(-A_2\upsilon)}{(\upsilon+1)^2}\right\}\right]d\upsilon,$$

where $A_2 = \frac{N_0}{(1-\theta)P_S\Omega_{S_1R_i}}$. The final expression of $I$ in Eq. (12.26) can be reexpressed as

$$I = I_3\left\{1-\left(1+\frac{P_R}{\rho\beta P_S\Omega_{R_iSS_2}}\right)\exp\left(-\frac{P_R}{\rho\beta P_S\Omega_{R_iSS_2}}\right)\right\}$$
$$+ I_4\left\{\left(1+\frac{P_R}{\rho\beta P_S\Omega_{R_iSS_2}}\right)\exp\left(-\frac{P_R}{\rho\beta P_S\Omega_{R_iSS_2}}\right)\right\}, \quad (12.34)$$

where $I_3$ and $I_4$ are given in Eq. (12.28) and in Eq. (12.33), respectively. The value of $I$ is put into Eq. (12.25), which gives the final expression of the SOP:

$$P_{out}^{SEC} = \left[1-(1-I)^2\right]^N \qquad (12.35)$$

### 12.2.3 Numerical Results Based on the SOP in Two-Way Communication with AF Relays

The numerical parameters are given in Table 12.1. The MATLAB-based simulation results have been found. The simulation results verify the analytical results based on Eq. (12.35).

Figure 12.3 presents the results of the SOP versus PS factor of energy harvesting. As $\theta$ increases, the harvesting energy increases as per Eq. (12.10). The selected relay uses the large power corresponding to harvested energy to transmit the amplified information signal. Received signals at both the sources have high strength after the cancellation of self-interference. The information channel capacity for both the sources increases due to high signal strength, but due to the jamming performance of the information signals at the untrusted relays and signals transmitted by both the sources with equal transmit power, the relay channel capacity for any of the signals remains unchanged. Thus the SOP decreases with increase in secrecy capacity before the optimal point of $\theta$. The obtained optimal values are 0.6 for $P_{PK} = -5$ dBW 0.6 for $P_{PK} = 0$ dBW and 0.65 for $P_{PK} = 5$ dBW. Beyond the optimal point, harvested energy increases with increase in $\theta$, but the part of the information signal for signal processing becomes very poor, i.e., signal becomes noisy. Because the amplification process of the relay does not increase the information signal strength up to a decodable strength due to its poor strength and the noise is amplified, the resulting signal becomes noisier. Thus, after the optimal value, the signal strength at both the sources decreases, and therefore the main channel capacity decreases due to noisier of the signals. The relay channel capacity remains constant with respect to $\theta$. As a result,

### TABLE 12.1
### Numerical Parameters with Their Values

| Numerical Parameters | Numerical Values |
| --- | --- |
| Maximum limit of transmit power of SNs ($P_{PK}$) | −5 dBW, 0 dBW, 5 dBW |
| Transmit power of the PT ($P_P$) | 5 dBW |
| Threshold outage rate of PR ($R_P$) | 0.2 b/s/Hz |
| Outage probability of PR ($P_{Out}^P = \Delta$) | 0.2 |
| Energy conversion efficiency ($\eta$) application | 0.7 |
| AWGN power ($N_0$) | $10^{-2}$ W |
| Threshold secrecy rate ($R_{TH}^{sec}$) | 1 b/s/Hz |
| Channel mean power of all links ($\Omega_j$) | 1 |
| Number of relays ($N$) | 5 |

# Physical Layer Security in Two-Way Wireless Communication System 223

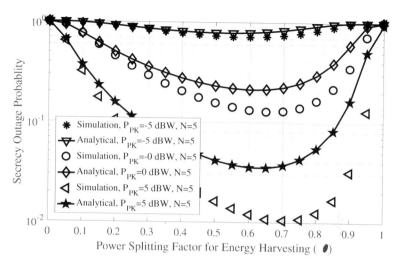

**FIGURE 12.3** SOP versus PS factor for harvesting energy for different values of maximum limit of transmit power of SNs.

the SOP increases with a decrease in secrecy capacity beyond the optimal point of $\theta$. Due to approximate expression of the SOP, there is some mismatching between analytical and simulation, but the nature of the curves are same, and optimal values are also almost the same in both analytical and simulation.

Figure 12.4 shows the SOP versus transmit power of primary network ($P_P$) under the different values of maximum limit of transmit power ($P_{PK}$) of SNs. An increase in $P_P$ increases the assigned power to the cognitive node to transmit the signal as per Eq. (12.4) and Eq. (12.6). The RF information signal is received by the selected relay with large energy, and the relay harvests more energy. The relayed signal, which is in high strength, is received by both the sources. Thus, the capacities of main channel increase with an increase in $P_P$, but due to the jamming performance of the information signal at the relay, the channel capacities of relay for any information signal remains constant, and thus the global secrecy capacity increases. An increase in secrecy capacity decreases the SOP, and performance is improved. Next, the device gets limited in power, such as $P_{PK}$, the cognitive nodes are allowed to transmit at a constant power for the signal transmission. This constant power keeps the secrecy capacity almost constant. Therefore the SOP remains constant. Further, the SOP decreases with an increase in $P_{PK}$, which allows cognitive nodes to transmit the signal with large power.

Figure 12.5 presents the SOP versus threshold outage rate of primary network for several values of $P_{PK}$. As per Eq. (12.4) and (12.6), with an increase in the threshold outage rate of the primary network, the assigned power to cognitive node decreases. Decrease in transmit power according to cognitive constraint decreases the strength of the information signals at corresponding to its each destination source, and the performance degrades in terms of SOP as shown in Figure 12.5.

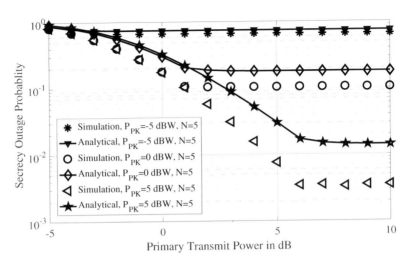

**FIGURE 12.4** SOP versus transmit power of primary network in dBW for different values of maximum limit of transmit power of SNs.

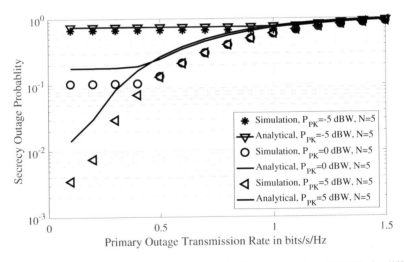

**FIGURE 12.5** SOP versus threshold outage rate of primary network in bits/s/Hz for different values of maximum limit of transmits power of SNs.

Figure 12.6 presents the SOP versus the number of relays for different values of $P_{PK}$. Increment in the number of relays increases the path diversity of information signal to send the message to respective destinations. As the number of relays increases, there is a higher chance to select the best relay, and therefore the respective destination receives the signal in with better strength. This in turn increases the capacity of main channel while the relay channel capacity remains unchanged due to the jamming performance of each information signal at the relay. Thus the secrecy

# Physical Layer Security in Two-Way Wireless Communication System

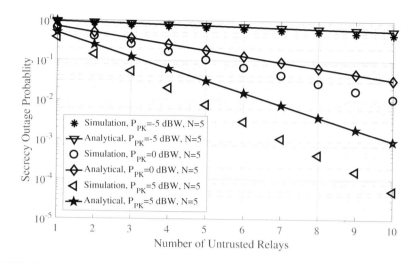

**FIGURE 12.6** SOP versus number of untrusted relays for several values of maximum limit of transmits power of SNs.

capacity increases and the SOP decreases with an increase in a number of relays. The decrease in SOP increases the secrecy performance.

In the next section, we present secrecy analysis in a TWC with two half-duplex DF relays when an external eavesdropper is present.

## 12.3 SECRECY AT THE PHYSICAL LAYER IN TWC WITH TWO HALF-DUPLEX DF RELAYS IN THE PRESENCE OF AN EXTERNAL EAVESDROPPER

### 12.3.1 SYSTEM MODEL OF TWO-WAY COMMUNICATION WITH DF RELAYS

In this model, two sources share their information signal with each other through two half-duplex DF relays as in Sharma et al. (2019). The complete system model of this TWC network has been shown in Figure 12.7. In this communication, an external eavesdropper efforts to eavesdrop the message of both the sources, but not simultaneously at a time. Each of the relays has two directional antennas, one is used for receiving the information signal, while the other is for transmitting the information signal. The directional antennas at each relay assist the relay for two-way communication. The rest of the communication nodes has omnidirectional antennas. The EAV keep its position in such a way that it presents at almost equal distance from both the sources, and this position provides the proper receiving of the information signal from both the sources. We have found this particular position of EAV on a line that is perpendicular to another line that connects to both the sources. Due to directional antennas, the EAV is unable to receive the relayed signal. The communication completes in two phases; first is a broadcasting phase, and another is the relaying phase as shown in Figure 12.8. In the broadcast phase, both the sources transmit the signals. In broadcasting phase, the information signal of source one (S1) is received by the

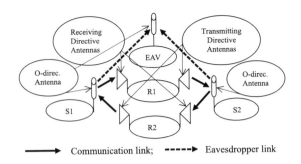

**FIGURE 12.7** System model of two-way communication with two half-duplex DF relay.

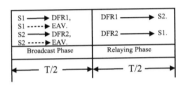

**FIGURE 12.8** The time frame of two-way communication with two half-duplex DF relay.

relay one (R1), and the information signal of source two (S2) is received by the relay two (R2). In the relaying phase, the S2 receives the signal from R1, and S1 receives the signal from R2.

All channels are Rayleigh fading. The channel has Gaussian noise with zero mean and variance $N_0$. The channel coefficients are denoted by $h_{ij}$, the channel gain is denoted by $g_{ij}$, and mean channel gains are denoted by $\Omega_{ij}$, where $i = (S1, R1, S2, R2, E)$, here subscript E indicates for EAV, and $j = (S1, R1, S2, R2, E)$. All the channels are independent and nonidentically distributed random variables.

### 12.3.1.1 Signal Strength

The signal strength at any receiving node is represented in signal-to-noise ratio (SNR) or signal-to-interference plus noise ratio (SINR) parameter. In the broadcasting phase, the information signal of the S1 is received by the R1, and the information signal of the S2 is received by the (R2). The received signal at the R1 is

$$y_{S1R1} = \sqrt{P_{S1}} h_{S1R1} x_{S1} + n_0 \qquad (12.36)$$

where $P_{S1}$ is the S1's transmit power, $x_{S1}$ message of S1, and $n_0$ is the Gaussian noise. From Eq. (12.36), SNR at the R1 is

$$\gamma_{S1R1} = \frac{P_{S1} g_{S1R1}}{N_0}. \qquad (12.37)$$

In the same way, the SNR at the $R2$ is

$$\gamma_{S2R1} = \frac{P_{S2}g_{S2R2}}{N_0}. \tag{12.38}$$

The SINR of information signal $S1$ at the EAV is

$$\gamma_{S1E} = \frac{P_{S1}g_{S1E}}{P_{S2}g_{S2E} + N_0}. \tag{12.39}$$

The SINR of $S2$ at the EAV is

$$\gamma_{S2E} = \frac{P_{S2}g_{S2E}}{P_{S1}g_{S1E} + N_0}. \tag{12.40}$$

In the relaying phase, the information signal strength of $S1$ at the $S2$ via $R1$ is

$$\gamma_{R1S2} = \frac{\alpha P_R g_{R1S2}}{N_0}, \tag{12.41}$$

where $\alpha$ is the fraction of relay power, i.e., $\alpha P_R$ has assigned to $R1$ and $(1-\alpha)P_R$ has assigned to $R2$. The information signal strength of $S2$ at the $S1$ via $R2$ is

$$\gamma_{R2S1} = \frac{(1-\alpha)P_R g_{R2S1}}{N_0}. \tag{12.42}$$

The end-to-end information signal strength of $S1$ at the $S2$ is

$$\gamma_{S1S2} = \min(\gamma_{S1R1}, \gamma_{R1S2}), \tag{12.43}$$

and end-to-end information signal strength of $S2$ at $S1$ is

$$\gamma_{S2S1} = \min(\gamma_{S2R2}, \gamma_{R2S1}). \tag{12.44}$$

### 12.3.1.2 Global Secrecy Capacity

Using the Shannon channel capacity formula, the channel capacity of information signal one is

$$C_{S1S2} = \left[\frac{1}{2}\log_2(1+\gamma_{S1S2})\right]^+, \tag{12.45}$$

and channel capacity of information signal two is

$$C_{S2S1} = \left[\frac{1}{2}\log_2(1+\gamma_{S2S1})\right]^+ \tag{12.46}$$

The channel capacity of the information signal one at the EAV is

$$C_{S1E} = \left[\frac{1}{2}\log_2(1+\gamma_{S1E})\right]^+. \quad (12.47)$$

The channel capacity of the information signal two at the EAV is

$$C_{S2E} = \left[\frac{1}{2}\log_2(1+\gamma_{S2E})\right]^+. \quad (12.48)$$

Now, the secrecy capacity of information signal one is defined as

$$C_{S1S2}^{SEC} = \frac{1}{2}\log(1+\gamma_{S1S2}) - \frac{1}{2}\log(1+\gamma_{S1E}). \quad (12.49)$$

The secrecy capacity of information signal two is defined as (Pan et al. 2015; Kalamkar and Banerjee 2017)

$$C_{S2S1}^{SEC} = \frac{1}{2}\log(1+\gamma_{S2S1}) - \frac{1}{2}\log(1+\gamma_{S2E}). \quad (12.50)$$

The global secrecy capacity is expressed as (Zhang et al. 2017; Sharma et al. 2019)

$$C_{S1S2}^{G\_SEC} = \left[\min\left(C_{S1S2}^{SEC}, C_{S2S1}^{SEC}\right)\right]^+, \quad (12.51)$$

## 12.3.2 SOP Formulation with DF Relays and Optimality of Source Power and Fraction of Relay Power

### 12.3.2.1 SOP Formulation

We have defined the SOP in the previous Section 12.2.2. Here, we express the SOP as

$$P_{OUT}^{SEC} = P\left(C_{S1S2}^{G\_SEC} < C_{TH}^{G\_SEC}\right) = P\left(\min\left(C_{S1S2}^{SEC}, C_{S2S1}^{SEC}\right) < C_{TH}^{G\_SEC}\right) \quad (12.52)$$

Equation (12.52) is further simplified as

$$P_{OUT}^{G\_SEC} = 1 - \left\{1 - P\left(C_{S1S2}^{SEC} < C_{TH}^{G\_SEC}\right)\right\}\left\{1 - P\left(C_{S2S1}^{SEC} < C_{TH}^{G\_SEC}\right)\right\}. \quad (12.53)$$

The term $P\left(C_{S1S2}^{SEC} < C_{TH}^{G\_SEC}\right)$ (let $P\left(C_{S1S2}^{SEC} < C_{TH}^{G\_SEC}\right) = I_1$) in closed form is expressed as

$$\begin{aligned}I_1 &= 1 - \exp(-A_7)\left[\int_0^\infty A_4 \frac{\exp(-A_8 y)}{(A_5 y + 1)} dy + \int_0^\infty A_5 \frac{\exp(-A_8 y)}{(A_5 y + 1)^2} dy\right] \\ &= 1 - \exp(-A_7)\left[1 - \frac{A_4 - A_8}{A_5}\left\{\exp\left(\frac{A_8}{A_5}\right) Ei\left(-\frac{A_8}{A_5}\right)\right\}\right],\end{aligned} \quad (12.54)$$

# Physical Layer Security in Two-Way Wireless Communication System

where all constants are given as

$$A_1 = \frac{P_{S1}\Omega_{S1R1}}{N_O}; A_2 = \frac{P_{R1}\Omega_{R1S2}}{N_O}; A_3 = \frac{A_1 A_2}{A_1 + A_2}; A_4 = \frac{N_O}{P_{S1}\Omega_{S1E}};$$
$$A_5 = \frac{P_{S2}\Omega_{S2E}}{P_{S1}\Omega_{S1E}}; A_6 = 2^{2C_{TH}^{G\_SEC}}; A_7 = \frac{A_6 - 1}{A_3}; A_8 = \frac{A_6}{A_3} + A_4$$

(12.55)

The term $I_2$ in closed form is expressed as

$$I_2 = 1 - \exp(-B_7)\left[\int_0^\infty B_4 \frac{\exp(-B_8 y)}{(B_5 y + 1)} dy + \int_0^\infty B_5 \frac{\exp(-B_8 y)}{(B_5 y + 1)^2} dy\right],$$
$$= 1 - \exp(-B_7)\left[1 - \frac{B_4 - B_8}{B_5}\left\{\exp\left(\frac{B_8}{B_5}\right) Ei\left(-\frac{B_8}{B_5}\right)\right\}\right]$$

(12.56)

where all constants are given as

$$B_1 = \frac{P_{S2}\Omega_{S2R2}}{N_O}; B_2 = \frac{P_{R2}\Omega_{R2S1}}{N_O}; B_3 = \frac{B_1 B_2}{B_1 + B_2}; B_4 = \frac{N_O}{P_{S2}\Omega_{S2E}};$$
$$B_5 = \frac{P_{S1}\Omega_{S1E}}{P_{S2}\Omega_{S2E}}; B_6 = 2^{2C_{TH}^{G\_SEC}}; B_7 = \frac{B_6 - 1}{B_3}; B_8 = \frac{B_6}{B_3} + B_4$$

(12.57)

We put $I_1$ from Eq. (12.54) and $I_2$ from Eq. (12.56) in Eq. (12.53); we get the SOP in final closed-form expression as

$$P_{OUT}^{G\_SEC} = 1 - \left[\begin{array}{l}\left\{\exp(-A_7)\left[1 - \frac{A_4 - A_8}{A_5}\left\{\exp\left(\frac{A_8}{A_5}\right) Ei\left(-\frac{A_8}{A_5}\right)\right\}\right]\right\} \\ \times \left\{\exp(-B_7)\left[1 - \frac{B_4 - B_8}{B_5}\left\{\exp\left(\frac{B_8}{B_5}\right) Ei\left(-\frac{B_8}{B_5}\right)\right\}\right]\right\}\end{array}\right].$$

(12.58)

### 12.3.2.2 Optimality of Source Power and Fraction of Relay Power

The optimal value of source power and fraction of relay power is obtained by differentiating the SOP with respect to source power and fraction of relay power, respectively Sharma et al. 2019). First, the SOP is differentiated with respect to $P_{S1}$ keeping the fixed value of $P_{S2}$ and $\alpha$. For finding the optimal value, we equate this differentiation to zero. The expression is given as

$$\frac{\partial P_{OUT}^{G\_SEC}}{\partial P_{S1}} = \frac{\partial I_1}{\partial P_{S1}}(1 - I_2) + (1 - I_1)\frac{\partial I_2}{\partial P_{S1}} = 0.$$

(12.59)

The expression of $\frac{\partial I_1}{\partial P_{S1}}$ can be expressed as

$$\frac{\partial I_1}{\partial P_{S1}} = -\frac{(A_6-1)N_0}{\Omega_{S1R1}P_{S1}^2}\exp(-A_7)\left[1-\frac{A_4-A_8}{A_5}\left\{\exp\left(\frac{A_8}{A_5}\right)Ei\left(-\frac{A_8}{A_5}\right)\right\}\right]$$
$$-\exp\left(\frac{A_8}{A_5}-A_7\right)Ei\left(-\frac{A_8}{A_5}\right)\left[\left\{\frac{A_5A_6N_0}{\Omega_{R1S2}P_{S1}^2}+(A_4-A_8)\frac{P_{S2}\Omega_{S2E}}{\Omega_{S1E}P_{S1}^2}\right\}/A_5^2\right]$$
$$+\left[\exp(-A_7)\left[\left\{\left(\frac{A_6}{\Omega_{S1R1}}+\frac{1}{\Omega_{S1E}}\right)\frac{A_5N_0}{P_{S1}^2}+\frac{A_8P_{S2}\Omega_{S2E}}{\Omega_{S1E}P_{S1}^2}\right\}/A_5^2\right]\right.$$
$$\left.\times\left[\exp\left(\frac{A_8}{A_5}\right)Ei\left[-\frac{A_8}{A_5}\right]\left(\frac{A_4-A_8}{A_5}\right)+\frac{A_5}{A_8}\right]\right], \quad (12.60)$$

and the expression of $\frac{\partial I_2}{\partial P_{S1}}$ can be expressed as

$$\frac{\partial I_2}{\partial P_{S1}} = -\frac{P_{S2}\Omega_{S2E}}{\Omega_{S1E}P_{S1}^2}\exp(-B_7)\left[\exp\left(\frac{B_8}{B_5}\right)Ei\left(-\frac{B_8}{B_5}\right)(B_4-B_8)\left(1+\frac{1}{B_5}\right)+B_5\right]. \quad (12.61)$$

We can find the final expression of Eq. (12.59) by putting $\frac{\partial I_1}{\partial P_{S1}}$ from Eq. (12.60) and $\frac{\partial I_2}{\partial P_{S1}}$ from Eq. (12.61) in Eq. (12.59). The obtained expression is a transcendental equation, which the bisection method of the numerical method is used to solve. The solution of the optimal value is shown in the next subsection.

Next, the optimality of $\alpha$ can be found following the same procedure as we have followed to find the optimality of $P_{S1}$. The derivatives of the SOP with respect to $\alpha$ is expressed as

$$\frac{\partial P_{OUT}^{G\_SEC}}{\partial \alpha} = \frac{\partial I_1}{\partial \alpha}(1-I_2)+(1-I_1)\frac{\partial I_2}{\partial \alpha} = 0 \quad (12.62)$$

where $\frac{\partial I_1}{\partial \alpha}$ can be expressed as

$$\frac{\partial I_1}{\partial \alpha} = -\frac{(A_6-1)N_0}{P_R\Omega_{R1S2}\alpha^2}\exp(-A_7)\left[1-\frac{A_4-A_8}{A_5}\left\{\exp\left(\frac{A_8}{A_5}\right)Ei\left(-\frac{A_8}{A_5}\right)\right\}\right]$$
$$+\left[\frac{A_4}{A_5}+\left\{\frac{1}{A_5}-\frac{A_4-A_8}{A_5^2}\right\}\frac{A_6N_0}{P_R\Omega_{R1S2}\alpha^2}\right]\exp\left(\frac{A_8}{A_5}-A_7\right)Ei\left(-\frac{A_8}{A_5}\right)$$
$$-\exp(-A_7)\frac{A_6N_0}{A_8P_R\Omega_{R1S2}\alpha^2}, \quad (12.63)$$

and $\dfrac{\partial I_2}{\partial \alpha}$ can be expressed as

$$\dfrac{\partial I_2}{\partial \alpha} = \dfrac{(B_6-1)N_0}{P_R \Omega_{R2S1}(1-\alpha)^2} \exp(-B_7)\left[1-\dfrac{B_4-B_8}{B_5}\left\{\exp\left(\dfrac{B_8}{B_5}\right)Ei\left(-\dfrac{B_8}{B_5}\right)\right\}\right]$$

$$+\left[\dfrac{B_4}{B_5}+\left\{\dfrac{B_4-B_8}{B_5^2}-\dfrac{1}{B_5}\right\}\dfrac{B_6 N_0}{P_R \Omega_{R2S1}(1-\alpha)^2}\right]\exp\left(\dfrac{B_8}{B_5}-B_7\right)Ei\left(-\dfrac{B_8}{B_5}\right) \quad (12.64)$$

$$+\exp(-B_7)\dfrac{B_6 N_0}{B_8 P_R \Omega_{R2S1}(1-\alpha)^2}.$$

From the expression in Eqs. (12.63) and (12.64), we obtain the final expression of Eq. (12.62) as a transcendental equation, which the bisection method of the numerical method is used to solve. The solution of the optimal value is shown in the next subsection.

### 12.3.3 Numerical Results Based on the SOP and Observation of Optimal Value

The numerical values for the two-way communication with DF relays are given in Table 12.2.

From Figure 12.9, we take the help of the graph and consider the two initial points from the graph in which one point from the left side and another is the right side of a ponit where curves cross $\dfrac{\partial P_{OUT}^{G\_SEC}}{\partial P_{S1}}=0$. Using the bisection method, which gives the roots for the different $P_{S2}$, the obtained roots are 5.5590, 8.4280, 11.7803, and 15.8188 dBW, for 0, 5, 10, and 15 dBW of $P_{S2}$, respectively. These obtained roots are the approximate optimal values for the corresponding power. In Figure 12.9, the points, where the derivative with respect to $P_{S1}$ crosses the zero line, are the optimal point of the $P_{S1}$ at which minimum SOP is obtained and performance is high. The obtained optimal values are approximately equal to obtained optimal points by the graphical method.

In the same way, we find the optimal point of $\alpha$ by the graphical method and the bisection method. By the graphical method, these roots are 0.5 for the equal power

### TABLE 12.2
### Numerical Parameters with Their Values

| Numerical Parameters | Values of Numerical Parameter |
|---|---|
| Transmit power of sources and relays ($P_{S1} = P_{S2} = P_R$) | 0 dBW, 5 dBW, 10 dBW, 15 dBW |
| Channel mean power of all links ($\Omega_{ij}$) except EAV links | 1 |
| Channel mean power of EAV links ($\Omega_{ij}$) | 0.5 |
| Threshold secrecy rate ($C_{TH}^{G\_SEC}$) | 0.5 b/s/Hz |
| AWGN power | $10^{-2}$ |

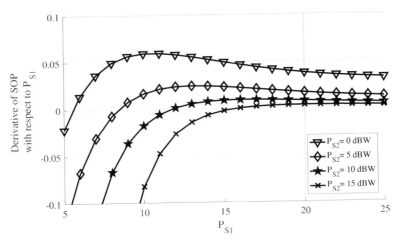

**FIGURE 12.9** Finding the optimal values of $P_{S1}$ with fixed values of $P_{S2}$ and $\alpha$.

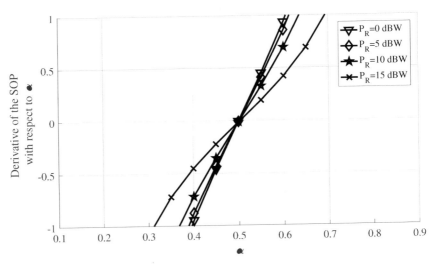

**FIGURE 12.10** Finding an optimal value of $\alpha$ for equal "transmit power of the source" (TPS).

of both the source as shown in Figure 12.10 and 0.4 for the unequal power (i.e., $P_{S1} = 5$ dBW and $P_{S2} = 10$ dBW) of both the sources as shown in Figure 12.11. With the bisection method, we find the approximate optimal points of $\alpha$ for equal transmit power of both the sources are 0.5006, 0.5014, 0.5021, and 0.5041, for 0, 5, 10, and 15 dBW of $P_R$, respectively. Further, we also find the approximate optimal points of $\alpha$ for unequal transmit power (i.e., $P_{S1} = 5$ dBW and $P_{S2} = 10$ dBW) with the bisection method. These optimal values are 0.3787, 0.3936, 0.3830, and 0.3732 for 0, 5, 10, and 15 dBW of $P_R$, respectively, which are nearly equal to the optimal values obtained with the graphical solution.

# Physical Layer Security in Two-Way Wireless Communication System

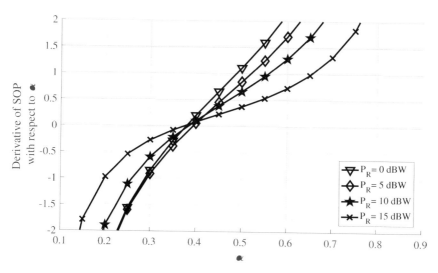

**FIGURE 12.11** Finding an optimal value of $\alpha$ for unequal TPS.

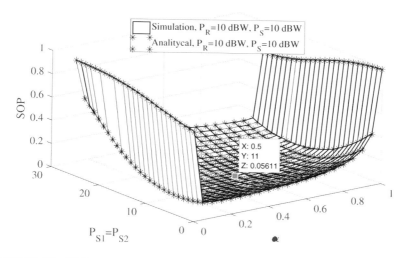

**FIGURE 12.12** SOP versus fraction of relay power ($\alpha$) and TPS one ($P_{S1}$).

Figure 12.12 shows the relation of SOP with respect to $\alpha$ and $P_{S1}$. We observe that for a fixed value of $P_{S1}$, with an increase in $\alpha$, SOP decreases and increases again. It is more visible in Figure 12.13, in which both the sources use the equal transmit power to transmit their own signal. Before the optimal values, as an increase in $\alpha$, increases in which is assigned to R1 and that for R2 decreases, but R2 still has sufficient power because fraction $(1 - \alpha)$ decreases from 1 to (1-optimal value). Resulting, the R1 to S2 channel capacity increases, but due to the constant power of the sources, the channel capacities of EAV remain unchanged. Therefore, the SOP decreases with increase in secrecy capacity, and thus performance improves. On the other hand, beyond the optimal values, as $\alpha$ increases, the fraction $(1 - \alpha)$ decreases

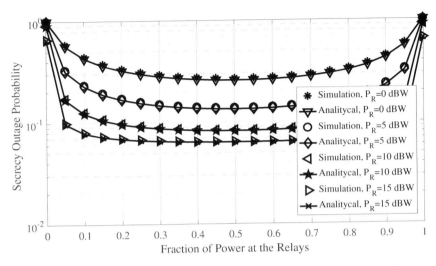

**FIGURE 12.13** SOP versus fraction of relay power ($\alpha$) and TPS one ($P_{S1}$).

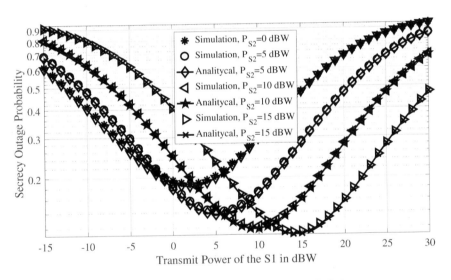

**FIGURE 12.14** SOP versus fraction of relay power ($\alpha$) and TPS ($P_{S1}$).

more, and corresponding assigned power to R2 decreases significantly. The end-to-end channel capacity decreases significantly, and the secrecy capacity also reduces. Thus, SOP increases, and performance reduces.

Next, in Figures 12.12 and 12.14, we also observe that for a constant value of $\alpha$ and $P_{S2}$, $P_{S1}$ also has the optimal value at which the SOP is minimum. We see in a particular curve of Figure 12.14, before the optimal value, $P_{S1}$ is less than $P_{S2}$. In this case, EAV is able to eavesdrop the message of S2 due to low value of $P_{S1}$. As $P_{S1}$ increases, interference at the EAV also increases, and thus secrecy capacity increases. The SOP decrease with increase in secrecy capacity before the optimal point. After the optimal point, the value of $P_{S1}$ becomes greater than the value of $P_{S2}$. In this case, the message

## Physical Layer Security in Two-Way Wireless Communication System 235

of S1 is eavesdropped by the EAV while the information signal of S2 creates interference, which is almost constant due to a fixed transmit power of the S2. Increase in $P_{S1}$ causes increase in eavesdropping by the EAV. Thus, with decrease in the secrecy capacity, performance degrades as the SOP increases beyond the optimal point of $P_{S1}$. The global optimal value $\alpha$ and $P_{S1}$ has been observed, which is $\alpha = 0.5$ and $P_{S1} = 11$ dBW.

Figures 12.15 and 12.16 show the results of the SOP versus relay power when both the sources transmit the signal with equal and unequal transmit power, respectively.

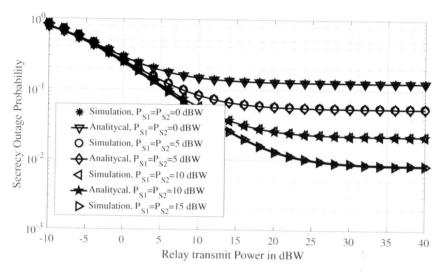

**FIGURE 12.15** SOP versus fraction of relay power ($\alpha$) for different equal TPS ($P_{S1} = P_{S2}$).

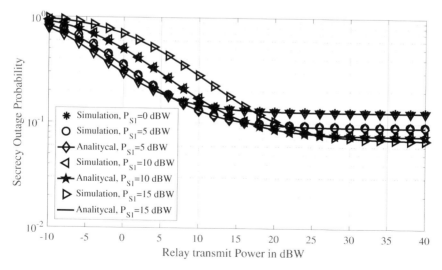

**FIGURE 12.16** SOP versus fraction of relay power ($\alpha$) for different transmit power of the S1 ($P_{S1}$) with constant transmit power of S2 ($P_{S2}$).

In both the figures, we observe that with increases in the relay power, the SOP decreases before the flooring point. The relay power only increases the channel capacities from the relay to the destination. The EAV channel capacities for both the information signal remain almost unchanged due to the fixed TPSs. Thus, up to a particular value of the relay transmit power, SOP decreases. After that, the end-to-end main channel capacity becomes unchanged due to the DF relay protocol as per Eqs. (12.43) and (12.44). Now, the EAV channel capacity and the main channel capacity are also almost constant, then the secrecy capacity remains unchanged and the floor SOP curve is obtained. Further in Figure 12.16, before the flooring point, we observe that the SOP increases for a particular value of $P_{S2} = 0$ dBW when $P_{S1}$ increases from 0 dBW to upward values. If $P_{S1}$ increases, then EAV eavesdrops the message of the S1 from the S1's information signal, while at the EAV, the S2's information signal, which is almost constant, jams the EAV in eavesdropping of the S1's information signal. Thus, with an increase in $P_{S1}$, the SOP increases before its flooring point and performance degrades, but after the flooring point, the performance increases slightly, i.e., increase in $P_{S1}$ with constant $P_{S2}$ causes more eavesdropping.

## 12.4 CONCLUSION

TWC becomes feasible with half-duplex "energy harvesting AF relays" or "DF relays." TWC is possible with half-duplex DF relays, which are assisted with directional antennas. The intermediate AF relay may be untrusted, or an external eavesdropper may eavesdrop the signal in DF relaying. Secrecy in such networks is accessed in terms of the secrecy capacity and secrecy outage probability. The PLS is investigated in TWC with half-duplex AF or DF relays. With multiple untrusted AF relays in CRN, security has been investigated under relay selection with maximum secrecy capacity. The CRN with AF relays secures the message under cognitive constraint and energy harvesting relay nodes. We have investigated the optimal point of the PS factor for energy harvesting, which maximizes secrecy outage. The transmit power of cognitive nodes increases with an increase in primary transmit power, and secrecy performance increases, while the increase in threshold outage rate of the primary network decreases the transmit power of the cognitive nodes and secrecy performance decreases. Further, the information security increases with an increase in number of relays.

In TWC with DF relays, we find an optimal value of transmit power of a particular source, which depends on the transmit power of the other source. It is also seen that the optimal value of a particular source is almost equal to the transmit power of another source. We have also searched the optimal point of the fraction of relay power. The global optimal points of the fraction of relay power and the source transmit power for a particular source is investigated. Now, the secrecy performance increases with an increase in relay power, while the increase in the transmit power of a particular source does not always increase the performance. The study presented is useful in designing a TWC network with the help of intermediate relay under eavesdropping by untrusted relay or external eavesdropper.

# REFERENCES

Jameel, Furqan, Zheng Chang, and Tapani Ristaniemi. 2018. "Intercept Probability Analysis of Wireless Powered Relay System in Kappa-Mu Fading." In *2018 IEEE 87th Vehicular Technology Conference (VTC Spring)*, Porto, Portugal, 1–6.

Kalamkar, Sanket S, and Adrish Banerjee. 2017. "Secure Communication via a Wireless Energy Harvesting Untrusted Relay." *IEEE Transactions on Vehicular Technology* 66 (3): 2199–2213.

Khandaker, Muhammad R A, Kai-Kit Wong, and Gan Zheng. 2017. "Truth-Telling Mechanism for Two-Way Relay Selection for Secrecy Communications with Energy-Harvesting Revenue." *IEEE Transactions on Wireless Communications* 16 (5): 3111–3123.

Pan, Gaofeng, Chaoqing Tang, Tingting Li, and Yunfei Chen. 2015. "Secrecy Performance Analysis for SIMO Simultaneous Wireless Information and Power Transfer Systems." *IEEE Transactions on Communications* 63 (9): 3423–3433.

Sakran, Hefdhallah, Mona Shokair, Omar Nasr, S El-Rabaie, and A A El-Azm. 2012. "Proposed Relay Selection Scheme for Physical Layer Security in Cognitive Radio Networks." *IET Communications* 6 (16): 2676–2687.

Salem, Abdelhamid, Khairi Ashour Hamdi, and Khaled M Rabie. 2016. "Physical Layer Security with RF Energy Harvesting in AF Multi-Antenna Relaying Networks." *IEEE Transactions on Communications* 64 (7): 3025–3038.

Sharma, Shashibhusham, Sanjay Dhar Roy, and Sumit Kundu. 2019. "Secrecy Performance of a Two-Way Communication Network with Two Half-Duplex DF Relays." *IET Communications* 13 (5): 620–629.

Sharma, Shashibhushan, Sanjay Dhar Roy, and Sumit Kundu. 2020. "Two-Way Secure Communication with Multiple Untrusted Half-Duplex AF Relays." *Wireless Personal Communications* 110 (4): 2045–2064.

Son, P.N. and Kong, H.Y. 2015. "Cooperative Communication with Energy-harvesting Relays under Physical Layer Security." *IET Communication* 9 (17): 2131–2139

Sun, Li, Taiyi Zhang, Yubo Li, and Hao Niu. 2012. "Performance Study of Two-Hop Amplify-and-Forward Systems with Untrustworthy Relay Nodes." *IEEE Transactions on Vehicular Technology* 61 (8): 3801–3807.

Tran, Hung, Hans-Jurgen Zepernick, and Hoc Phan. 2013. "Cognitive Proactive and Reactive DF Relaying Schemes under Joint Outage and Peak Transmit Power Constraints." *IEEE Communications Letters* 17 (8): 1548–1551.

Wang, Chao, and Hui-Ming Wang. 2014. "On the Secrecy Throughput Maximization for MISO Cognitive Radio Network in Slow Fading Channels." *IEEE Transactions on Information Forensics and Security* 9 (11): 1814–1827.

Wang, Hui-Ming, Miao Luo, Qinye Yin, and Xiang-Gen Xia. 2013. "Hybrid Cooperative Beamforming and Jamming for Physical-Layer Security of Two-Way Relay Networks." *IEEE Transactions on Information Forensics and Security* 8 (12): 2007–2020.

Wyner, Aaron D. 1975. "The Wire-Tap Channel." *Bell System Technical Journal* 54 (8): 1355–1387.

Xu, Hongbin, Li Sun, Pinyi Ren, and Qinghe Du. 2015. "Securing Two-Way Cooperative Systems with an Untrusted Relay: A Constellation-Rotation Aided Approach." *IEEE Communications Letters* 19 (12): 2270–2273.

Zhang, Chensi, Jianhua Ge, Jing Li, Fengkui Gong, and Haiyang Ding. 2017. "Complexity-Aware Relay Selection for 5G Large-Scale Secure Two-Way Relay Systems." *IEEE Transactions on Vehicular Technology* 66 (6): 5461–5465.

Zhang, Rongqing, Lingyang Song, Zhu Han, and Bingli Jiao. 2012. "Physical Layer Security for Two-Way Untrusted Relaying with Friendly Jammers." *IEEE Transactions on Vehicular Technology* 61 (8): 3693–3704.

Zou, Yulong, Xuelong Li, and Ying-Chang Liang. 2014. "Secrecy Outage and Diversity Analysis of Cognitive Radio Systems." *IEEE Journal on Selected Areas in Communications* 32 (11): 2222–2236.

# 13 Design and Simulation of Bio-Inspired Algorithm
## Based Cognitive Radio for 5G Networks

*Sadaf Ajaz Khan, Javaid A. Sheikh,
Tanzeela Ashraf, and Mehboob-ul-Amin*
University of Kashmir

## CONTENTS

| | | |
|---|---|---|
| 13.1 | Introduction | 240 |
| 13.2 | Cognitive Radio | 242 |
| 13.3 | Introduction to Genetic Algorithm | 243 |
| | 13.3.1 GA Nomenclature | 244 |
| | 13.3.2 Classification of GA | 244 |
| | 13.3.3 Algorithm Outline | 245 |
| | 13.3.4 Fitness Value | 246 |
| | 13.3.5 Mutation | 247 |
| | 13.3.6 Crossover | 247 |
| 13.4 | System Model | 247 |
| 13.5 | Optimization Problem | 250 |
| 13.6 | Framework for Optimal Solution | 250 |
| 13.7 | Results and Discussions | 251 |

SCENARIO 1: P A AAA: When Only one PU is in Attendance and All Other Slots are Vacant .................................................. 252
SCENARIO 2: P P AAA: When One PU and One Secondary User (SU) are in Attendance and the Remaining Three Slots are Vacant ............ 252
SCENARIO 3: P PPA A: When One PU and Two SUs are in Attendance, and the Remaining Two Slots are Vacant ........................ 253
SCENARIO 4: P PPP A: When One PU and Three SUs are in Attendance, and the Remaining One Slot is Vacant ......................... 254
SCENARIO 5: P PPP P: When All Slots are Occupied ........................ 254

13.8 Conclusion and Future Scope ...................................................................256
References ........................................................................................................257

## 13.1 INTRODUCTION

Wireless communications are evolving at a rapid pace, and the quest for new applications and services is driving researchers toward the rapid introduction of new technologies (such as 5G) in the marketplace. In the information and communications technology (ICT) industry, 5G arrangements are thought to be the prime facilitator and infrastructure contributor by means of presenting a diversity of services with miscellaneous requirements (Li et al. 2017). Three types of usage scenarios are recommended by the International Telecommunication Union (ITU) for the fifth generation (5G) of mobile communication systems. First, broadband multimedia to humancentric use cases is endowed by the enhanced mobile broadband (eMBB) (Prathisha 2018). The eMBB addresses bandwidth-famished applications such as video streaming and aggrandized reality. Second, to account for more assorted and resource famished applications the ultrareliable low-latency service (URLLC) is presented. URLLC has stern requirements in terms of latency (of the order of ms) and reliability for applications like self-governed driving, drones, and tangible Internet. The third scenario includes massive machine-type communications (mMTC). The mMTC is a service class primarily used for connecting enormous figures of devices and broadcasting the petite amount of timely susceptible information. The mMTC is used for facilitating substantial sensing, scrutinizing, and metering in support of enormous deployments of Internet of Things (IoT) (Yao et al. 2019). Communication networks, for over a century now, have been devised through the primary goal of performance optimization (for criterion such as data rate, throughput, and latency). In the most recent decade, energy efficiency has turned out as an eye-catching figure of merit owing to monetary and functioning problems and environmental apprehensions. Therefore, the design of emerging 5G arrangements will need to reckon the energy consumption and efficiency as the prime facilitator necessarily (Buzzi et al. 2016.). These next generation networks will have to serve an unparalleled number of devices, while providing ubiquitous connectivity and pioneering services. According to the forecast, by 2020 the number of connected devices will surpass 50 billion, which in turn indicate six connected devices per person, including both human and machine-type communications. The intention is to create an associated society where all sensors, cars, and wearable devices will make use of already existing cellular structures and give rise to various new and innovative services like smart cities, smart cars, etc. Therefore, if the future networks have to deliver to such huge numbers of devices, then these networks will call for expanded capacities in contrast with capacity provided by current networks. It is estimated that in 5G networks, the volume of traffic will increase per month to exabytes ($1000^6$ bytes). This means that the capacity provided by these networks must be at least 1,000 times larger than what is provided by the current networks. In order to achieve the above-mentioned goal of increased capacity, depending on the architecture and theory of current networks is unbefitting as it brings about an energy crisis. There is a shift toward a heterogeneous paradigm in wireless networks so as to entertain the escalating demand for information transfer. A futuristic 5G network comprises numerous layers of cells such as macro, femto, pico, and relays. The network coverage and capacity is improved by efficient reuse of radio resources. For the purpose of accomplishing the increased capacity, the cells in a 5G network must be densely deployed, i.e., almost 40–50 times denser than the current 4G networks. However, new challenges are posed toward the networks due to this dense deployment of nodes. A classic 5G node has to configure about 2,000 parameters subject to the application of the latest air interface methods, wide-ranging services, and terminals. According to an estimate, the operational complexity in 5G will increase 53–67 times compared to 4G. Configuring the network entirely by humans is an extremely difficult and time-consuming task for the operators. Therefore, increasing automation in wireless mobile networks has now become obligatory. A potential solution for this involves initiating artificial intelligence (AI) for the control and management of the network, as an alternative of labor-intensive optimization practice. The concept of self-operation network (SON) has previously evolved in second-, third-, and fourth-generation networks, but instead of interacting with the environment and smart decisions, the computerization is realized by predefined policies. The cellular networks in the 5G era have different choices for access and facility providing methods and therefore achieve the groundwork for applying preliminary intelligence. For realizing the self-organizing features including self-configuration, self-optimization, and self-healing, 5G networks are required to enhance their intelligence. With frequently changing existing services and constantly evolving new types of services, 5G cellular networks still fall short in functionalities for the purpose of automatically recognizing an innovative form of service, deduce suitable prerequisite method, and set up the necessary network slice. Fortunately, AI provides solutions regarding how to gain knowledge of deviations, categorization of the problems, anticipation of potential challenges, and finding possible explanations by collaborating with the surroundings. Consequently, for hastening the evolution toward an intelligent 5G era, the cellular networks may possibly exploit cognitive radio approach and collaborating with the setting using AI. AI is the intellect established by machines compared to natural human intelligence. AI enables the machines to resolve the problems intelligently similar to human beings (Russel and Norvig 2013). AI facilitates imitating of human intellect by machines and "gradual learning" from the surroundings, to decipher the issues by augmenting the success probability. By including AI, the networks are self-governing and administered more proficiently, thereby improved performance is attained. AI has advanced to multidisciplinary techniques, for instance, machine learning, optimization theory, game theory, control theory, and meta-heuristics (Long et al. 2017). Among these, the most essential subfield of AI is machine learning. This field is inspired from nature, and its mechanism also relies on it. Nowadays various machine learning systems hugely rely on cognitive technology. Another field that depends on nature's bounty is the field of genetic algorithms (GAs) and genetics-based machine learning. There have been significant advancements in GA from theory as well as application point of view such that genetics-based systems are finding their way into everyday commercial use (Badoi et al. 2011).

## 13.2 COGNITIVE RADIO

In order to cater to the requirements of 5G (like high data rates and system capacity, lower latencies, low cost, etc.) current approaches are not sufficient. Therefore, besides the existing technologies, some novel technologies (like FBMC, NOMA, massive MIMO, mmWaves) (Andrews et al. 2014; Mitra, and Agrawal. 2015) must be adopted to meet the demands of 5G networks. 5G systems call for new network management schemes that are improved by machine learning techniques to extort facts from the system and steady learning in the attendance of intrinsic uncertainties. The demand for higher data rates is increasing tremendously and hence will require more spectrums, which is a scarce resource. So, efficient utilization of the spectrum is an important factor to meet the demands of the customers. This results in amalgamation of technology known as cognitive radio (CR) in 5G. CR was proposed by Joseph Mitola at a conference in 1999. The CR technology appeared in reaction to the sparse spectrum resource crisis with the objective of optimum spectrum utilization with the intention of enhancing the efficiency of the spectrum (Goldberg and Holland 1988). The U.S. Federal Communications Commission (FCC) defines cognitive radio as an aware adaptive radio that senses the radio environment and adapts to it to enable more efficient use of the available spectrum. In CR nomenclature, users who have possession of a license that officially authorizes them to exploit the frequency band for the intention indicated via license are referred to as the primary user (PU). A user that acquires the primary user's spectrum in their absence is known as a secondary user (SU). The approach for such an access must be controlled in such a manner that there is no destructive interference to the PU due to the SU (Mehboob et al. 2016). CR technology is established on the actuality that the spectrum is not always being used by the primary users (i.e., certified systems) and hence remains underutilized. The secondary user has to hand over the spectrum whenever the primary user requires it back. For facilitating bandwidth prerequisites, the SUs opportunistically exploit the spectrum of numerous PUs. Aforesaid opportunistic use of the spectrum is referred as dynamic spectrum access (DSA) and is vital for solving the spectrum scarcity issue. DSA has the following components:

1. *Spectrum sensing:* It is attributed to the capability of cognitive radio to identify the obtainable vacant channels contained by the preexisting licensed bands. The CR must be able to perceive the PU's actions and proceed according to their objectives. Spectrum-sensing operation consists of signal detection, signal classification, and availability of channel.
2. *Spectrum sharing:* Following the identification of available channels, the subsequent move is to share the holes in the spectrum. This provides a notion regarding sharing of spectrum and its exploitation by the radio devices. The schemes for spectrum sharing maybe categorized into horizontal, vertical, and hierarchical spectrum sharing.
3. *Spectrum handover:* This refers to the change in frequency for the purpose of maintaining transmission. Whenever a PU wishes to exploit the same channel again or whenever the transmission is degrading, SU changes its frequency and hands over the spectrum.

# Design and Simulation of Bio-Inspired Algorithm

In 5G networks, CR can be seen as a new solution for interference management. Spectrum-sensing quality CR permits the users to discover the existence or nonexistence of other users so that the interference induced to the erstwhile users in attendance is insignificant.

## 13.3 INTRODUCTION TO GENETIC ALGORITHM

Lately by virtue of effortlessness of installation and operation of untethered wireless medium, wireless communication technology is becoming conventional. However, the main challenge faced is the design of wireless networks because the environmental conditions are highly dynamic, due to which parameter optimization is becoming even more difficult and complex (Mata 2017). Thus the standards for wireless networking are highly dependent on the techniques of machine learning and AI algorithms. . The strategies that employ the computational illustrations of the natural development processes as a means for elucidating problems are called evolutionary algorithms (Guvencet al. 2012). The study of nature has guided researchers who work in the field of AI and machine learning. During the 1950s and 1960s, quite a lot of computer researchers inspected these natural evolutionary systems independently with the insight that the natural process of progression may possibly be used as a tool for optimization in engineering problems. In the 1970s, John Holland invented genetic algorithms and perhaps may be contemplated as a subdivision of the evolutionary algorithms. Genetic algorithms (Mitola et al. 1999) are search algorithms that imitate natural genetic operations artificially (Walters, and Sheble 1993). The genetic algorithm (GA) is an evolutionary algorithm that provides a complete framework by which tasks of AI such as classification, learning, and optimization can be implemented. These are methods for solving optimization problems (constrained as well as unconstrained). The GA mimics the natural evolution process by imitating various procedures by which numerous organisms adapt to their environment. The search for the most favorable solution to an optimization problem is accomplished by moving between the individuals of a previous population to latest population (old to new) using genetics-like operators. Due to their versatile nature and remarkable generality, GA has been adapted in an extensive assortment of applications in wireless networks. GA has been used in various research fields as diverse as the aircraft industry, chip design, computer animation, and drug design, in telecommunications, software design, and financial markets. This algorithm introduced the concept of population. The basic procedure for optimization simply involves processing of the fit individuals in the population in order to generate better offspring intended for the next generation. Thus genetic algorithms can conclude which of the solutions (individuals) are fittest and then make a decision on what individuals will replicate and send forward their genetic code to the later generations. Therefore, genetic algorithm is based on the principle of "survival-of-the-fittest" given by Darwin.

Genetic algorithms do not always endow with the most favorable result when the algorithm runs, and perhaps is considered a population-based meta-heuristic. Although this algorithm endows with a good solution, there is no guarantee to attain the finest solution for that problem. It is a corollary of this method because it provides operations based on randomness. Although it does have random components, to guide the search it engages the knowledge of the current situation despite having random components. Genetic algorithms are basically motivated because of their general and versatile nature. They have adaptive nature in addition to online problem-solving capability and are capable of discovering good building blocks. Genetic algorithms are scalable and parallel in nature and possess multiobjective optimization capability. GAs are simply implemented and provide ease of global optimization.

### 13.3.1  GA Nomenclature

Because this evolutionary algorithm has been derived from the natural process of evolution, quite a few metaphors related to that are used here. The following terms are used:

1. Organism: The unit that is to be optimized (a radio parameter, a wireless resource etc.).
2. Population: It consists of collection of possible solutions.
3. Chromosome: It represents the string of binary symbols that are used to encode the solution for the problem under study. They are submitted to genetic operations.
4. Fitness: In natural process the fit individuals pass on their characteristics to the next generation. It is a measure of how good a solution is.
5. Gene: The components that make up the chromosomes.
6. Allele: Alternative forms taken up by a gene.
7. Locus: It is the location of a gene of interest on a chromosome.
8. Mutation: It represents random changes to an individual from the current population to create children for the next generation.
9. Selection: The process of choosing the most fit individuals so that they can send forward their genes to the later generation and weed out the weaker ones.

### 13.3.2  Classification of GA

Various classifications of genetic algorithms are presented based on number-of-objectives, applications, network type, etc. The categorization for GA is shown in Figure 13.1 (Mata 2017).

1. *Single-objective GAs:* To optimize a single objective that has a scalar value of fitness function, Single-objective GAs (SOGA) is accepted. These SOGAs are easier in computation. In case of multiobjective GAs, these can be made to run several times by means of diverse weight vectors.
2. *Multiple-objective GAs:* Various networking parameters such BER or SNR may need optimization. For such scenarios, multiobjective optimization is more suitable because a single solution may be fit for one objective but may worsen the others. Thus, multiobjective GAs (MOGAs) optimize the overall system simultaneously in all directions using multiple solutions.

# Design and Simulation of Bio-Inspired Algorithm

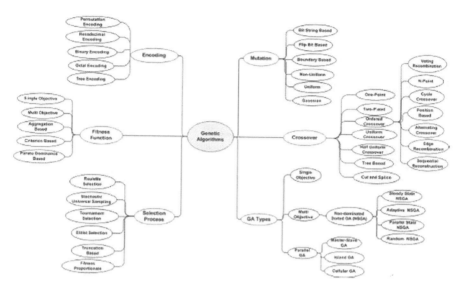

**FIGURE 13.1** Genetic algorithm categorization.

3. *Nondominated sorted GA (NSGA):* An occurrence of MOGA that aims to develop the fitness value for a population to the Pareto front inhibited through a set of objective functions. This method sorts the population into groups called subpopulations derived from the configuration of Pareto dominance. This method calculates the major resemblance between the members of each level with regard to the Pareto front. The ensuing subdivision in conjunction with its resemblance count is used to endorse the solution front that is nondominated.
4. *Parallel GA (PGAs):* There are three main categories of parallel GAs: (a) Master-slave PGAs that require stable communication between master and slave; (b) cellular PGAs that require all agents within each generation of PGA to communicate locally; and (c) island PGAs, which are the best choice for wireless networks as they permit migration policy to be controlled.
5. *Distributed GA or coarse-grained GA:* In this technique also referred to as island GA, the population is divided into several subpopulations referred to as islands that have a weaker interaction with each other. The aforementioned islands communicate among one another through the relocation of individuals. The island GAs are most widely used in wireless networking.

## 13.3.3. Algorithm Outline

The flowchart of the algorithm is shown in Figure 13.2 (Mata 2017). The algorithm primarily begins by creating an arbitrary initial population of candidate solutions called individuals. The performance and efficiency of GA is mostly influenced by the size of population. For small population size, the performance of GA is poor. Each individual in the population is represented by a chromosome that is a sequence of

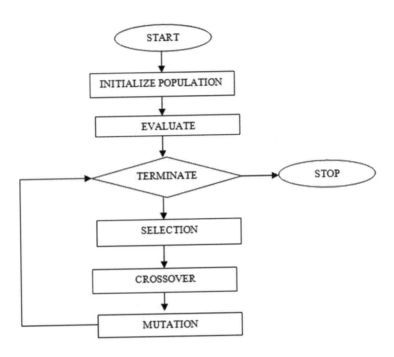

**FIGURE 13.2** Flowchart for GA.

binary symbols typically taken from the binary alphabet. In the next step, a sequence of new population is created by making use of the individuals of the current generation. During every step, the algorithm selects the individuals in the current generation (known as parents) who impart their genes (vector entries) to the children in the new generation. Every member of the current population is scored by computing its fitness score.

The individuals with the best fitness value are selected as parents for next generation. The current population is then exchanged with the next generation's children. A particular individual can be chosen as a parent more than once. This means that this individual will contribute its genes to more than one child. The algorithm stops when one of the stopping criteria is met.

### 13.3.4 Fitness Value

The fitness function, also referred to as the objective function, provides a method for evaluating the individual chromosomes (that represents the design solution). Fitness function can comprise single or multiobjectives. The individual chromosome is the variable input factor for the fitness function. Objective function makes sure that the chromosomes that are to be sent forward to the future generation do not defy any constrictions. A fitness score is assigned to every individual based on the eminence of the individual in respect of the objectives and optimization objective it provides. During fitness estimation, each design solution is simulated or tested. After each

# Design and Simulation of Bio-Inspired Algorithm

phase of concluded simulation, the worst $N$ design solutions are deleted, and new $N$ ones are created from the finest design solutions. Therefore, a figure of merit needs to be applied to each design solution to assess how close it has come to satisfy the overall specification.

### 13.3.5 MUTATION

The mutation operator is used to determine new possibilities and traits by arbitrarily varying genes in chromosomes. Mutation leads to the diversification of population and in so doing enhances the likelihood that individuals with improved fitness values are generated. Mutation enables GA to steer clear of local optima and hence lead them in the direction of global optima. Mutation techniques have numerous types such as bit-string mutation, flip bit mutation, boundary mutation, nonuniform mutation, uniform mutation, and Gaussian mutation.

### 13.3.6 CROSSOVER

Crossover facilitates the algorithm to extract the superlative genes from different individuals and recombine them into potentially better-quality children. Fundamentally, the process is an endeavor in taking advantage of the paramount traits of the existing chromosomes and to blend them with the intention of improving their fitness. A locus is arbitrarily chosen by the crossover operator. To generate an offspring pair, the subsequences prior to and following that locus are swapped among two parent chromosomes. To support further investigation there can be numerous crossover points. Various methods are available for crossover of chromosomes such as one-point crossover, two-point crossover, uniform crossover and half uniform crossover, cut and splice, three-parent crossover, and ordered chromosome crossover.

## 13.4 SYSTEM MODEL

Think of a future (5$^{th}$) generation cellular network comprising of primary and secondary users denoted by PUs and SUs, respectively. The system model for the said network is shown in Figure 13.3.The location of PU fluctuates with respect to its $K^{th}$ position within the cell. By the side of every $K^{th}$ location, a macro base station (MBS) services a PU. Assume that m ϵM users in totality (PU and SUs) dispersed within the cell, while SUs pursue Gaussian-type distribution for sufficiently huge numeral of variables.

The $h_{PU}$ is used to designate the medium (channel) linking MBS and a PU at $K^{th}$ position, and $h_{CU}$ is used to designate the channel linking PU at $K^{th}$ position and $m^{th}$ SU. Presume that channel state information (CSI) is completely acknowledged both at MBS and as well as at PU. A link called backhaul link connects MBS and PU. Another link called access link connects PU and channel. For the purpose of downlink transmission, time division duplexing mode is employed. For the duration of initial (first) time slot, the received signal at $K^{th}$ position of PU is given by:

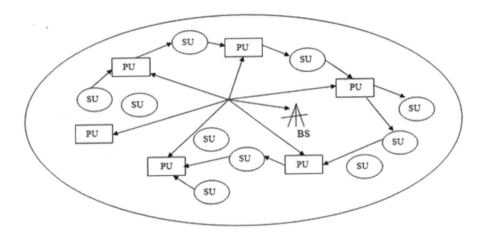

**FIGURE 13.3** System model.

$$Y_{mk} = \sum_{m=1}^{M}\sum_{K=1}^{N} h_{PU}^{m} X + n_m \qquad (13.1)$$

where $n_m$ correspond to AGWN, such that $K$ takes values from 1 to $N$ (i.e., 1, 2, 3, 4, 5 .... $N$) and entails that $N$ different positions are intended for PU within the system. The received symbol is linearly processed by the PU at the $K^{th}$ spot via exploiting a weighting matrix designated as "$w_{mk}$." During the second time slot, the said PU transmits: $T_{mk=} Y_{mk} G_{mk}$ in the direction of access link, where $G_{mk}$ symbolizes the genetic algorithm vector that broadcasts the signal in the required direction and curbs whichever interference is present amid PU and SU.

Therefore, for the duration of subsequent (second) time slot, the arriving signal can be represented as:

$$Z_{mk} = \sum_{m=1}^{M}\sum_{k=1}^{N} h_{SU}^{m} T_{mk} + n_m. \qquad (13.2)$$

The cumulative power limitation at $K^{th}$ PU and $m^{th}$ SU is specified by:

$$\sum_{k=1}^{Kmax} E\{E_r(Y_{mk}Y_{mk}^*)\} \le P_{PU}^{k} \qquad (13.3)$$

$$\sum_{m=1}^{m} E\{E_r(Z_{mk}Z_{mk}^* \le P_{SU}^{m})\}. \qquad (13.4)$$

# Design and Simulation of Bio-Inspired Algorithm

The cumulative channel capacities of the contour, i.e., PU and SUs, are to be deliberated for the purpose of defining the optimization problem. Total system throughput will be same as

$$C_{Tout} = \sum_{m=1}^{M}\sum_{k=1}^{N} R_{m,k}\left(G_m^-\right), \quad (13.5)$$

where $R_{m,k}\left(G_m^-\right) = \sum_{m=1}^{M}\sum_{n=1}^{k}(C_{PU}, C_{SU})$.

Here, throughput of PU and SU is indicated as $C_{PU}$ and $C_{SU}$, respectively. Both $C_{PU}$ and $C_{SU}$ follow the normalized and modified Shannon Hartley capacity formulae in the unit of capacity known as Shannon.

$$C_{PU} = \log_2\left(1 + \frac{h_{PU}^m \left(h_{PU}^m\right)^H Y_{mk}\left(Y_{mk}\right)^H G_{PU}^k \left(G_{PU}^k\right)^H}{n_m}\right) \quad (13.6)$$

$$C_{SU} = \log_2\left(1 + \frac{h_{SU}^m \left(h_{SU}^m\right)^H Z_{mk}\left(Z_{mk}\right)^H G_{SU}^k \left(G_{SU}\right)^H}{n_m}\right), \quad (13.7)$$

where $G_{PU}$, $G_{SU}$ are genetic algorithm-based power allocation coefficients for primary (PU) and secondary user (SU), respectively.

The cumulative power expenditure for PU and SUs is as follows:

$$P_{PU} = \sum_{m=1}^{M}\sum_{k=1}^{N} p_{PU} \quad (13.8)$$

Subject to $\sum p_{PU} \leq 1$

and

$$P_{SU} = \sum_{m=1}^{M}\sum_{k=1}^{N} p_{SU}. \quad (13.9)$$

subject to $\sum p_{CU} \leq 1$ where $p_{PU}$ and $p_{SU}$ represent the power allocation coefficients for PU and SU, respectively. Thus cumulative power expenditure in the aforementioned contour is given as:

$$P_{Total} = p_{PU} + p_{SU}. \quad (13.10)$$

The energy efficiency will thus be same as:

$$EE = \frac{C_{Tout}}{P_{Total}}. \qquad (13.11)$$

## 13.5 OPTIMIZATION PROBLEM

The optimization problem is formulated as follows:

$$\max_{p_{PU},\, p_{SU}} C_{Tout} \qquad (13.12)$$

$$\max_{p_{PU},\, p_{SU}} EE \qquad (13.13)$$

subject to following constraints: P1:

$$\sum_{m=1}^{M}\sum_{k=1}^{N} p_{PU} \leq P_{PU} \qquad (13.14)$$

$$\sum_{m=1}^{M}\sum_{k=1}^{N} p_{SU} \leq P_{SU} \qquad (13.15)$$

$$p_{PU} \leq 1 \qquad (13.16)$$

$$p_{SU} \leq 1 \qquad (13.17)$$

$$\sum_{m=1}^{M}\sum_{k=1}^{N} p_{PU} h_{PU} \leq J_{th}^{PU} \qquad (13.18)$$

$$\sum_{m=1}^{M}\sum_{k=1}^{N} p_{SU} h_{SU} \leq J_{th}^{SU}, \qquad (13.19)$$

where $C_{Tout} = \sum_{m=1}^{M}\sum_{k=1}^{N} R_{m,k}\left(G_m^{-}\right)$ is the aggregate system throughput, and $J_{th}^{PU}$ and $J_{th}^{SU}$ are interference threshold for $K^{th}$PU and $m^{th}$SU link.

## 13.6 FRAMEWORK FOR OPTIMAL SOLUTION

Suppose the fractional programming (FP) problem with nonconcave fractional objective function and nonlinear constraints is represented as P1, and it's difficult to find the optimal solution. Thus P1 is transformed into a subtractive problem as:

$$P2: \max\left(C_{Tout} - \lambda\left(p_{PU} + p_{SU} + \mu_{mk} C_{Tout}\right)\right) \qquad (13.20)$$

$$\text{S.T.} \quad \left(C_{PU}, C_{SU}\right), \qquad (13.21)$$

where $\lambda = (\lambda_{1k}, \lambda_{2k}, \lambda_{3k}, \ldots \lambda_{mk})$ is the Lagrangian multiplier vector, and $\mu_{mk}$ is a symbol of Lagrangian multiplier associated with $m^{th}$ power constant on both links. We can represent the Lagrangian function as:

$$L = \sum_{m=1}^{M}\sum_{k=1}^{N}\left[\log_2\left(1 + \frac{h_{PU}^m\left(h_{PU}^m\right)^H Y_{mk}\left(Y_{mk}\right)^H G_{PU}^k\left(G_{PU}^k\right)^H}{n_m}\right)\right.$$

$$+ \log_2\left(1 + \frac{h_{SU}^m\left(h_{SU}^m\right)^H Z_{mk}\left(Z_{mk}\right)^H G_{SU}^k\left(G_{SU}^k\right)^H}{n_m}\right) \quad (13.22)$$

$$\left. + \mu_{mk} P_{Total} \sum\left(p_{PU} - P_{PU}\right)\sum\left(p_{SU} - P_{SU}\right)\right].$$

For the purpose of attaining an optimal solution, Karush Kuhn Tucker condition (KKT) can be employed to solve the concavity of the problem.

After solving $\dfrac{\partial L}{\partial p_{PU}} = 0$ and $\dfrac{\partial L}{\partial p_{SU}} = 0$,

KKT can be used to attain optimal solutions as:

$$p_{PU} = \mu_{mk}\left[\sum_{m=1}^{M}\sum_{k=1}^{N}\frac{\left(P_{Total}\mu_{mk} + G_{PU}^k - n_m\right)}{\sum P_{PU}}\right] \quad (13.23)$$

$$p_{SU} = \mu_{mk}\left[\sum_{m=1}^{M}\sum_{k=1}^{N}\frac{\left(P_{Total}\mu_{mk} + G_{SU}^k - n_m\right)}{\sum P_{SU}}\right]. \quad (13.24)$$

## 13.7 RESULTS AND DISCUSSIONS

It is a well-established fact that 5G systems form the base of the next generation networks. Numerous techniques are being proposed for their implementation. This work aims at studying the implementation of future-generation networks using GA. We have incorporated the concept of cognitive radio where five cognitive signals act as five primary users. We are considering five cases starting from when only one slot in the channel is being used to the case where all slots are being filled by the users. We are aiming to increase the channel capacity and spectral efficiency of the network. We have tried to increase the spectral efficiency using eNB, AF relaying, and DF coding, which give satisfactory results. To further increase the spectral efficiency, proposed genetic algorithm is incorporated. Spectral efficiency has been calculated at both the 5th and 50th percentiles of cdf. This has been done using the bio-inspired GA in Matlab software, and the corresponding simulation results are discussed.

### SCENARIO 1: P A AAA: WHEN ONLY ONE PU IS IN ATTENDANCE AND ALL OTHER SLOTS ARE VACANT

As shown in Figure 13.4, the eNB (enhanced node B) provides the least value of spectral efficiency. When amplify and forward relaying (AF relaying) is applied, the spectral efficiency somehow increases. A higher value for spectral efficiency can be obtained by using decode and forward coding (DF coding). The genetic algorithm provides the superior value at numerous percentiles of cdf for spectral efficiency. The maximum spectral efficiency provided by DF coding on the 5th percentile cdf is 0.5 bits/s/Hz, while the value for GA is 1.6 bits/s/Hz. At the 50th percentile cdf, DF coding provides a spectral efficiency 2 bits/s/Hz, while the value of spectral efficiency using GA is 15 bits/s/Hz. The net increase in the value of spectral efficiency at the 50th percentile of cdf is 650%.

### SCENARIO 2: P P AAA: WHEN ONE PU AND ONE SECONDARY USER (SU) ARE IN ATTENDANCE AND THE REMAINING THREE SLOTS ARE VACANT

The SU senses the spectrum and after finding an empty slot occupies it. As shown in Figure 13.5, the eNB (enhanced node B) provides the least value of spectral efficiency. When amplify and forward relaying (AF relaying) is applied, the spectral efficiency somehow increases. A higher value for spectral efficiency can be obtained by using decode and forward coding (DF coding).

The genetic algorithm provides the superior value at numerous percentiles of cdf for spectral efficiency. The maximum spectral efficiency provided by DF coding on the 5th percentile of cdf is 6 bits/s/Hz, while the value for GA is 32 bits/s/Hz. At the

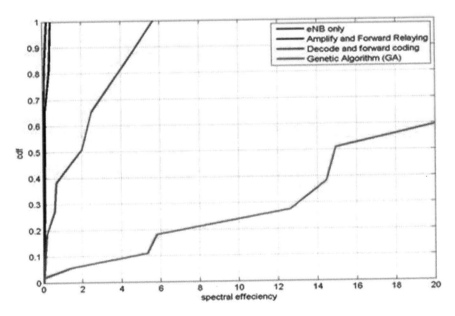

**FIGURE 13.4** Plot of cdf vs. spectral efficiency when only a single primary user is present, and the rest of slots are vacant.

# Design and Simulation of Bio-Inspired Algorithm

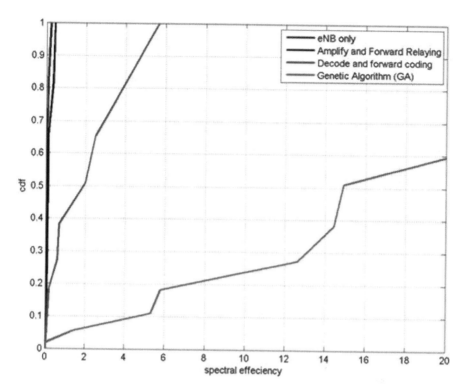

**FIGURE 13.5** Plot of cdf vs. spectral efficiency when one PU and one SU are present, and the rest of the slots are vacant.

50th percentile of cdf, DF coding provides a spectral efficiency 18 bits/s/Hz, while the value of spectral efficiency using GA is 200 bits/s/Hz. The net increase in the value of spectral efficiency at the 50th percentile of cdf is 1011%.

## SCENARIO 3: P PPA A: WHEN ONE PU AND TWO SUs ARE IN ATTENDANCE, AND THE REMAINING TWO SLOTS ARE VACANT

The SUs sense the spectrum and after finding an empty slot occupy it. As shown in Figure 13.6, the eNB (enhanced node B) provides the least value of spectral efficiency. When amplify and forward relaying (AF relaying) is applied, the spectral efficiency somehow increases.

A higher value for spectral efficiency can be obtained somehow increases. A higher value for spectral efficiency can be obtained by using decode and forward coding (DF coding). The genetic algorithm provides the superior value at numerous percentiles of cdf for spectral efficiency. The maximum spectral efficiency provided by DF coding on the 5th percentile of cdf is 40 bits/s/Hz, while the value for GA is 90 bits/s/Hz. At the 50th percentile of cdf, DF coding provides a spectral efficiency 300 bits/s/Hz, while the value of spectral efficiency using GA is 1600 bits/s/Hz. The net increase in the value of spectral efficiency at the 50th percentile of cdf is 433%.

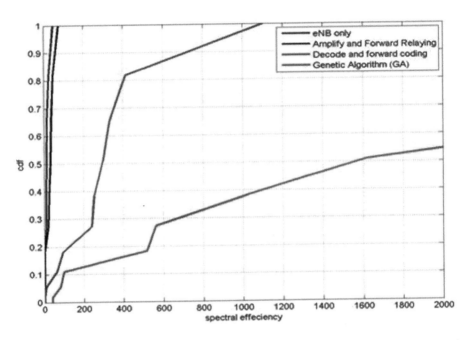

**FIGURE 13.6** Plot of cdf vs. spectral efficiency when one PU and two SUs are present, and the rest of the slots are vacant.

### SCENARIO 4: P PPP A: WHEN ONE PU AND THREE SUs ARE IN ATTENDANCE, AND THE REMAINING ONE SLOT IS VACANT

The SUs sense the spectrum and after finding an empty slot occupy it. As shown in Figure 13.7, the eNB (enhanced node B) provides the least value of spectral efficiency. When amplify and forward relaying (AF relaying) is applied, the spectral efficiency somehow increases.

A higher value for spectral efficiency can be obtained by using decode and forward coding (DF coding). The genetic algorithm provides the superior value at numerous percentiles of cdf for spectral efficiency. The maximum spectral efficiency provided by DF coding on the 5th percentile of cdf is 300 bits/s/Hz, while the value for GA is 500 bits/s/Hz. At the 50th percentile of cdf, DF coding provides a spectral efficiency 800 bits/s/Hz, while the value of spectral efficiency using GA is 1,700 bits/s/Hz. The net increase in the value of spectral efficiency at the 50th percentile of cdf is 112%.

### SCENARIO 5: P PPP P: WHEN ALL SLOTS ARE OCCUPIED

The SUs sense the spectrum and after finding an empty slot occupy it. Hence all the slots are occupied. As shown in Figure 13.8, the eNB provides the least value of spectral efficiency. When amplify and forward relaying (AF relaying) is applied, the

# Design and Simulation of Bio-Inspired Algorithm

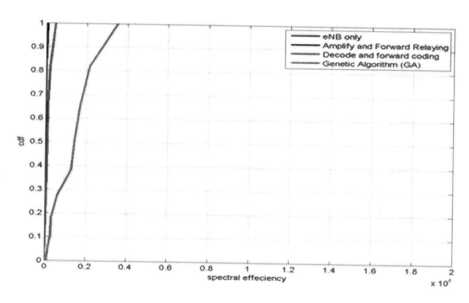

**FIGURE 13.7** Plot of cdf vs. spectral efficiency when one PU and three SUs are present, and one slot is vacant.

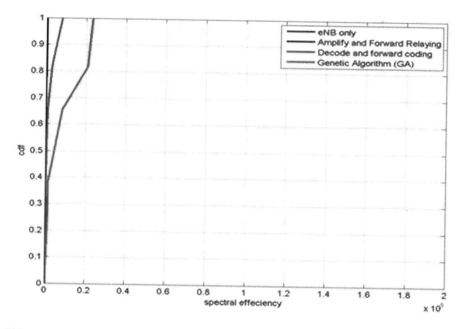

**FIGURE 13.8** Plot of cdf vs. spectral efficiency when all the slots are occupied.

## TABLE 13.1
### Simulation Results for Increase in Spectral Efficiency at the 5th and 50th Percentiles w.r.t. DF Coding

| Case | 5th Percentile | 50th Percentile |
| --- | --- | --- |
| P A AAA | 220% | 650% |
| P P AAA | 433% | 1011% |
| P PP A A | 125% | 433% |
| P PPP A | 66.6% | 112% |
| P PPPP | 150% | 50% |

spectral efficiency somehow increases. A higher value for spectral efficiency can be obtained by using decode and forward coding (DF coding). The genetic algorithm provides the superior value at numerous percentiles of cdf for spectral efficiency. The maximum spectral efficiency provided by DF coding on the 5th percentile of cdf is 2,000 bits/s/Hz, while the value for GA is 5,000 bits/s/Hz. At the 50th percentile of cdf, DF coding provides a spectral efficiency of 8,000 bits/s/Hz, while the value of spectral efficiency using GA is 12,000 bits/s/Hz. The net increase in the value of spectral efficiency at the 50th percentile of cdf is 50%. The results are summarized in Table 13.1.

## 13.8 CONCLUSION AND FUTURE SCOPE

In this chapter, we discussed various aspects of future 5G networks. In order to be aware of the network coverage and capacity for futuristic 5G networks, various measures have been undertaken. The efficient resource utilization and improved spectral efficiency in all kinds of deployment settings in futuristic wireless cellular systems can be achieved by development of new and innovative resource allocation methods. In this chapter, we have tried to enhance the spectral efficiency of multi-user networks by using CR technique. Further, bio-inspired GA was used to assign resources and for solving the optimization problem. The GA mimics the natural evolution process by imitating various procedures by which numerous organisms adapt to their environment and is based on the theory of survival of the fittest by Darwin. It comprises an initial population called parent chromosomes, which may undergo mutation or cross over to create the child chromosomes. The fitness function is calculated, and the process of pairing of chromosomes continues until we get the best fit. The search for the most favorable solution to an optimization problem is accomplished by moving between the individuals of a previous population to latest population (old to new) using genetics-like operators. After applying GA, the results show that the spectral efficiency increases by many fold. The results were obtained using the MATLAB software. The results obtained using the MATLAB software have been supported by a mathematical model that shows the spectral efficiency of multi-user network with GA is better as compared to eNB, AF relaying, or DF coding. It is seen that when there is only one PU, the efficiency at the node is 650%. If we increase the number of users along with CR and GA algorithm, the efficiency rises to 50%.

The future Scope of proposed work is that some other bio-inspired algorithm like particle swarm optimization, game theory, etc. can be applied to allocate the resources and to solve the optimization problem and see its effect on the spectral efficiency of various multi-user networks.

## REFERENCES

Andrews, Jeffrey G, Stefano Buzzi, Wan Choi, Stephen Hanly, Angel Lozano, Anthony CK Soong, and Jianzhong Charlie Zhang. 2014. "IEEE JSAC Special Issue on 5G Wireless Communication Systems What Will 5G Be?," *IEEE Journal on Selected Areas in Communications*, 32 (6): 1065–1082.

Badoi, Cornelia-Ionela, Neeli Prasad, Victor Croitoru, and Ramjee Prasad. 2011. "5G Based on Cognitive Radio." *Wireless Personal Communications* 57 (3): 441–464.

Buzzi, Stefano, Chih-Lin I, Thierry E Klein, H Vincent Poor, Chenyang Yang, and Alessio Zappone. 2016. "A Survey of Energy-Efficient Techniques for 5G Networks and Challenges Ahead." *EEE Journal on Selected Areas in Communications* 34 (4): 697–709. *Ieeexplore.Ieee.Org*.

Goldberg, DE, and JH Holland. 1988. "Genetic Algorithms and Machine Learning." *Machine Learning* 3, 95–99. doi:10.1023/A:1022602019183.

Guvenc, Ugur, Bekir Emre Altun, and Serhat Duman. 2012. "Optimal Power Flow Using Genetic Algorithm Based on Similarity Power System Optimization View Project Optimal Power Flow Using Genetic Algorithm Based on Similarity." *Energy Education Science and Technology Part A: Energy Science and Research* 29: 1–10.

Li, R, Z Zhao, X Zhou, G Ding, Y Chen. 2017. "Intelligent 5G: When Cellular Networks Meet Artificial Intelligence." *EEE Wireless Communications* 24 (5): 175–183. *Ieeexplore. Ieee.Org*.

Long, F, N Li, and Y Wang. 2017. "Autonomic Mobile Networks: The Use of Artificial Intelligence in Wireless Communications." *2017 2nd International Conference on Advanced Robotics and Mechatronics (ICARM)*, Hefei, pp. 582–86, *Ieeexplore.Ieee. Org*.

Mata, SH. 2017. "A New Genetic Algorithm Based Scheduling Algorithm for the LTE Uplink."

Mehboob, Usama, Junaid Qadir, Salman Ali, and Athanasios Vasilakos. 2016. "Genetic Algorithms in Wireless Networking: Techniques, Applications, and Issues." *Soft Computing* 20 (6): 2467–2501.

Mitola, J, and GQ Maguire. 1999. "Cognitive Radio: Making Software Radios More Personal." *IEEE Personal Communications* 6 (4): 13–18. *Ieeexplore.Ieee.Org*.

Mitra, RN, DP Agrawal, ICT Express. 2015. "5G Mobile Technology: A Survey." *ICT Express* 1 (3): 132–7.

Prathisha, R Raj. 2018. "A Study on Use of Artificial Intelligence in Wireless Communications." *Asian Journal of Applied Science and Technology (AJAST) (Open Access Quarterly International Journal)* 2 (1): 354–360.

Russel, S, and P Norvig. 2013. *Artificial Intelligence: A Modern Approach*. Harlow, UK: Pearson.

Walters, DC, and GB Sheble. 1993. "Genetic Algorithm Solution of Economic Dispatch with Valve Point Loading." *IEEE Transactions on Power Systems* 8 (3): 1325–1332. *Ieeexplore.Ieee.Org*.

Yao, Miao, Munawwar Sohul, Vuk Marojevic, and Jeffrey H Reed. 2019. "Artificial Intelligence-Defined 5G Radio Access Networks." *EEE Communications Magazine* 57 (3): 14–20. *Ieeexplore.Ieee.Org*.

# 14 Evaluating the Performance of Quasi and Rotated Quasi OSTBC System with Advanced Detection Techniques for 5G and IoT Applications

*Priyanka Mishra*
Noida International University
*Mehboob-ul Amin*
University of Kashmir

## CONTENTS

14.1 Introduction .................................................................................................259
14.2 Multiplexing Gain and Its Relation with Diversity ....................................262
14.3 System Model ..............................................................................................263
    14.3.1 K-Best Algorithm for Sphere Decoder ............................................265
    14.3.2 Capacity of Rotated Quasi OSTBC Channels ................................267
    14.3.3 Decoding of R-QOSTBC .................................................................269
14.4 Results and Discussion ...............................................................................270
14.5 Conclusion ...................................................................................................273
References ............................................................................................................274

## 14.1 INTRODUCTION

Wireless Internet, cellular video, e-mail, etc. requires tremendous speed. For such high-speed applications, wireless networks have become the promising field of modern engineering. High data rates and capacity requirement for long-term evolution (LTE) and 5G networks are still a challenging task for researchers due to severe restriction in the wireless communication channels. These restrictions are caused due to an enormous number of users getting access to the network. This results in inter-user interference and intersymbol interference. Use of additional nodes between base stations and mobile users to decrease the path loss further leads to additional interference known as intermodal interference. The challenge for researchers is to resolve these issues to get smooth communication between base station and users. Thus, researchers are involved in solving various problems arising in a communication channel like multipropagation fading, intersymbol interference, and distortion of one's own signal.

In spatial multiplexing, the signal is transmitted in such a way that a receiver receives numerous copies of an original signal in three dimensions viz space, frequency, and time. Thus, it increases the number of paths between transmitter and receiver and probability that two or more paths will experience the fade at same time. If some path is experiencing a fade, a signal can be switched to another route. This increases the spatial diversity gain and reduces multipath fading. If a multiple input multiple output (MIMO) system is using $N_t$ and $N_r$ transmit and receive antennas, respectively, the total number of independent fading links available will be $N_t N_r$. The spatial multiplexing gain thus means a linear increase in data rate.

To approach the need of a MIMO wireless channel system is to exploit space–time coding, constructed with multiple antennas at the transmitter side. This system provides an efficient technique for utilizing the diversity provided by the time and space for fading suppression purpose increasing capacity, data rate, and spectral efficiency. However, it is not simple to implement because complexity rises in an exponential manner by incrementing the antenna elements either at the transmitting or receiving end.

MIMO schemes are mainly divided into spatial multiplexing, spatial diversity, and beamforming. Beamforming technique is related to the concept of the smart antenna. Spatial multiplexing schemes gain data bit transmission speed due to parallel transmission and are used to get the high data rate, although spatial diversity technique decreases the signal fading effect by reducing BER. It is clear from the above discussion that it is impossible to achieve both the benefits simultaneously. This is done with the help of two-dimension coding, that is, in space and time domain proposed by (Tarokh et al. 1998) to generate core relation between the signals achieved from numerous transmitting antennas at different time slots called space–time codes. Basically, space–time codes are classified as space–time block codes and space–time trellis codes, and in this chapter, essential emphasis is on orthogonal space–time block codes (OSTBC).

Space–time block codes are the orthogonal codes, that is, columns from the channel matrix are orthogonal to each other with a simple and optimal linear decoding scheme at the front end. The only standard STBC that can achieve full transmission rate and full diversity was introduced by (Alamouti 1998). In this phenomenon, two antennas are deployed at the back end of system. Further, Alamouti space–time block codes (STBC) are modified by employing more antennas at the transmitter side, but achieving 100% diversity and code rate becomes too difficult. Tarokh et al. (1999) proved that using the calculations and examples of the set of STBCs with more than two transmit antenna cannot achieve rate 1 code; it can be a maximum of ¾. In order to obtain the 100% diversity gain, Jafarkhani (2001) brought in a generalized method known as quasi orthogonal space–time block code (QOSTBC) arrangement utilizing four antennas at the transmitting side, by pairing symbols. But the limitation of this pairing was that the matrix columns at the transmitter end did not follow the orthogonality principle, hence full diversity was not achieved. Most of the literature so far has not been able to achieve the full diversity, while selecting the symbols from the same constellation phases using various phase modulations. Thus, we have to rotate the symbols using various rotation factors to minimize the hamming distance in both spatial and temporal domain. These rotation-based codes are termed as rotated Q-OSTBC (Ahmadi, Talebi, and Shahabinejad 2014).

Typically, it was realized that a new scheme known as maximum-likelihood (ML) technique can be implemented to decode these codes, giving superior performance among all conventional decoders (Alabed, Paredes, and Gershman 2011). This chapter first reviews the realization of different STBC MIMO systems with ML detector with respect to BER and SNR. Alamouti (1998) generated complex orthogonal codes for two antennas at transmitter employing a ML detection algorithm at the receiver. The simplicity is due to the orthogonal codes. Similarly, it was performed with higher-order STBCs for more than two transmit antennas, and the literature suggests that computational complexity of the decoding algorithm rises at 4G and 5G modulation schemes like 256 QAM ,1024 QAM, and OQAM for newly generated codewords. Finally, some different decoding methods have been proposed to reduce the complexity.

An alternative approach was proposed (Wolniansky et al. 1998) by employing the V-BLAST algorithm at the receiver. It is a simple detection technique to exploit the spectral efficiency and capacity of the next generation wireless system using a well-defined antenna system employed at back and front end. Spectral efficiency close to 40 bits/sec/Hz in space research centers with interference suppression and successive interference cancellation techniques are achieved using this algorithm. Foschini (1996) proposed a Diagonal Bell Laboratories Layered Space–Time architecture specified as D-BLAST. This further enhances the capacity and information rate. This architecture provides the benchmark for MIMO wireless communications. V-BLAST architecture is the simplest version designed to reduce the computational complexity inherently occurring in D-BLAST systems. Practical implementation of this architecture in MIMO wireless systems can achieve spectral efficiencies more than 40 bits/sec/Hz. V-BLAST finds practical application in MIMO systems because of its immense spectral efficiency, simplicity, and ease to implement on any test bed or VLSI kits. In V-BLAST algorithm researchers introduced numerous concepts based on coding theory and system model to design the BLAST system. This included space–time block coding. There is ordered successive interference cancellation (OSIC) and many decoding schemes used at receivers such as ML decoding (Azzam and Ayanoglu 2009). This process involves much improved performance for V-BLAST systems. Two important receiver decoding steps based on detecting and decoding of symbols is performed in a layered way successively. To avoid the interference and to nullify it, researchers introduce successive interference cancellation mechanism at the receiver site.

This chapter mainly focuses on decoding algorithm based on sphere decoder, also termed the Fincke–Pohst algorithm, proposed by Pohst and later advanced by Fincke and Pohst (1985). This algorithm was introduced in STBC-based MIMO systems at higher modulation techniques like 256 QAM to reduce the computational complexity of ML detectors. The main idea behind the algorithm was to choose the shortest vector path in order to reduce the hamming errors. In the algorithm we introduce certain code words using various encoders and encoding techniques and constrain these codewords within the inner boundary of the sphere. The sphere radius should be equal to received signal vector. Thus, in order to get a ML solution vector, we use a minimum metric method to select the transmitted signal vector (Agrell et al. 2002).

## 14.2 MULTIPLEXING GAIN AND ITS RELATION WITH DIVERSITY

Consider a system employed with $M_T$ transmitting antennas and $M_R$ receiving antennas (Chang, Lin, and Chung 2012), so that the received vector is given as

$$R = Hs + n \tag{14.1}$$

The MIMO system channel can be represented as $N_R \times N_T$ matrix,

$$H = \begin{bmatrix} H_{1,1} & H_{1,2} & \cdots & H_{1,N_T} \\ H_{2,1} & H_{2,2} & \cdots & H_{2,N_T} \\ \vdots & \vdots & \ddots & \vdots \\ H_{N_R,1} & H_{N_R,2} & \cdots & H_{N_R,N_T} \end{bmatrix}. \tag{14.2}$$

Multiple transmitting and receiving antennas are used to obtain the diversity gain. As a result, one is able to go for higher spectral efficiencies by increasing the information rates. Because the diversity is directly proportional to the min or max of quantity of transmit or receive antennas, increasing the quantity level of these elements at the either end of the system will result in increased bit rates (Dalton and Georghiades 2005). We define the spatial multiplexing gain (SMG) using the following equation:

$$SMG = \lim_{\gamma \to \infty} \frac{R}{\ln \gamma}. \tag{14.3}$$

$R$ represents rate. The units used are bits per channel. From the equation, it can be followed that SMG, information rate $R$, along with diversity gain $D$ relate to signal to noise ratio (SNR) using following relation:

$$\lim_{S_{nr} \to \infty} R(SNR) \ln SNR = R. \tag{14.4}$$

The probability of error $P_e$ will be given as:

$$\lim_{snr \to \infty} Pe \ln SNR = -D. \tag{14.5}$$

Because $P_e(snr)$ and logsnr are related, so are $D$ and $r$.

Multiplexing Gain: For the multiplexing gain, we use the asymptotic analysis method to derive the slope of the outage capacity. Plotting is done on a linear–log scale using modified Shannon capacity analysis as:

$$R_{max} = \lim_{\rho \to \infty} \frac{C_{out,\rho}(\rho)}{\log_2 \rho}. \tag{14.6}$$

$R_{max} = \min(\max\{N_r, N_t\})$. Thus we can conclude that forward transmission rate increases by the factor $\min(N_r, N_t)$ for 3 dB increase in SNR.

Accordingly, maximum diversity gain can be achieved by taking the negative values of asymptotes in terms of signal to noise ratio slope of FER,

$$D_{max} = -\lim_{\rho \to \infty} \frac{\log_2 P_e(\rho, R)}{\log_2 \rho} \tag{14.7}$$

$D = M_R M_T$. Thus, the error rate decreases by $2^{-M_R M_T}$.

## 14.3 SYSTEM MODEL

Considering a system with $N_t$ antennas at transmitter side and $N_r$ antennas at the receiving end. Figure 14.1 depicts a diagram in the form of blocks of the space–time coded model where RQSTBC is applied at the transmitter, while lattice point-based sphere decoding algorithm is applied at the receiver. Lattice coding is used for interleaving the information bits at the transmitter. $K$ antennas are being deactivated using $v = \log_2 \binom{n_t}{k}$ bits. OQAM is used to activate remaining $n_t - k$ antennas (Damen,

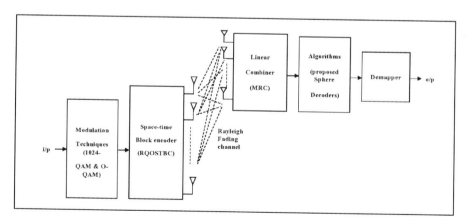

**FIGURE 14.1** System model for RQOSTBCs with advance detection scheme.

Chkeif, and Belfiore 2000). In the proposed model we design the detector, in which the modulated signals are based on the rotational quasi space–time block coded antennas, position of symbol draw from block to block. The arrangement of decoding algorithm corresponding to the transmitting antenna order is variable, i.e., the number of nodes created by various root nodes are not steady (upper triangular matrix will be different from lower triangular matrix). The proposed detection algorithm becomes different from the conventional algorithm, full antenna diversity where the number of nodes is created by every root node depending on constellation size.

We are using a space–time coded MIMO system in which we map a bit stream into a symbol stream $\{\tilde{x}_i\}_{i=1}^{N}$. Figure 14.1 shows a stream of symbols of size $N$. This stream is encoded into $\{x_k^{(t)}\}_{k=1}^{N_t}$, $k = 1, 2, \ldots, T$, where antenna index is represented by $k$ and time index by $t$.

The quantity of symbols in a coded matrix gets constrained to $N_t T$ (i.e., $N = N_t T$). This can also be represented by $\{x_k^{(t)}\}_{k=1}^{N_t}$, $t = 1, 2, \ldots, T$, thus forming a space–time codeword. The symbol rate is defined as:

$$R = \frac{N}{T}\left[\frac{\text{symbol}}{\text{channel}} \text{ used}\right]. \tag{14.8}$$

The receiver side estimates the symbol stream $\{\tilde{x}_i\}_{k=1}^{N}$ using the received signal vectors $\{y_j^{(t)}\}_{j=1}^{N_r}$, $t = 1, 2, \ldots, T$. Assume a Rayleigh-distributed channel gain $H_{jk}^{(t)}$ from the $K_{th}$ transmitted antenna to the $J_{th}$ receive antenna. This happens over the $T_{th}$ symbol period ($K = 1, 2, \ldots N_t$, $J = 1, 2, \ldots, N_r$, and $T = 1, 2, \ldots, T$). The channel gains are considered to be static over $T$ duration of time, which allows us to bypass symbol time index. Thus allowing transmitter and receiver antennas at larger, distances apart, we can achieve $N_r \times N_t$ fading gains.

If $x_k^{(t)}$ is a signal that gets transmitted from the $k_{th}$ transmit antenna over $t_{th}$ symbol duration, the signal at the $j$th receiving antenna (Ding et al. 2016) at the front end will be:

$$y_j^{(t)} = \sqrt{\frac{E_x}{N_0 N_t}} \left[ H_{j1}^{(t)} H_{jj2}^{(t)} \ldots H_{jN_t}^{(t)} \right] \begin{bmatrix} x_1^{(t)} \\ x_2^{(t)} \\ \vdots \\ x_{N_t}^{(t)} \end{bmatrix} + n_j^{(t)}, \tag{14.9}$$

where $n_j^{(t)}$ is the noise vector. The noise vector is modeled by AWGN. The noise vector is having a unit variance with some definite positive energy. The energy is averaged over the $t$th duration, and average value is represented as $E_x$. Thus, we constraint the total transmitted power as (Kostina and Loyka 2011):

$$\sum_{i=1}^{N_t} \left| x_i^{(t)} \right|^2 = N_t, \quad t = 1, 2, \ldots \ldots T. \quad (14.10)$$

### 14.3.1 K-Best Algorithm for Sphere Decoder

At the receiver side $N$ separate substreams are generated, and each is transmitted from different transmitting antennas from the input side; we represent the output in the form of vector O as:

$$O = X.H + \mathcal{N}. \quad (14.11)$$

$H$ is channel gain matrix and $\mathcal{N}$ is AWGN noise matrix with $1 \times M$. In Equation (2.13), we divide the reals and imaginaries of the received vector as:

$$\left( \mathfrak{R}\{O\} \mathfrak{I}\{O\} \right) = \left( \mathfrak{R}\{X\} \mathfrak{I}\{X\} \right) \cdot \begin{pmatrix} \mathfrak{R}\{X\} & \mathfrak{I}\{H\} \\ -\mathfrak{I}\{O\} & \mathfrak{R}\{H\} \end{pmatrix} + \left( \mathfrak{R}\{\mathcal{N}\} \mathfrak{I}\{\mathcal{N}\} \right). \quad (14.12)$$

Equation (14.12) involves only reals and is written in terms of real matrices as

$$O' = X'.H' + \mathcal{N}' \quad (14.13)$$

where $O' = \left( \mathfrak{R}\{R\} \mathfrak{I}\{R\} \right), X' = \left( \mathfrak{R}\{X\} \mathfrak{I}\{X\} \right), \mathcal{N}' = \left( \mathfrak{R}\{\mathcal{N}\} \mathfrak{I}\{\mathcal{N}\} \right)$, and

$$H' = \begin{pmatrix} \mathfrak{R}\{H\} & \mathfrak{I}\{H\} \\ -\mathfrak{I}\{O\} & \mathfrak{R}\{H\} \end{pmatrix}. \quad (14.14)$$

The lattice generated for the given set of symbols is done using the following equation (Hassibi and Hochwald 2002):

$$\wedge = \{(X|X) = X'.H'\}, \quad (14.15)$$

and taking $N = M$ for the ease of calculation.

The lattice $\Lambda$ will now be defined by matrix $G : K_n \rightarrow K_n$ for $n$ dimensional lattice with $n = 1, 2\ldots\ldots n$. A search is initiated for the shortest length vector $k$ in the translated lattice $y - \Lambda$ in the $n$ dimensional space $K_n$. This problem is thus summarized as

$$\min_{X \in \Lambda} \|Y - X\| = \min_{K \in y-\Lambda} \|K\|. \tag{14.16}$$

The original sphere is now translated into a new sphere having squared radius $d$ and centered at the receiving point. The new system will now be defined by $K_n$. Therefore,

$$\|K\|^2 = Q(K_n) = K_n G G^T K_n^T = K_n G K_n^T \tag{14.17}$$

Where $G$ will denote the Cholesky's factorization of Gram matrix $G = GG^T$. The Equation (14.17) can be formulated as following constraint

$$\sum_{i=1}^{n} G_i K_i \leq d. \tag{14.18}$$

If $K$ can be written as upper triangular matrix, $G$ can be decomposed as

$$G = K_n^T K_n. \tag{14.19}$$

Then,

$$Q(K_n) = K_n K_n^T K_n K_n^T = \|K_n K_n^T\|^2 = \sum_{i=1}^{n} \left( Y_{ii} K_i + \sum_{i=j+1}^{n} Y_{ij} K_j \right)^2 \leq d. \tag{14.20}$$

Substituting $r_{ii} = Y_{ii}^2$ for $i = 1$ to $n$, we can write

$$Q(K_n) = \sum_{i=1}^{n} r_{ii} \left( \left( K_i + \sum_{i=j+1}^{n} r_{ii} K_{j+1} \right) \right)^2 \leq d. \tag{14.21}$$

Starting from, $K_n$, the domain for components $v_n$, and $v_{n-1}$ exist between:

$$\left[ -\sqrt{\frac{d}{r_{nn}}} + \rho_n \leq v_n \leq \sqrt{\frac{d}{r_{nn}}} + \rho_n \right] - \left[ \sqrt{\frac{d - r_{nn} K_{nn}^2}{r_{n-1,n-1}}} + \rho_{n-1} + r_{n-1,n} K_n \right]$$

$$\leq v_{n-1} \leq \left[ \sqrt{\frac{d - r_{nn} K_{nn}^2}{r_{n-1,n-1}}} + \rho_{n-1} + r_{n-1,n} K_n \right], \tag{14.22}$$

where $\rho_n = \{\rho_1, \rho_2, \ldots \ldots \rho_3\}$ and $\rho_{n-1}$ are the Lagrangian coefficients associated with $n_{th}$ and $(n-1)_{th}$ iteration. The solution can be extended to the $i_{th}$ integer component as:

$$\left[-\sqrt{\frac{1}{r_{ii}}\left(d-\sum_{l=i+1}^{n}r_{ll}\left(K_l+\sum_{j=l+1}^{n}r_{lj}K_j\right)^2\right)}+\rho_i+\sum_{j=i+1}^{n}r_{ij}K_j\right]\leq v_i$$

$$\leq\left[\sqrt{\frac{1}{r_{ii}}\left(d-\sum_{l=i+1}^{n}r_{ll}\left(K_l+\sum_{j=l+1}^{n}r_{lj}K_j\right)^2\right)}+\rho_i+\sum_{j=i+1}^{n}r_{ij}K_j\right]. \quad (14.23)$$

Variable updating for the bounds for sphere decoder S is done using recursive formulae as:

$$S_i = S_i(K_{i+1},\ldots\ldots K_n) = \rho_i + \sum_{l=i+1}^{n} r_{il}K_l$$

$$T_{i-1} = T_{i-1}(K_i,\ldots\ldots K_n) = d - \sum_{l=1}^{n} r_{ll}\left(K_l + \sum_{j=l+1}^{n} r_{lj}K_j\right)^2 = T_i - r_{ii}(S_i - v_i)^2, \quad (14.24)$$

where $T_i$, $T_{i-1}$ represent the partial Euclidean distance (PED) of the digits $v_i$ and $v_{i-1}$ at the $i_{th}$ and $(i-1)_{th}$ level. Thus, bounds of the sphere decoder change for every carry operation from one digit to another. The sphere decoder vector inside the sphere is obtained by taking the square of distance R from the received point as:

$$R^2 = D - T_i + r_{ii}(S_i - v_i)^2 \quad (14.25)$$

If $R^2 < D$, then we have to search for closest point candidate and terminate the process only, when testing of all vectors inside a sphere is done. Figure 14.2 shows the flowchart for the decoding algorithm RQOSTBC.

### 14.3.2 Capacity of Rotated Quasi OSTBC Channels

The SNR at the receiver is given by $\left(\frac{\rho}{M_T}\right)\|H\|_F^2$, and the capacity is given by

$$C_{ROSTBC} = r_s(1+\rho/M_T)\|H\|_F^2 \quad (14.26)$$

where $r_s$ is the code rate. The capacity analysis is given as provided no channel state information at transmitter (Hochwald et al. 2000; Hochwald and Marzetta 2000):

$$C = \log_2 \det\left(I_{M_R} + \frac{E_s}{M_T N_0} HH^H\right) \quad (14.27)$$

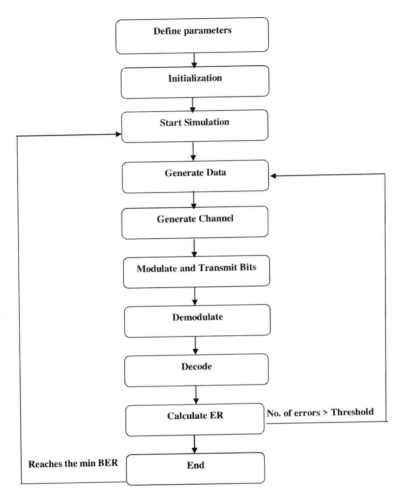

**FIGURE 14.2** Flowchart for decoding algorithm.

$$= \log_2 \prod_{k=1}^{r} \left(1 + \rho/M_T \lambda_k\right) \tag{14.28}$$

$$= \log_2 \left(1 + \frac{\rho}{M_T} \|H\|_F^2 + \frac{\rho^2}{M_T}(\cdot) + \ldots\right) \geq C_{\text{RQOSTBC}}, \tag{14.29}$$

where $\lambda_k$ are the eigenvalues of $HH^H$. The capacity of orthogonal space–time block code (OSTBC) channels is inferior to the channel with optimal coding except for Alamouti's scheme, wherein the code rate $r_s = 1$, causing $C = C_{\text{RQOSTBC}}$ (He and Ge 2003). However, the outage properties of RQOSTBC will be superior to the outage obtained with optimal coding for a given transmission rate because RQOSTBC fundamentally improves the link.

### 14.3.3 DECODING OF R-QOSTBC

The decoding of R-QOSTBC is formulated using the following mathematical analysis. The decoder for $G_3$ minimizes the decision metric (Leuschner and Yousefi 2008a, 2008b),

$$\left[\sum_{j=1}^{m}\left(r_1^j a_{1,j}^* + r_2^j a_{2,j}^* + r_3^j a_{3,j}^* + r_4^j a_{4,j}^* + \left(r_5^j\right)^* a_{1,j} + \left(r_6^j\right)^* a_{2,j} + \left(r_7^j\right)^* a_{3,j}\right) - s_1\right]^2$$
$$+ \left(-1 + 2\sum_{j=1}^{m}\sum_{i=1}^{3}|a_{i,j}|^2\right)|s_1|^2 \tag{14.30}$$

for decoding $s_1$, the decision metric

$$\left[\sum_{j=1}^{m}\left(r_1^j a_{2,j}^* - r_2^j a_{1,j}^* + r_4^j a_{3,j}^* + \left(r_5^j\right)^* a_{2,j} - \left(r_6^j\right)^* a_{1,j} + \left(r_8^j\right)^* a_{3,j}\right) - s_2\right]^2$$
$$+ \left(-1 + 2\sum_{j=1}^{m}\sum_{i=1}^{3}|a_{i,j}|^2\right)|s_2|^2 \tag{14.31}$$

for decoding $s_2$, the decision metric

$$\left[\sum_{j=1}^{m}\left(r_1^j a_{3,j}^* - r_3^j a_{1,j}^* - r_4^j a_{2,j}^* + \left(r_5^j\right)^* a_{3,j} - \left(r_7^j\right)^* a_{1,j} - \left(r_8^j\right)^* a_{2,j}\right) - s_3\right]^2$$
$$+ \left(-1 + 2\sum_{j=1}^{m}\sum_{i=1}^{3}|a_{i,j}|^2\right)|s_3|^2 \tag{14.32}$$

for decoding $s_3$, the decision metric

$$\left[\sum_{j=1}^{m}\left(-r_2^j a_{3,j}^* + r_3^j a_{2,j}^* - r_4^j a_{1,j}^* - \left(r_6^j\right)^* a_{3,j} + \left(r_7^j\right)^* a_{2,j} - \left(r_8^j\right)^* a_{1,j}\right) - s_4\right]^2$$
$$+ \left(-1 + 2\sum_{j=1}^{m}\sum_{i=1}^{3}|a_{i,j}|^2\right)|s_4|^2 \tag{14.33}$$

for decoding $s_4$.

The decoder for $G_4$ minimizes the decision metric

$$\left| \sum_{j=1}^{m} \left( \begin{array}{c} r_1^j a_{1,j}^* + r_2^j a_{2,j}^* + r_3^j a_{3,j}^* + r_4^j a_{4,j}^* + \left(r_5^j\right)^* a_{1,j} \\ + \left(r_6^j\right)^* a_{2,j} + \left(r_7^j\right)^* a_{3,j} + \left(r_8^j\right)^* a_{4,j} \end{array} \right) - s_1 \right|^2 + \left( -1 + 2 \sum_{j=1}^{m} \sum_{i=1}^{3} |a_{i,j}|^2 \right) |s_1|^2 \quad (14.34)$$

for decoding $s_1$, the decision metric

$$\left| \sum_{j=1}^{m} \left( \begin{array}{c} r_1^j a_{2,j}^* - r_2^j a_{1,j}^* - r_3^j a_{4,j}^* + r_4^j a_{3,j}^* + \left(r_5^j\right)^* a_{2,j} \\ - \left(r_6^j\right)^* a_{1,j} - \left(r_7^j\right)^* a_{4,j} + \left(r_8^j\right)^* a_{3,j} \end{array} \right) - s_2 \right|^2 + \left( -1 + 2 \sum_{j=1}^{m} \sum_{i=1}^{3} |a_{i,j}|^2 \right) |s_2|^2 \quad (14.35)$$

for decoding $s_2$, the decision metric

$$\left| \sum_{j=1}^{m} \left( \begin{array}{c} r_1^j a_{3,j}^* + r_2^j a_{4,j}^* - r_3^j a_{1,j}^* - r_4^j a_{2,j}^* + \left(r_5^j\right)^* a_{3,j} \\ + \left(r_6^j\right)^* a_{4,j} - \left(r_7^j\right)^* a_{1,j} - \left(r_8^j\right)^* a_{2,j} \end{array} \right) - s_3 \right|^2 + \left( -1 + 2 \sum_{j=1}^{m} \sum_{i=1}^{3} |a_{i,j}|^2 \right) |s_3|^2 \quad (14.36)$$

for decoding $s_3$, the decision metric

$$\left| \sum_{j=1}^{m} \left( \begin{array}{c} r_1^j a_{4,j}^* - r_2^j a_{3,j}^* + r_3^j a_{2,j}^* - r_4^j a_{1,j}^* + \left(r_5^j\right)^* a_{4,j} - \left(r_6^j\right)^* a_{3,j} \\ + \left(r_7^j\right)^* a_{2,j} - \left(r_8^j\right)^* a_{1,j} \end{array} \right) - s_4 \right|^2 + \left( -1 + 2 \sum_{j=1}^{m} \sum_{i=1}^{3} |a_{i,j}|^2 \right) |s_4|^2 \quad (14.37)$$

for decoding $s_4$.

## 14.4 RESULTS AND DISCUSSION

Consider simulation setup for MIMO system with $n_t = 6$ and $n_r = 6$ antennas. Initially, we fix $K = 0$ for the conventional sphere decoder. Then, we vary the value of $K$ in

### TABLE 14.1
### Optimum Rotation for Different Modulation Techniques

| Modulation Techniques | Optimum Rotation |
| --- | --- |
| 1024-QAM | $\Pi/4$ |
| O-QAM | $\Pi/4$ |

### TABLE 14.2
### Simulation Parameters for System Model

| S. no. | Parameters | Values |
| --- | --- | --- |
| 1 | No. of transmitters | 4 |
| 2 | No. of receivers | 1 |
| 3 | Max. Doppler shift (fm) | 200 Hz |
| 4 | Sampling frequency (fs) | 8000 Hz |
| 5 | Career modulation | 64 QAM |
| 6 | Bandwidth | 20 MHz |
| 7 | Sampling time ($ts$) | 1/fs |
| 8 | No. of Doppler shift ($N_0$) | 8 |
| 9 | Doppler frequency ($f_d$) | 926 Hz |
| 10 | Number of pilot subcarriers | None |
| 11 | Window type | No windowing used |

accordance to sphere decoding algorithm for different proposed sphere decoders. We put $K = 0$ for SD, $K = 1, 2, 3, 4$ for K-Best Sphere decoders. Table 14.1 shows the optimum rotation for the 1024 QAM and OQAM modulation, respectively, and Table 14.2 shows the values of the simulation parameters used in the system model for both the modulations.

Figure 14.3 demonstrates the system performance graph of different combination of SD's in terms of Bit Error Rate for 1024 QAM. The original SD ($K = 0$) is having the BER of 0.15. For $K = 1$, k1 SD reduces the BER to 0.08, for $K = 2$, BER reduces to 0.07, and for $K = 3, 4$ BER reduces to 0.065 and 0.06, respectively.

Figure 14.4 demonstrates the spectral efficiency (S.E.) graph for different combinations of SDs using RQOSTBC at the transmitter for 1024 QAM modulation. The results obtained are in accordance with the analytical approach discussed in Sections 14.3.1 and 14.3.4. Conventional SD is having least spectral efficiency, whereas spectral efficiency increases at higher K values. The 5th and 50th percentile level CDF is used as performance metric for calculating the efficiencies. The 5th percentile corresponds to cell edge users while 50th percentile reflects the performance of central cell users. At 5th percentile CDF level SD is having S.E. of 0.95 (bits/sec/Hz). At $K = 1$, SD is having S.E. of 1.15. For $K = 2, 3, 4$ S.E. of 1.25, 1.35, 1.45 is achieved, respectively. Similarly, for the 50th percentile, conventional SD is having SE of 1. For $K = 1, 2, 3, 4$ S.E. of 1.2, 1.3, 1.4, 1.5 is achieved, respectively.

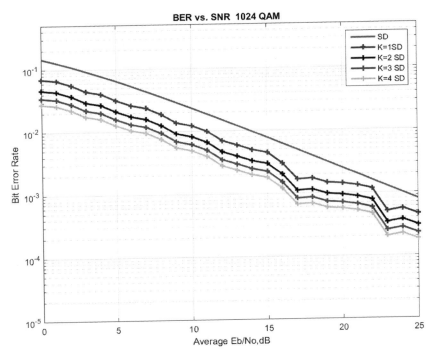

**FIGURE 14.3** Spectral efficiency for 1024 QAM modulation.

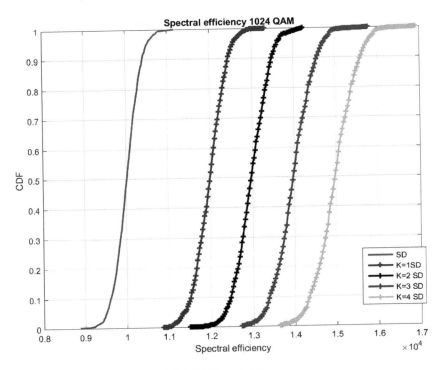

**FIGURE 14.4** BER vs. SNR for 1024 QAM modulation.

# Evaluating the Performance of Quasi and Rotated Quasi OSTBC System

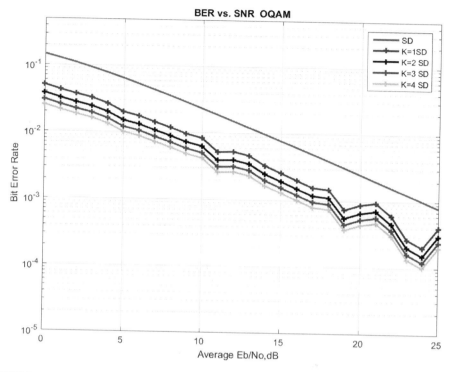

**FIGURE 14.5** BER vs. SNR for O-QAM modulation.

Figure 14.5 demonstrates the system performance graph of different combination of SDs in terms of bit error rate for OQAM. The original SD ($K = 0$) is having the BER of 0.13. For $K = 1$, k1 SD reduces the BER to 0.06, for $K = 2$, BER reduces to 0.05, and for $K = 3$, 4 BER reduces to 0.04 and 0.0356, respectively.

Figure 14.6 demonstrates the spectral efficiency (S.E.) graph for different combination of SDs using RQOSTBC at the transmitter for OQAM modulation. At the 5th percentile CDF level SD is having S.E of 1.15 (bits/sec/HZ). At $K = 1$, SD is having SE of 1.35. For $K = 2, 3, 4$ SE of 1.45, 1.55, 1.65 is achieved, respectively. Similarly, for the 50th percentile, conventional SD is having S.E. of 1.2. For $K = 1, 2, 3, 4$ S.E. of 1.4, 1.5, 1.6, 1.7 is achieved, respectively.

## 14.5 CONCLUSION

In this chapter we have evaluated the performance of RQOSTBC using various decoding techniques for 1024 QAM and OQAM modulation. A novel K best sphere decoding algorithm is used for improving the performance of conventional sphere decoder. The algorithm gives much improved results at higher K values. Simulated results supported by analytical analysis proved the effectiveness of the proposed techniques. Performance matrix used in the proposed work is bit error rate and spectral efficiency. Bit error rate gets decreased at higher values of K, whereas spectral efficiency is increased at higher value. Graphical results demonstrate that OQAM

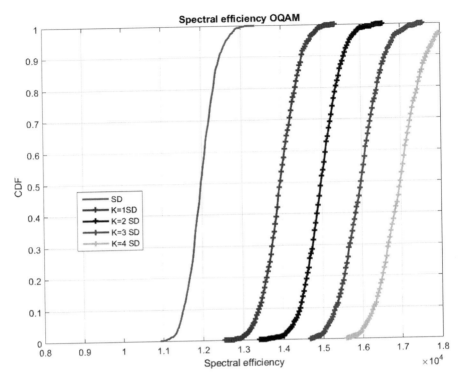

**FIGURE 14.6** Spectral efficiency for O-QAM modulation.

performed reasonably better than 1024 QAM. Thus, the modified sphere decoders can be effectively used in 5G and IOT-enabled devices.

## REFERENCES

Agrell, Erik, Thomas Eriksson, Alexander Vardy, and Kenneth Zeger. 2002. "Closest Point Search in Lattices." *IEEE Transactions on Information Theory* 48 (8): 2201–2214.

Ahmadi, Adel, Siamak Talebi, and Mostafa Shahabinejad. 2014. "A New Approach to Fast Decode Quasi-Orthogonal Space-Time Block Codes." *IEEE Transactions on Wireless Communications* 14 (1): 165–176.

Alabed, Samer J, Javier M Paredes, and Alex B Gershman. 2011. "A Low Complexity Decoder for Quasi-Orthogonal Space Time Block Codes." *IEEE Transactions on Wireless Communications* 10 (3): 988–994.

Alamouti, Siavash M. 1998. "A Simple Transmit Diversity Technique for Wireless Communications." *IEEE Journal on Selected Areas in Communications* 16 (8): 1451–1458.

Azzam, Luay, and Ender Ayanoglu. 2009. "Real-Valued Maximum Likelihood Decoder for Quasi-Orthogonal Space-Time Block Codes." *IEEE Transactions on Communications* 57 (8): 2260–2263.

Chang, Ronald Y, Sian-Jheng Lin, and Wei-Ho Chung. 2012. "Efficient Implementation of the MIMO Sphere Detector: Architecture and Complexity Analysis." *IEEE Transactions on Vehicular Technology* 61 (7): 3289–3294.

Dalton, Lori A, and Costas N Georghiades. 2005. "A Full-Rate, Full-Diversity Four-Antenna Quasi-Orthogonal Space-Time Block Code." *IEEE Transactions on Wireless Communications* 4 (2): 363–366.

Damen, Oussama, Ammar Chkeif, and J-C Belfiore. 2000. "Lattice Code Decoder for Space-Time Codes." *IEEE Communications Letters* 4 (5): 161–163.

Ding, Yuehua, Nanxi Li, Yide Wang, Suili Feng, and Hongbin Chen. 2016. "Widely Linear Sphere Decoder in MIMO Systems by Exploiting the Conjugate Symmetry of Linearly Modulated Signals." *IEEE Transactions on Signal Processing* 64 (24): 6428–6442.

Fincke, U. and M. Pohst. 1985. "Improved Methods for Calculating Vectors of Short Length in a Lattice, Including a Complexity Analysis." *Mathematics of Computation*, 44 (170): 463–471.

Foschini, G. J. 1996. "Layered Space-time Architecture for Wireless Communication in a Fading Environment When Using Multi-element Antennas." *Bell Labs Technical Journal* 1: 41–59.

Hassibi, Babak, and Bertrand M Hochwald. 2002. "High-Rate Codes That Are Linear in Space and Time." *IEEE Transactions on Information Theory* 48 (7): 1804–1824.

He, Lei, and Hongya Ge. 2003. "A New Full-Rate Full-Diversity Orthogonal Space-Time Block Coding Scheme." *IEEE Communications Letters* 7 (12): 590–592.

Hochwald, Bertrand M, and Thomas L Marzetta. 2000. "Unitary Space-Time Modulation for Multiple-Antenna Communications in Rayleigh Flat Fading." *IEEE Transactions on Information Theory* 46 (2): 543–564.

Hochwald, Bertrand M, Thomas L Marzetta, Thomas J Richardson, Wim Sweldens, and Rüdiger Urbanke. 2000. "Systematic Design of Unitary Space-Time Constellations." *IEEE Transactions on Information Theory* 46 (6): 1962–1973.

Jafarkhani, Hamid. 2001. "A Quasi-Orthogonal Space-Time Block Code." *IEEE Transactions on Communications* 49 (1): 1–4.

Kostina, Victoria, and Sergey Loyka. 2011. "Optimum Power and Rate Allocation for Coded V-BLAST: Average Optimization." *IEEE Transactions on Communications* 59 (3): 877–887.

Leuschner, Jeff, and Shahram Yousefi. 2008a. "A New Sub-Optimal Decoder for Quasi-Orthogonal Space-Time Block Codes." *IEEE Communications Letters* 12 (8): 548–550.

Leuschner, Jeff, and Shahram Yousefi. 2008b. "On the ML Decoding of Quasi-Orthogonal Space-Time Block Codes via Sphere Decoding and Exhaustive Search." *IEEE Transactions on Wireless Communications* 7 (11): 4088–4093.

Tarokh, Vahid, Nambi Seshadri, and A Robert Calderbank. 1998. "Space-Time Codes for High Data Rate Wireless Communication: Performance Criterion and Code Construction." *IEEE Transactions on Information Theory* 44 (2): 744–765.

Tarokh, Vahid, Hamid Jafarkhani, and A Rovert Calderbank. 1999. "Space-Time Block Codes from Orthogonal Designs." *IEEE Transactions on Information Theory* 45(5), 1456–1467.

Wolniansky, Peter W, Gerard J Foschini, Glen D Golden, and Reinaldo A Valenzuela. 1998. "V-BLAST: An Architecture for Realizing Very High Data Rates over the Rich-Scattering Wireless Channel." In *1998 URSI International Symposium on Signals, Systems, and Electronics. Conference Proceedings, Pisa, Italy (Cat. No. 98EX167)*, 295–300.

# Index

## A

adaptive power allocation, 58
additive White Gaussian noise, *see* AWGN
advanced metering infrastructure, 39
AI, 10, 16, 24, 25, 27, 137, 241
AI-assisted surgeries, 143
AIoT, 138
Alamouti code, 198
Allele, 244
all-IP networks, 4, 18, 20
amplify and forward, 188
AMPS, 2
anchor-based localization, 128
anchor-free localization, 129
anchor guiding mechanism, 129
anechoic chamber, 166
angle of arrival, 121
ANN, 65
antenna gain, 155
antenna in package, 152
antenna on-chip, 152
Apple, 21
artificial intelligence, *see* AI
artificial intelligence of things, *see* AIoT
artificial neural network, *see* ANN
atmospheric absorption, 10, 170
augmented machine vision, 144
automation, 7, 19, 34, 142, 241
autonomous vehicle, 6
AWGN, 54, 92, 216

## B

bandwidth, 80
beam-forming, 4, 9, 162, 200
beamforming vector, 179
beam steering, 162
biasing effect, 123
big-data, 37
bisection method, 230
bit error rate, 91, 196
blind channel estimation, 112
broadcast, 3
BST, 196

## C

capacity, 4, 7, 9, 15, 46, 72, 167, 240, 249, 267
capacity gain, 9
carrier-to-noise ratio, 196
CDMA, 3, 8, 18, 48
CDMA-2000, 3

channel capacity, *see* capacity
channel estimation, 11, 73, 84, 92, 101, 106, 183
channel modeling, 11, 170
channel sensing, 11
channel state information, *see* CSI
chromosome, 244
circular convolution, 105
cloud computing, 35
cloud jammers, 212
cognitive radio, 33, 177, 240, 242
coherence interval, 95
combined and differentiated localization, 127
connected intelligent vehicles, 142
covariance matrix, 110
coverage, 7, 10, 81, 86, 89, 158
COVID-19, 7, 140
CP, *see* cyclic prefix
*Cramer-Rao lower bound*, 123
cryptography, 37
CSI, 73, 82, 179, 215
cumulative distribution function, 216
cyber physical system, 35
cyclic prefix, 105

## D

data analytics, 5, 25
D-BLAST, 261
decode and forward, 188
deep learning, 2, 11, 39, 133
DEMUX, 104
DFLAR, 133
diffraction, 170
*digital platforms*, 21
direct current, 52
direct line-of-sight, *see* DLOS
directivity, 155
dirty paper coding, 85
distributed intelligence, 140
diversity gain, 9, 89, 181, 260
DLOS, 49
D2D, 4
dynamic spectrum access, 180, 242

## E

EDGE, 3, 20
edge computing, 36
effective beam-scanning efficiency, 159
e-health, 5, 7
8K resolution, 144
EIRP, 156
e-learning, 7

277

energy efficiency, 96, 185
enhanced mobile broadband, 138, 240
enhanced node B, 252
envelope correlation coefficient, 166
*epoch*, 132
equal gain combining, 199
Euclidean distance, 60, 130
evolutionary algorithms, 243
expectation maximization, 114

**F**

Facebook, 20
far-field, 156
FBMC, 242
FDD, 4, 84, 161
femtocells, 5
FFT, 102
fitness value, 246
5G, 6–10, 15, 143, 145, 177, 185
flat fading, 101, 104
flip-error, 131
fog computing, 35
Fog of Everything, 35
force-directed graphs, 130
4G, 3, 4, 14, 15, 196
4G LTE, 4
free space loss, 10
frequency division duplexing, *see* FDD
FTTX, 23

**G**

gain ratio power allocation, 62
gene, 244
genetic algorithm, 243
global positioning system, 120
Google, 24, 144
GPRS, 2, 3, 18, 20
GSM, 2, 3, 8, 17, 18, 20

**H**

healthcare, 38, 142
high gain arrays, 160
HSPA, 2, 3, 5
HSPA+, 4

**I**

IDFT, 104
IFFT, 102
Incremental Redundancy, 3
independent component analysis, 113
Industrial Internet of Things, 141
industrial manufacturing, 139, 141

Industry 4.0, *see* smart factories
information rate, 261
insertion losses, 151
Instagram, 21
Intelligent Surveillance, 143
intelligent transportation systems, 47
interference, 8, 9, 46, 64, 82, 91, 112
interleaving, 263
Internet of Everything, 33
Internet of Things, *see* IoT
inverse power allocation, 55
IoT, 2, 5, 19, 33, 46, 137
IPv6, 5, 33
ISI, 82, 113
isotropic antenna, 155

**J**

jamming, 9, 141, 212, 222
joint power allocation, 65

**K**

Kalman filter, 128
Karush Kuhn Tucker condition, 187, 251
K-best algorithm, 265

**L**

Lambertain coefficient, 66
Lambertain order, 66
Lambertian radiation pattern, 52
large scale MIMO, 9, 79
latency, 2, 4–7, 9, 10, 18, 36, 138, 142, 150, 240
Lattice coding, 263
least square estimation, 106
LED, 9, 46, 49, 52
LiFi, 9
light emitting diodes, *see* LED
linear least-squares, 123
line-of sight, 49
Link Adaptation, 3
localization, 119
locus, 244
long term evolution, *see* LTE
low density spreading-CDMA, 48
low power wide area network, 34
LTE, 4, 8
LTE-A, 5, 8
LTE-Advanced, *see* LTE-A

**M**

machine learning, 2, 11, 16, 36, 145, 241
macrocells, 5
massive machine type communication, 138

# Index

massive MIMO, 11, 19, 91, 103, 104, 114, 163, 169
matched filter, 87
maximal ratio combining, 199
maximizing minimal rate, 69
maximizing sum rate, 69
maximum likelihood estimation, 108
mean squared error, 109
MIMO, 1, 2, 4, 9, 79, 102, 195
MIMO-CDMA, 196
MIMO-OFDM, 101
minimum mean squared error estimation, 109
MISO, 97, 102
MMSE, 109
mmWave, 2, 8, 10, 19, 86, 152, 160, 169, 178
MNO, 15
*mobile robots*, 128
modulation, 3, 4, 196
Monte Carlo Localization, 123
*motes*, 119
moving beacon, 128
M-PSK, 196
    8PSK
    QPSK, 196
    16-PSK, 196, 199
    64-PSK, 203
MRC, 199
M2M, 4, 33, 36
multi-agent system, 38
multicast, 3
multi-cell, 93
multilayer precoding, 89
multiple-input multiple-output, *see* MIMO
multiuser massive MIMO, 91
MU-MIMO, 5, 96
mutation, 247
muti-user MIMO, *see* MU-MIMO
MUX, 104
MVNO, 21

## N

network slicing, 10, 11, 19, 140
NFC, 38
NMT, 2
noise impedance, 152
NOMA, 46
non-LOS, 82
non-orthogonal multiple access, 46

## O

OBST, 196
ODQ, 196
OFDM, 5, 48, 101, 104, 196
OFDMA, 4, 8, 49
omnidirectional, 156

1024 QAM, 271
1G, 14, 15, 150
online gaming, 7, 144
optimal power allocation, 50
OQAM, 261
ordered successive interference cancellation, 261
orthogonal frequency division multiple access, *see* OFDMA
orthogonal multiple access, 48
orthogonal space-time block codes, *see* OSTBC
OSTBC, 196
OTT, 15, 19, 23
outage capacity, 263
over-the-top, *see* OTT

## P

P2M, 34
P2P, 34
partial Euclidean distance, 267
path planning algorithm, 128
pay-as-you-go, 14
PDC, 2, 18
PDMA, 38
people-to-machine, *see* P2M
people-to-people, *see* P2P
phased zero forcing, 88
photo-detectors, 9
photodiode, 46, 62
picocells, 5
pilot symbols, 106
polarization, 8, 165
power density, 170
power domain multiplexing access, *see* PDMA
Precision Medicine, 143
precoding, 86
primary receiver, 212
primary transmitter, 212
primary user, 242
probability distribution function, 130, 216
16-PSK, *see* M-PSK

## Q

QAM, 4, 5, 261
QoE, 2, 3, 20
QoS, 3, 7, 20, 81, 196
QOSTBC, 269
QPSK, *see* M-PSK
quality of experience, *see* QoE
quasi orthogonal space time block code, *see* QOSTBC

## R

radiation pattern, 156
radio access technologies, *see* RAT

radio frequencies, *see* RF
radio signal strength indicator, *see* RSSI
range-based localization, 121
range-free localization, 124
RAT, 8
ratiometric vector iteration, 123
Rayleigh fading, 112
real time services, 5
reflection coefficient, 156
regulatory signature distance, 124
remote healthcare services, 39
renewable energy, 39
RF, 9
Rician gain factor, 178
robust extended Kalman filter, 128
root-mean squared error, 123
RSSI, 121

**S**

scalability, 40, 47, 86, 140
SDN, 19
secondary user, 252
secrecy capacity, 212, 217
secrecy outage probability, 212
selection combining, 199
self-operation-network, 241
semi blind channel estimation, 113
semidefinite relaxation, 113
SeRLoc, 124
shadowing, 86, 170
Shannon channel capacity, 219, 227
Shannon Hartley capacity, 249
signal to interference noise ratio, 64
SIMO, 102
SINR, 181, 226
SISO, 102
6G, 7, 8, 28, 154
64-PSK, *see* M-PSK
skin effect, 170
smart cities, 5, 7, 144, 240
smart embedded system, 39
smart factories, 5
smart grid, 34
smart homes, 5, 39
SNR, 49, 80, 167
software defined networking, *see* SDN
software defined radio, 5, 133
sparse code multiple access, 48
spatial diversity, 198
spatial multiplexing, 198
spatial multiplexing gain, 260, 262
spectral efficiency, 82
spectrum, 2, 8, 141
spectrum handover, 242
spectrum sensing, 242
spectrum sharing, 242

sphere decoder, 265
sphere decoding algorithm, 271
successive interference cancellation, 48, 51, 54
survival-of-the-fittest, 243
systems on chips, 138

**T**

TDMA, 2, 8
TDoA, 121
3G, 3, 4, 14, 15
3.5G, 3
3.75G, 3, 4
3GPP, 2, 4
thermal effects, 151
throughput, 4, 86
THz, 2, 9, 10, 46, 151
time-based positioning scheme, 122
time-difference of arrival, 122
time division multiple access, *see* TDMA
TOPSIS, 58
transceiver, 151
transformed least-squares, 122
transportation & logistics, 142
trilateration, 122
256 QAM, 261
2G, 2, 14
two-way communication, 212

**U**

UE, 5, 154, 160
ultra reliable low latency communication, 138, 150
ultra-dense small cell networks, 10
UMTS, 3, 4, 8, 18
underwater communication, 46
under-water sensor networks, 121
Unmanned Aerial Vehicles, 142
URLLC, 19, 150, 240
user equipment, *see* UE

**V**

V2I, 47
V2V, 46, 47
value-added services, *see* VAS
VAS, 18, 21
V-BLAST, 197, 261
vehicle-to-everything, 142
vehicle-to-infrastructure, *see* V2I
vehicle-to-vehicle, *see* V2V
video conferencing, 7
virtual reality, *see* VR
visible light communication, *see* VLC
VLC, 9, 49

# Index

VL-NOMA, 49
VoIP, 15
VR, 5, 19

## W

WCDMA, 2, 3, 5, 8
WhatsApp, 20
WiFi, 19, 47
WiMAX, 3, 8, 19, 79, 196
wireless acoustic sensor network, 120

wireless broadband, 4
wireless high definition, 178
wireless sensor network, 119
WISP, 23
WLAN, 79, 178
WSN, 119

## Z

zero forcing precoding, 87
zero padding, 112

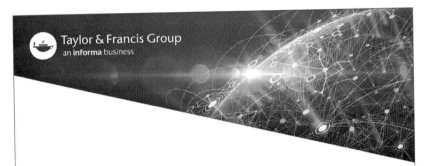

Printed in the United States
By Bookmasters